London
November 2007

A SHEARWATER BOOK

The Remarkable Life of William Beebe

The Remarkable
Life *of*
William Beebe

Explorer and Naturalist

Carol Grant Gould

Island Press / SHEARWATER BOOKS
Washington · Covelo · London

A Shearwater Book
Published by Island Press

Copyright © 2004 Carol Grant Gould

SHEARWATER BOOKS is a trademark of
The Center for Resource Economics.

Library of Congress Cataloging-in-Publication data
Gould, Carol Grant.
The remarkable life of William Beebe : explorer and naturalist /
Carol Grant Gould.
p. cm.
"A Shearwater book."
Includes bibliographical references (p.   ) and index.
ISBN 1-55963-858-3 (cloth : alk. paper)
1. Beebe, William, 1877-1962. 2. Zoologists—Biography
3. Explorers—Biography.   I. Title.
QL31.B37G68 2004
508'.092—dc22   2004012082

British Cataloguing-in-Publication data available

Printed on recycled, acid-free paper ✪

Design by David Bullen Design

Manufactured in the United States of America

10 9 8 7 6 5 4 3 2 1

To Jim, with love

❧ CONTENTS ❧

The beauty and genius of a work of art may be reconceived, though its first material expression be destroyed; a vanished harmony may yet again inspire the composer; but when the last individual of a race of living beings breathes no more, another heaven and another earth must pass before such a one can be again.

William Beebe, *The Bird,* 1906

William Beebe did not want his biography written. Or at least that's what he told his wife Elswyth Thane, a writer of historical romances. In fact, though, he had been telling the story of his own life in detailed journals since childhood. As he grew older, he added reams of scientific data and observations to the chronicle. He mined his diaries extensively for the hundreds of articles and books he wrote during a long and productive life; when he died, Beebe left his papers and the journals to his longtime companion and colleague, Jocelyn Crane, to keep unpublished until Elswyth's death.

I was incredibly lucky to have been given access to these materials, which Jocelyn placed in the Department of Rare Books and Special Collections at Princeton University's Firestone Library, inaccessible without her permission. Other scholars had attempted to use them, but had been effectively shut out by friends and colleagues of Beebe who were determined to protect his privacy. After Elswyth died in 1984, however, Jocelyn, fearful that all memory of this amazing man would be lost, began to look for a biographer, and her eye lighted on me.

Over the course of many years' friendship, Jocelyn tried repeatedly to tempt me to take on Beebe's long-overdue biography. She bubbled over with humorous stories about Beebe whenever we met. She surprised me with books or cartoons or greeting cards Beebe had written or designed, or provocative clippings from yellowing newspapers that would arrive unexpectedly in Christmas boxes or in large brown envelopes addressed in Jocelyn's bold, eager scrawl. To my assertions that I was uniquely unqualified, she was blithely indifferent. In her eyes I had enough science background to do a proper job, but most important, I knew firsthand the difficulties inherent in getting research funded. In those early days before the government stepped in, scientists with large or costly aspirations had to look to the private sector, no matter how critically important their work. Beebe was keenly aware that his reputation as a scientist had been compromised by having to play the funding game; Jocelyn wanted to make sure that his biographer would tell the story fairly.

For better or worse, Jocelyn persuaded me at last. I felt—and still feel—unequal to the task: how can any writer recreate the life of a man who had been a legend in his time, a naturalist par excellence, an expert in ornithology, ichthyology, entomology, marine biology, tropical ecology—fields few individual scientists know as broadly or as well?

With Jocelyn's encouragement and the support of colleagues who knew and loved Beebe's writings, I launched into the daunting task of tracing the life of this charismatic, driven, insightful man, in love with nature, voracious for knowledge, enraptured by beauty. I have tried to be fair both to his work and to the critics of it, to the loyalty and humor that made him a beloved and true friend, and to the depression, exacting standards, and driven personality that made him difficult to work with. My research was especially rewarding when I could talk with people who had known Beebe, who could provide insights into the character of the man and into life with him in the field. My conversations with Jocelyn after I had agreed to write this book introduced me to new facets of her lively and generous spirit, and illuminated her own adventurous life.

Born four years before the electric light bulb, Beebe lived to see the beginning of space exploration. Reading through his letters and journals, a labor of several years, was a rich and rewarding experience both because of the sense of history as it happened and because of William

Beebe himself. It was a wonder and a privilege to eavesdrop on the development of a man of vision and character from eager boyhood, through impetuous adolescence and dreamy young adulthood, and on to his attainment of his lifelong goals. It was a pleasure to meet his friends and to feel the warmth and affection of his close-knit family.

Jocelyn Crane died suddenly while I was still in the early phases of research. So much invaluable knowledge and eager inspiration passed on with her that it was hard to continue. Fortunately, before her death she changed her mind and left me a trove of private correspondence and memorabilia that she had planned to have destroyed. In conjunction with what she had communicated personally and the large archival holdings, her generous gift allowed the completion of this biography.

No book can do full justice to William Beebe; a single volume can only begin. But the natural world desperately needs the excitement that Beebe inspired with his innovative work and passionate writing. If this story of his remarkable life can reignite some of that fire, it will have succeeded.

*Beebe and the
bathysphere.*

At three o'clock Eastern Standard Time on September 22, 1932, a fair part of the world sat breathless beside the radio, straining to hear the voice of William Beebe speaking from a cramped position in his tiny bathysphere nearly half a mile beneath the surface of the ocean. As ship and radio crews and apprehensive colleagues watched, Beebe and the sphere's designer, Otis Barton, had been lowered on a fragile tether over the side of their ship and down, through infinite gradations of translucent greens and blues, into the pitch-black depths, where they hoped to view the life of the deep ocean for the first time. Through a thick window of fused quartz, Beebe glimpsed unearthly creatures with strange shapes and ghostly lights, deep-sea animals totally unknown to the science of his day.

> While we hung in mid-ocean at our lowest level . . . a fish poised just to the left of my window, its elongate outline distinct and its dark sides lighted from sources quite concealed from me. . . . I saw it clearly and knew it as something wholly different from any deep-sea fish which

had yet been captured by man. It turned slowly head-on toward me,
and every ray of illumination vanished, together with its outline and
itself—it simply was not, yet I knew it had not swum away.[1]

The bathysphere's descent through depths no human had plumbed,
from which one tiny miscalculation would make return impossible,
shattered only one of many barriers Will Beebe breached in his long
career as naturalist, explorer, and writer. In a life that spanned eight
decades and hundreds of thousands of miles, he opened previously
inaccessible worlds to the light of scientific exploration, and to the
eager eyes of armchair travelers. Driven by an insatiable curiosity, he
explored places that more circumspect scientists avoided, delving
deeper and farther in quest of knowledge about the natural world,
which he worshiped above all things.

Born at the end of the nineteenth century—when Victorian science,
with its obsession with collecting and naming every organism, was
beginning to shift into the narrow specializations that would characterize the twentieth—Beebe lived to see natural science shift yet again,
into a comprehensive field that embraced the whole of life on earth.
When the Bronx Zoo opened in New York in 1898, Beebe was its first
curator of birds. He circled the globe researching the lives of pheasants,
slipping and sliding down icy Himalayan slopes, paddling up plague-
ridden rivers in Malaysia, and bushwhacking through malarial jungles
in Borneo. To add weight to the elegant concept of natural selection,
so compelling yet so frustratingly elusive, he explored the Galápagos
Islands—still terra incognita in the 1920s—in Darwin's footsteps, and
dredged the oceans for undiscovered creatures that might bridge gaps
in the patchy fossil record. His own intensive researches, especially his
studies of the complex ecosystems of jungles and coral reefs, became
cornerstones of the nascent fields of tropical and marine ecology.

The bathysphere was Beebe's most daring attempt to discover more
about life on earth, and it captured the imagination of a world weary of
internecine bickering and economic blight. People had mounted on
wings with Lindbergh, escaped the leaden pull of gravity with Piccard
in his balloon, and flown and trudged hundreds of frigid, backbreaking
miles with Peary and Amundsen and Byrd en route to the poles. With
William Beebe, they plunged to depths only fantasy writers had imag-

ined, and even heard his voice piped up a cable and sent pulsing along the airwaves.

The alien world he traversed had not even the cold stolidity of the Antarctic, or the undependable but familiar clarity of the air. Dark, deep, unremittingly hostile, the sea had always kept its secrets to itself. Interlopers died of drowning, of asphyxiation, of cold, and of the "bends," their veins frothing with nitrogen gas. The most advanced submarine of the time had not descended past 383 feet and had no windows. An armor-suited diver had made it to 525 feet, but he could barely move or see. Beebe's desire to see what was there overcame caution; his need spawned a belief in the plausibility of the endeavor that had as much to do with faith as with science.

The twenty-four books he wrote during his life of discovery describe his varied experiences with childlike wonder. They were wildly popular and enthusiastically reviewed, making the best-seller lists for months at a time. His essays graced the pages of *Scribner's* and *The Atlantic Monthly* and supplied examples of clear, graceful prose to countless rhetoric texts. The press covered his adventures with the same loving zeal they

Beebe collecting bromeliads
from the forest canopy.

gave Byrd, Lindbergh, and Piccard, and he was feted and lionized. By the time the bathysphere made headlines around the world, Beebe's bald head and thin, eager features were already as well known as any celebrity's.

Through all the publicity, however, Beebe remained a very private person. He allowed just enough exposure to generate funds for his work and for the zoo, and to raise public awareness of the dangers that threatened the natural world. His personal life he kept almost fanatically to himself. His secretive personality has kept him unknown to successive generations, who never experienced the thrill of his radio broadcasts or the excitement of his firsthand reports of amazing discoveries.

But despite his desire to live on only through his work, Beebe kept meticulous diaries and journals from childhood throughout his eventful life. Recently released, they provide a compelling portrait of a man whose love for and curiosity about the natural world, and concern for its preservation, opened new fields of research, entertained and educated a wide public, and inspired future generations of biologists, ecologists, and conservationists. The breadth of William Beebe's vision opened a panoramic view of the interdependence of life that has become crucial to the survival of our shrinking planet.

PART I

Naturalist

CHAPTER I

Counting Crows

To be a Naturalist is better than to be a King.
William Beebe, journal, 12.31.1893

GROWING UP IN THE LATE 1800s with a keen awareness of the natural world was easy. The climate, whether hot or cold, penetrated every building and shelter; accident and disease raged unabated by effective medical intervention; animals were ever present as transportation, labor, and food. Outside even the densest cities wilderness pressed in, offering adventure, sport, and danger. Without television or radio, people were alert to any aspect of the world that could provide diversion, and nature was the most universally accessible form of entertainment. Even city children grew up watching, collecting, and playing with whatever they could find or catch.

What set someone like Will Beebe apart from other childish collectors was that he persisted until he knew everything there was to know about every creature, stone, or shell he found. What set the adult Beebe apart from explorers such as Byrd, Shackleton, or Hillary was that he was a scientist first. His explorations were for the sole end of discovering more about the natural world, whether in the Himalayas, the jungles of British Guiana, or the depths of the sea. He never explored unknown regions unless he felt a pressing need to find out how the creatures that lived there developed, survived, and interacted. Doing anything just to have done it was, to his way of thinking, the depth of folly.

That said, Will Beebe never shirked danger. He always believed, with good reason, that he led a charmed life. Born in 1877 to educated, hardworking parents devoted to each other and to their only child, he grew up self-confident and outgoing, driven by a nervous energy and a work ethic that was daunting to friends and colleagues alike. His father, Charles, was one of four sons of a prosperous Brooklyn paper merchant. Roderick Beebe Sr. had been one of the first to recognize the possibilities of acting as a middleman between the mammoth upstate paper and pulp companies, and manufacturers who could use their output. After college he put his sons to work in the family business; Charles and his identical twin brother, Clarence, remained with it throughout their lives.

The picturesque town of Glens Falls, New York, in the foothills of the Adirondack Mountains, was a frequent stop on the route of any young paper salesman. With its thriving twin pulp mills, the town was home to many well-to-do businessmen; the young sales representatives were constantly in and out of the houses of the mill owners and their customers. At one of these houses Charles Beebe met Henrietta Marie Younglove, daughter of John Younglove and Elizabeth Van Buren Geer.

Charles, left, and Clarence Beebe,
ca. 1900.

The Geers and Van Burens, both powerful forces in Glens Falls society and politics, were not entirely happy with the romance between their offspring and the son of a salesman from New York City. Nettie's mother, Elizabeth, was the oldest in a family of six girls, daughters of a state judge and cousins of President Martin Van Buren. All six had been educated at the elite Glens Falls Academy to display the era's requisite skills, polished with a comely veneer of gentility. Elizabeth's husband, Nettie's father, had never been a success, and had to bring his small family back to live with the Geers. Another one of the sisters had "run off" and made what in Geer eyes was a bad marriage. As a result, Nettie knew she needed to be tactful when she presented a Brooklyn salesman as her future husband.

The Beebes had their own share of family pride. Charles's twin, Clarence, eventually assembled and published a massive genealogy tracing the Beebe name back to John Beebe of Broughton, England, who had come to America in 1650, and even to Richard and Guillaume de Boebe, Norman knights in the retinue of William the Conqueror. And Charles himself was one of those knights of the newly prosperous middle class, a "college man" from Cornell.

Henrietta Younglove Beebe,
Will's mother, ca. 1875.

Charles's affability and courtly bearing combined with his promising business prospects to win the family's approval, and Nettie and Charles were married at Judge Geer's substantial home on a hill overlooking Glens Falls on June 28, 1876. The couple then moved to Brooklyn, where they lived with Clarence and the elder Beebes for a time before moving into their own home nearby. On July 29, 1877, Charles William, who would be known as Will, was born in Brooklyn. The Beebes' only other child, John Younglove, died before his second birthday.

When Charles had built up a strong customer base, he moved the family across the Hudson to East Orange, New Jersey. In those days, East Orange was an ideal place for a family that had business in New York but wanted to raise a child outside the city. To Nettie, who had always had access to the hills and forests of upstate New York, the city and even busy "suburban" Brooklyn were stifling. East Orange was miles away, across the Hudson River and the water meadows of New Jersey. It boasted modern, efficient train links to both the city and upstate, helpful because Charles's business had begun to take him more and more frequently to the big paper mills on Lake Ontario, near Gouverneur, New York. Most important to the education-conscious Beebes, East

The house at 73 Ashland Avenue.

Orange was being settled by well-to-do commuter families committed to cultural refinements and a fine school system.

During those early years at 73 Ashland Avenue in East Orange, Will took full advantage of the suburban neighborhood. His father, who had added talc to his list of sales products, was seldom at home. Nettie made sure Will did his schoolwork, and supplemented it with lessons in music and natural history. They were not wealthy, and Will helped at home by running errands and weeding the tiny yard, sometimes for small sums of money, which he hoarded zealously.

The house itself was typical of the Victorian row houses that were springing up along the rail lines across northern New Jersey and southern New York. Three stories tall and very narrow, it had three gable windows in the attic, which looked out over woods and hills—now completely built up—to a range of low mountains. As an adult, Will remembered clambering out of the gable windows onto the roof to watch the sunset, then transcribing his perceptions into his notebook.

The small house was stretched past its limits by the arrival of Nettie's Aunt Abby and Aunt Hetty, who moved in with their niece and her family in 1884. Nettie and Will had grown to enjoy their privacy, spiced deliciously by Charles's visits, which meant cozy evenings of talk about Will and his achievements, and stories of the wild and rugged forests of upstate New York where Charles spent much of his time. But the maiden aunts, the last of the female Geer household, were family, and the Beebes needed the small rent the ladies would pay. So the aunts

*The living room at
73 Ashland Avenue.*

descended, bringing their straitlaced, querulous ways along with their endless curiosity.

Because Charles was so often away, Will and his mother wrote letters to him almost daily, letters that Charles stored carefully in a box in his room in Gouverneur. Discolored and cracking today, they bear sketches of shells, animals, tracks, and birds' eggs along with descriptions, in a cramped, boyish hand, of football games, classroom pranks, and a bicycle he wistfully admired. They tell of Aunt Abby's fits and Aunt Hetty's snooping, and of the way Will and his mother would kick each other conspiratorially under the table when Aunt Hetty would say something excessively genteel or disparage the soft but pervasive Brooklyn accent that Will was acquiring from his father.

Perhaps most important, the act of writing these letters, which were lovingly read and meticulously answered by an absent father, established in the young boy a lifelong habit of recording every event, every detail he noticed. The earliest surviving letter, written when Will was eight, reported to his father that he had climbed the tree in the front yard to collect some robin's eggs, and had put them in his mouth for safekeeping while he descended. But in a rough landing, "I swallered them."[1]

At the end of the nineteenth century, the world was full of mysteries that seemed accessible to any determined seeker. Vast areas remained unexplored, skies rang with the songs of unknown birds, the oceans teemed with unimaginable forms. Anyone with curiosity could be a naturalist; any boy with a spirit of adventure could be an explorer. Most children of upwardly mobile parents were encouraged to learn about nature the way today's children are given computer classes: it was both education and entertainment.

The newly opened American Museum of Natural History in Central Park was one of New York City's greatest tributes to the growing public demand for knowledge about science. For years, groups of philanthropists and scientists had tried to erect a natural history museum in the New York region. Every great city in Europe boasted one, and many felt that New York could never aspire to cultural equality without its own. Harvard University boasted the great Agassiz Museum, and Washington, D.C., the Smithsonian, but New York's citizens seemed doomed to ignorance of the world beyond bricks and asphalt.

By 1869, however, a group of education-minded donors had braved

Boss Tweed's opposition to obtain a charter, as well as the deeds to a desolate sea of mud, squatters' shanties, and mangy goats. This wasteland would become the fashionable area surrounding Central Park, home to not only the American Museum but its sister institution, the Metropolitan Museum of Art. In 1874 President Grant laid the cornerstone for the first building of the massive museum complex. In 1877, the year Will Beebe was born, the American Museum of Natural History opened its doors, stocked with vast collections of beetles and bones and curiosities from all corners of the world.

The great mission of the natural history museum was at that time to act as guardian of nature's riches and educator of the people. Naturalists had been working steadily through the centuries to accumulate and catalog all the plants and animals in nature—an undertaking whose end continued to recede as new frontiers disclosed unimagined species, and as new concepts of an evolving world gained adherents over the old idea of a static creation. Museums housed collections of flora and fauna, current and extinct, from all corners of the world; they provided scholars—now as then—with research materials, and the public with exposure to exotic realms and ideas they were unlikely to experience elsewhere.

On Will's first awed visits, painters were still working on the landscapes that would be background for the grand dioramas, glass cases of stuffed specimens in which different habitats, meticulously reproduced down to the tiniest creatures and plants, would be displayed. In the warren of offices and laboratories of the upper floors, scientists hunkered over microscopes and trays of rocks or shells or preserved insects, studying them for the minute differences that identified them as to species or sex. Accommodating staff members answered the questions of curious children, and identified flowers or creatures or minerals they brought in. There was always a chance that a specimen clutched in a small grubby fist would be a new species, as yet uncataloged, which might add to the compendium of knowledge. On Saturdays noted scientists gave lectures, and Will attended as many as he could. To his wondering eyes, the life of a naturalist was the highest and most exciting calling anyone could aspire to.

For Nettie Younglove, the wild foothills of the Adirondacks had been a refuge from the repressive gentility of the Geer household, inspiring her to become knowledgeable about botany. Trained as a teacher, she entertained Will and his friends by taking them on nature walks, encouraging them to make collections and to learn the names and life histories of the things they collected. When they were older, she held natural history classes for Will and his best friend, Warrie Mountain, after school. "Warrie came over and Mama heard our lessons, facts & essays," Will wrote his father. "Our next essays will be as follows: Warrie's, Alligators, & mine: Snakes. (I love you.) I think I can find a great deal to write about them." Previous essays had been on vultures and storks; the next week's were to be on kangaroos and monkeys. In the first essay, Will wrote, Warrie had written *more,* but Will's had been more *interesting.*[2] "I want you to try to learn the orders, etc.," the twelve-year-old wrote to his father, "& when you come down you can recite them to me, & if you don't get them good, <u>you</u> get a spanking, & sent to bed."[3]

Nettie taught herself the techniques of preservation so she could teach the boys, and gave them small experiments to conduct. Spurred by magazines published by the burgeoning Audubon clubs, she started Will on a habit he never outgrew: counting crows as they wheeled overhead. This exercise helped him develop an ability to gauge flock size from a representative sample; the data from researchers and novice bird watchers around the country were used in almanacs to calculate and forecast seasonal weather, and by ornithologists to track migration patterns. Will mastered the technique and religiously recorded numbers and direction with date and time in his journals, feeling part of the scientific community as he wrote.

When Will was eleven, his parents took him to be assessed at Fowler and Wells Phrenological Cabinet on Broadway. Phrenologists claimed to be able to read a person's intellectual and moral character from his or her skull, and many nineteenth-century children were subjected to this practice until it was supplanted by psychological test batteries. Nettie put great stock in the reading. The learned man determined, after detailed skull measurements—and a lengthy consultation with the subject—that Will had great "executive" powers. "This boy is smart, quick, clear-headed, and he is a good talker and a critic. He reads strangers well and has strong preferences and prejudices. He is fond of music, and of mechanism and he is going to be a strong character if he can have the body—the handle, as it were,—well developed."[4] He cautioned against too much emphasis on "mental" education, which was already precociously developed, and stressed physical culture and a healthy diet. The Beebes were confirmed in their belief that their son was destined for great things, but for the rest of his life, Will would tell friends that he knew from this moment that he would never be a great intellectual force—his head was too small.

CHAPTER 2

Fledging

A cabinet of eggs is not only an interesting object, but if the owner has collected them himself, he must necessarily acquire an amount of scientific knowledge that will not only at once make him an authority upon ornithology, even among learned men, but at the same time put him ahead of all the boys in wood-craft.

Daniel Beard, *The American Boy's Handy Book*

ALTHOUGH CHARLES BEEBE'S work revolved around the clay and talc mines and pulp mills of northern New York, he lived in what seemed to his suburban son a paradise of unspoiled wilderness. In his spare time Charles wrote voluminously to Nettie and Will of hunting and fishing trips, birds he had seen, minerals he had collected. In the spring of 1891 he mentioned that he might send thirteen-year-old Will a stuffed owl. For the next six weeks every letter Will wrote to his father displayed the perseverance the boy's parents had worked so hard to implant in him: Had Papa found an owl yet? Was the taxidermist there good? When would it arrive? Was he aware that it had not come *yet?*

> My very dear Pop:—
> I have only got a few minutes to write you in, so I can't write a very long letter, as it is 9 o'clock. The <u>Owl</u> came this morning, so this noon I opened it. Isn't it <u>lovely</u>! It greatly exceeded my <u>expectations</u>. . . . The box in which the Owl came will make a lovely cabinet, & tomorrow I am going to get boards, etc. & Mama is going to help me make a nice one. . . . I have got a good beginning now, of 5 stuffed birds, & 1

stuffed mole, & I am going to have a <u>splendid</u> collection of stuffed
things. I have got about 50 shells named, & about 35 or 40 different
minerals (named). . . . P.S. Come over <u>early Saturday</u>. Lots of love,
Will.[1]

Charles encouraged Will in his reading and sent him books he had
finished. Will devoured James Fenimore Cooper, who had written
about the very area of upstate New York Charles worked in. He read
Balzac's eerie story "A Passion in the Desert" and declared it "lovely."
Will and Nettie read them together and discussed them with Charles
when he was able to visit. Like other children of his time, Will collected
colorful nature stamps, which he pasted assiduously into albums, and
waited eagerly for each issue of *St. Nicholas,* a popular monthly magazine read by generations of children from 1873 to 1943. He read Jules
Verne, and G. A. Henty's stirring tales of heroic British soldiers and
sailors — tales of adventure, empire, character, and ideals.

Faced with the heat and humidity of New York and New Jersey summers, everyone fled if they could manage it, or at least sent their families to the mountains. In 1889, when Will was twelve, the family summered at Pine Hill, in Pennsylvania's Pocono Mountains. Here Will
learned to be alone, with only the woods and their creatures as company. He fished, chased butterflies, stalked unsuspecting animals, and
camped out when Charles visited on weekends, dining on campfire-
cooked trout with chewing gum for dessert. His idol was Cooper's
Uncas, the noble Mohican whose stealthy, moccasin-clad tread allowed
him to elude pursuit or approach any animal. Will practiced until he
could sneak up on chipmunks and birds without alerting them, then
watch them without moving.

In school, Will's quickness assured him a place in the most advanced
classes, and in 1890 he was part of a group of children being "prepped"
for the opening of the new East Orange High School. At this time the
vast majority of children ended their academic careers before high
school; the opening of East Orange High indicates the importance of
education to that small community. As part of the prepping, those in
the rowdy all-boys "A" class had been moved into a classroom with the
"A" girls. The boys were appalled. For one thing, in Miss Femby's class
they had had only eighteen spelling words a day. In this accelerated
class, they began with fifty and quickly geared up to a hundred. And
there was worse:

> If you whisper in Miss Day's room you get 3 taken off your deportment
> card at the end of the month, but if you only do something not <u>very</u>
> bad, you have to write down "one mistake," which takes off 1. I got "1
> mistake" to-day. I will try and not get any more.[2]

There were compensations. Warrie sat right behind him, and poor
Miss Day had "never taught any boys, so she is kind of soft and we can
have lots of fun." Will's coiled-spring energy, in conjunction with a well-
knit, wiry frame, earned him spots on most sports teams, and his competitive spirit was keen. He played baseball and football, ran track, and
was fiercely devoted to tennis. By high school he played piano, guitar,
banjo, and mandolin in concerts, and took part eagerly in dramatic presentations.

It was in high school that Will began what was to become an endur-
ing relationship with a supply house called Lattin's. The catalog was to
young naturalists like Will both gospel and siren: its alluring pages
described every sort of collectible, from individual beetles or eggs to
comprehensive collections of everything from Indian artifacts to birds'
nests (with or without eggs) to preserved specimens of ordinary or
exotic creatures. Several of the collections offered grab bags of unspec-
ified wonders to lure the thrifty or thrill-seeking naturalist. One of
Will's first orders, eagerly anticipated with every day's mail and opened
with Pandora-like excitement, contained "three varieties of arrowheads
(named) a horse shoe crab, an iron sling-shot (a <u>dandy</u>! I can kill lots of
sparrows now.) & a drill & blowpipe. I am trying to blow a chicken's egg,
& I can drill a hole in it lovely, but it takes me a long while to blow the
insides out, but I will get used to it."[3]

Later one of Lattin's "Marvelous Collections" enriched the museum
that was rapidly forming in Will's room with "Yellow, pink, and organ-
pipe coral, 'Electric' stone, native lodestone, Petrified Wood, Alligator
tooth . . . Fossil Sea Urchin, Fossil Shark tooth, Crinoid Limestone,
Indian Pottery, Tapa cloth, Coralline, and Japanese leaf of book."
(11.30.1893)

The only animals Will killed during this period were the so-called
English sparrows, the "rats of the air," which were proliferating in suburban environments and driving the native songbirds out of their habitats. Unwisely introduced to North America from England in the mid-
nineteenth century, these noisy and gregarious birds were known to kill

nestlings and even adults of other species to take over a nest site. Their besetting sin seems to have been that they ate every seed people planted, but they were widely regarded as disease vectors as well. Will, like many boys of his time, was paid for every one he killed.

Disease was rife in these days before antibiotics, and people looked to nutrition and spirituality for protection. In addition to the patent or homeopathic medicines that newspaper ads and traveling salesmen touted, control of mind over body became a popular theme for lecturers, and "right thinking" became a byword in religion, particularly among the bluestocking Calvinists of the East.

Although the Beebe family attended the local Presbyterian church regularly, Nettie ventured on her own into the nascent Christian Science movement, which employed right thinking as a way to preserve health. She encouraged Will to keep his heart, mind, and body pure; for added insurance, she also insisted that he take "phosphates," widely regarded as a cure for all manner of ailments thought to originate in the digestive tract.

This was the time when John Harvey Kellogg, later of breakfast cereal fame, was establishing his sanitorium in Battle Creek, Michigan, which taught that a pure gut ensured a pure life. The dogma that the bowels needed to be cleansed by dietary, physical, or chemical means was promoted at the most fashionable dinner tables. The digestive problems that were to plague Will all his life may well have resulted from this early and thorough indoctrination.

In 1891, the summer before Will entered high school, Charles was finally able to bring his family up to Gouverneur. At last Will would be able to see for himself the wonders his father had been writing about for so long. Just after Christmas, Will had written his father a description of his ideal vacation, complete with sketches.[4] In May, with the prospect of Gouverneur alive in his active imagination, Will wrote "Only 65 more days now, or 1560 hours, 93600 min, or 5,616,000 seconds."

Probably the most evident benefit of the summer excursions, whether to Gouverneur or the Poconos, was the chance they offered for the small family to be alone together without the aunts. Back home, Will was reminded repeatedly of his position as head of a household of females. When he was eight, Charles had had him load his gun with nails in case tramps came into the yard and threatened the ladies. When

Nettie got one of her migraines, it was Will who had to run for the doctor. He took pride in the responsibility, and never complained if it was for his mother. But in the small East Orange house, two elderly great-aunts, however well-intentioned, were a strain. If Hetty's fastidious refinement wasn't enough, Will wrote Charles in January 1891 that "the Aunts have had their first row yesterday." Soon after, Aunt Abby had the first of what were to become all-too-common fits, which might have been some form of epilepsy. At two in the morning her sister Hetty screamed for help, and Will was sent to find a doctor. In April there were more fits, and the thirteen-year-old Will was again in charge:

> I have been having a lovely time again. About 5 o'clock this morning we were awakened by one of Aunt Abby's yells (enough to raise the hair on your head) & knew only too well what it meant, we jumped out of bed & ran downstairs, & found Aunt A. in a fit. Mama went for those little glass things & after losing one or two on the floor managed to break one, & let Aunt A. smell of it, but as they didn't do her any good, I went for Dr. Graves, but he was sick abed with the Grip *[sic]*. . . .

Then, shifting abruptly back to his favorite subject, "Yesterday Harry & I went bugging, & got 4 snakes, & I have got them downstairs." [5]

Will's increasing desire for arcana to fill out his varied collections made him eager to earn extra money. When he had saved enough to buy a few chicks, he began an egg- and chicken-raising business that gener-

Will's ideal vacation, ca. 1893: "Go to a place like this; and shoot things like this; and find insects like this; and have adventures like this."

ated the extra funds he needed to support his habit at Lattin's. He studied books on bird rearing and made painstaking entries in his journal, tabulating expenses and profits. He learned to care for the chicks, to treat their injuries and cure their diseases, and wrote detailed accounts of illness, treatment, and outcome. His skill increased with experience, as did his affection for the birds, and for several years he made good money selling eggs to local merchants as he learned.

Will's letters to his father, along with entries in his journals, show that he was becoming increasingly serious about his natural history interests. Still an eager boy who reveled in adventure stories, games, and pranks, he spent much time outdoors with friends, but also loved sitting immobile, quietly observing. He watched, fascinated, while a tadpole hatched in an aquarium in his room. He conducted crude experiments, tying a June bug to his toy cannon and finding that the beetle could pull it ten inches in seven minutes.

To give himself more reasons to be outside, he tried to increase his collections not so much by purchasing them from Lattin's tempting selection but by learning more about the organisms and collecting them himself from the fields and woods. He traded them with his friends, and other children brought him objects to identify and eggs to blow for their own collections: "Jeanette brought me 3 E. Sparrows eggs to blow, & Mary Dutton 4, so you see I am beginning to be a Celebrated Naturalist."[6]

Although Warrie remained Will's closest friend and they continued to go "bugging" and "fossiling" together, other boys began to be attracted to Will's energy and sunny disposition. At the new high school, good-natured Ferdinand Knolhoff shared his love of the outdoors and was willing to follow Will's lead.

Harry Macdonough, whose father was a noted tenor and vaudeville actor in New York, also began to figure in Will's escapades. Will loved evenings spent at the Macdonoughs' house, singing and playing his banjo with a raucous assortment of "Mr. Mac's" music hall friends from the city. The three boys often rode the train into the city to explore the reaches of Central Park or the cavernous halls of the American Museum of Natural History.

In 1893, the summer Will turned sixteen, the family journeyed again to Pine Hill. Will spent long days in the woods and fields, collecting insects—which he chloroformed and pinned carefully—and anything else that caught his eye. It was here that he collected and skinned his first bats. Will and his father built a tent hut, furnished it with blankets, a pistol, and an old double-barreled muzzle-loader, and camped out, surrounded by the dark wild woods. "I have had a very profitable summer as far as collecting goes," Will recorded on his return, "having got 234 specimens of insects, 4 bats, numerous crystals, etc, and 125 var. of woods." (9.2.1893)

That fall Will attended a series of talks given by Frank Chapman, a young naturalist and bird expert on the American Museum's staff, whose name was already appearing in ornithological literature. He watched Chapman pitch for the casual staff softball games at lunch hour ("he pitches almost as well as he writes!") and admired the easy camaraderie among the men of science. Years later Chapman remembered his introduction to this inquisitive youngster, who would become the greatest naturalist of his time.[7]

Chapman's talks were followed by several lectures on minerals, and then several more on other topics of interest to a scientific audience. Each time Will journeyed into the city to attend a lecture at the American Museum, he spent the day there researching the identities of beetles and shells he had collected or birds he had seen. Sometimes, if his father was working at the offices of the family company on Nassau Street, the two would attend the lecture together and then go home on the elevated railway. Every jaunt was an adventure. They went up to the Statue of Liberty's crown, watched silver bars entering the Treasury under guard, attended the unveiling of the statue of Nathan Hale at City Hall, and watched the Old Guard, his father's old militia unit, fire a salute from the Battery. With his friend Ferd Knolhoff, Will cadged a tour of a Navy ship; he spent hours at the docks, watching vessels with exotic names heading for tropical ports he could only dream of. "I will go some day," he vowed.

In the late fall came an important rite of passage: at the age of sixteen, he got his own gun.

Hurrah! Aunt Abby . . . has given me $5.00 and I gave it to Papa. . . . On my way home from school I met him with the gun, and it is a beauty. Mama was away so we went down in the cellar and fired two shots at the piece of paper in the back stoop. (11.8.1893)

Soon Will was able to shoot a penny from twelve feet with a ball car-
tridge. On a trip to the woods he shot his first live prey with his own gun,
an English sparrow. Excitedly, he skinned and stuffed the little bird—
the first specimen he had collected and prepared by himself. His second
bird was a chipping sparrow, which he stuffed "better than the first."

Will approached the art of taxidermy as he did every other self-
appointed task: as something he would work on until he got it right. He
evolved his own methods for wiring the tiny frames, and stuffed the
birds with all the cotton they could hold, finishing them off with beady
artificial eyes ordered from Lattin's. A curved needle (also from Lattin's)
worked well, he found, since he didn't have to press the delicate feathers down to get the point through. (12.8.1893)

Daniel Carter Beard's *American Boy's Handy Book* was a bible to many
children, with its uncomplicated instructions for dozens of hobbies
and adventurous projects, most of which Will tried. Beard details the
popular hobby of stuffing and mounting animals minutely, even advis-
ing his readers where they should go to get the arsenic powder required
to preserve skins. Along with instructions for making kites, aquariums,
boats, fishing poles, and snow forts, he tells how to trap rats and snare
birds and spear fish. There are instructions for blowguns modeled on
those of the "Dyaks of Borneo," and a spring shotgun powerful enough
to stun a small bird but "not injure the specimen by making ugly holes
in the skin and staining the feathers with blood." [8]

> Let us suppose that an owl has been lowering suspiciously near the
> pigeon house or chicken coop, and that you have shot the rascal. Do
> not throw him away. What a splendid ornament he will make for the
> library! How appropriate that wise old face of his will be peering over
> the top of the book-case! He must be skinned and stuffed! [9]

For Will, winters were always times of trial. He hated the cold, with
its endless ailments and confinement. He counted the vast irregular
flocks of crows that passed overhead with the cold winds of fall and
wondered how they kept warm; he read of the tropics, where the bright
warblers that enlivened New Jersey forests spent the hard winters, and
dreamed of going there himself someday, to the lands where winter
never came.

But although he wished the days of winter would go quickly, he was glad to have so much time to prepare for the next collecting season. "My insects especially have got to be classified and arranged in order, so . . . I can place an insect, as soon as it is caught, in its family and genus. I also want to make an aquarium, and hundreds of other things, which will occupy me so that I will find that the winter will pass quickly enough." (11.27.1893)

On a rare crisp, clear December day, sixteen-year-old Will wrote about a "red-letter day," a day in which he had done what he called a "good day's work in Natural History."

> At 8 o'clock this morning I started for the Mountain alone. I had Papa's old hunting cap and I also took my gun and plenty of cartridges. I had my thick overcoat on as the wind is cold and is blowing sharply. I took a buttered slice of bread, 2 doughnuts and an apple which with four chestnuts, and water from springs, made my dinner. . . . When about half-way through I saw a flock of Junco's and as I did not see any-one I fired and brought one down. I made a little cornucopia of news-paper and pinned it in. Then I went on to the top of the mountain. Here I got a drink at the spring, which we named "Uncas" spring. I heard a note of a bluebird or goldfinch, I think the latter. Then I went on for about two miles and came to the Evergreen Woods. I had seen no birds except crows, but here I found a flock of about 200 Gold-finches. I fired twice at them but got none, and they flew away. (12.26.1893)

On the last day of December, he tabulated the year's accomplish-ments and looked eagerly forward to the next.

> So with this I say good-by to Volume III of my notes, and the last words I write this year are: —
>   To be a Naturalist is better than to be a King.
>   C. Will Beebe

This 1893 declaration is not a bad manifesto for the rest of Will Beebe's life.

Portrait of the Naturalist as a Young Man

I am tired from my tramps to-day, and no wonder, for I have been hunter, fisherman, ornithologist, entomologist, conchologist, & taxidermist, & feel I have been fairly successful in each position.

William Beebe, journal, 8.26.1896

IN THE EARLY PAGES OF HIS new journal for 1894, sixteen-year-old Will began what would become a tradition of his, of reflecting on the previous year and making plans for the coming one. For several months, he wrote, he had been troubled about the direction he ought to go with his multifarious interests. His collecting ventures had gotten out of hand, and his mind—as well as his room—needed more specialization. His natural history hobby was becoming all-consuming, and his inherent love of order demanded that he make some hard choices. "The two main sciences to which I shall devote myself are, Ornithology and Entomology. Under the first, would come, birds (mounted and skins), 2. Nests and eggs. The second science would include all kinds of insects, but especially beetles. Of course when I have the chance I will collect fishes, reptiles, mammals and shells, but if I had note-heads, I would put 'Ornithologist and Entomologist' on them." (1.11.1894)

By this time Will had established many of the habits and preferences that would form his adult character. He observed the world around him with a trained eye and recorded his observations meticulously. His schoolwork seems to have cost him very little time or energy, and he devoted the rest of both to scientific pursuits, which to him were infinitely exciting and rewarding. Alone or with a friend he would hike into the countryside, or up to Orange Mountain, to observe or collect new birds and insects. Imagining himself an explorer or an Indian facing perilous odds, he worked to be as silent as Uncas, as observant as Darwin, as dauntless as Peary trekking through Arctic wastes.

Like all good amateur ornithologists of the time, Will subscribed to various natural history periodicals. *St. Nicholas* had given way to the *American Magazine of Natural History, Ibis, The Auk,* and *Oölogist,* a magazine devoted to the science of egg collecting. He pored over the articles and ads in each issue, and formed grand schemes for enhancing and displaying his collections. He was able to see how other collectors labeled their specimens, and learned better methods of mounting and preserving his own. In the magazines' reader exchange sections he advertised nests and eggs he had collected and blown, and traded them with other collectors. One notice netted him ten letters and five sets of eggs. "Yesterday I received 30 species of shells (63 specimens)," he wrote in March 1894, "of which 17 var. are of the genus Helix, and 1 of Cypraea. All but two are new to my collection. They were sent from Texas by a Mr. W. Westgate in exchange for $2.26 worth of eggs which I sent him. I now have 17 birds all of which I have shot (with the exception of the Red-Tailed Hawk) and skinned." (3.11.1894)

Will's collecting techniques were improving as he gained expertise. His aim was extraordinary, and he could shoot birds at a fair distance with very little damage to the skins. He also shot and stuffed unlucky snakes and any small mammals he ran across, such as chipmunks, squirrels, bats, and even one of the large brown rats that lived under his chicken coop and raided the feed supply. He stretched its hide on a board and mounted it as a trophy in his room—which, with its huge stuffed owl, cages of living insects and aquaria of freshwater creatures, its forest of branches with mounted birds and boxes of eggs and nests, and cabinets of pinned and stored specimens—was beginning to look very much like a natural history museum itself.

Nettie had encouraged Will's interest in flora as well as fauna, and he went on field trips with his high school botany class to nearby wild places where rare wildflowers and ferns, mosses, and lichens could be collected. Outside school Will and his biology teacher, Mr. Lottridge, took bikes and guns on increasingly ambitious collecting expeditions. With the help of an experienced teacher who could bring his encyclopedic observations into focus, Will saw these places with new eyes. At Eagle Rock, whose dizzying overlook offers a spectacular view of Manhattan, Mr. Lottridge directed Will's eye to the spot where he caught his first breathless glimpse of a bald eagle on the wing, swooping far out over the gorge below.

In 1894 the Beebes again summered in the Poconos, this time sharing a farmhouse with the Irwin family from Philadelphia. Their son Wallace shared Will's love of nature, and the two boys became fast friends, making Will's summer much more enjoyable. One day Charles brought home a tame young barred owl, which Will named Moses. His original plan was to study the bird for a while to learn its habits, then to kill and stuff it for his collection. Moses rapidly became such a character, however, that Will grew quite attached to him. For food, Moses was happy with the castoffs from Will's collecting—a molting chickadee, a mouse, some leftover fish. "To-day he had two trout which I caught this morning, a large piece of fat ham, and a Vesper sparrow this evening. He always swallows the birds whole except that he generally pulls 4 or 5 of

Will in his room in East Orange, New Jersey, age eighteen.

the longer primaries out, but when he is hungry he omits this even."
(3.11.1894)

This was the summer that Will taught himself to swim. There was no
pool in East Orange, and no formal instruction available. But in the
swimming hole near the house in the mountains, he worked with characteristic intensity until he "got the knack of it" and then was unstoppable. After mastering the breaststroke, he worked on the backstroke
and then on diving. Alone, because Wallace hated the icy water, he dove
repeatedly into the falls, determined that swimming against the current would build his stamina. Unfortunately for someone prone to
sinus and ear infections, the concentrated regimen of swimming and
diving was a recipe for disaster. One day he woke up ill and became
increasingly miserable as the days wore on. Without antibiotics, infections simply ran their sinister courses, and he suffered with earache and
terrible sinus pain that would punctuate his life with periods of misery.

To Will, whose happiness derived from accomplishment, the greatest curse of a summer illness was falling behind in his collecting. This
first time, he lay in a hammock watching a hummingbird, trying to make
up for lost time by writing his impressions of the scene around him.
Although the boy's style mimics the "passive-effusive" constructions of
much contemporary nature writing, the notes are lyrical in their own
way, and acutely observed. "From the other side of the creek near the
bridge comes the strange and not unmusical medley of the Catbird's
apology for a song. Its nest is in one of the four or five clumps of willows
or rhododendrons. It sings almost incessantly and can be heard every
hour of the day." (7.13.1894)

Feeling gradually better, and desperate to make the most of his
remaining days in the mountains, Will attacked the disorganization in
his room, so as to be "all ready for another weeks steady work, and I
hope to average five specimens a day including two birds, if I don't get
sick again." (7.22.1894) He began by collecting a fine large female kingbird and catching a green snake. "But it got out in my room and was so
quick in its motions that I had great difficulty putting it in alcohol. In
the kingbird's stomach I found many insects, but only two were recog-
nizable a bee and the wing-covers of a painted clytus." He stuffed a five-
and-a-half-foot king snake first with cotton, which came out lumpy,
and then with sawdust retrieved from the sawmill, which worked fine.
Toward the end of the season Will was once again put out of commis-

sion, this time by a terrible earache, "but I found to-day that a caterpillar had crawled up my ear I suppose the night I camped out & I have just succeeded in getting it out." (7.28.1894)

Will's last two years at East Orange High School marked him as an outstanding "all-around" student. He was captain of the football team, played a mean game of tennis, and took part in track meets at the Y, where he was known for his speed in the 110-yard dash. He was an enthusiastic participant in a banjo club with Harry Macdonough and several other friends and parents, and was active in staging dramatic presentations at the high school—at least one of them in German. The New Jersey suburbs were becoming noted for their progressive educational programs, as their public schools added courses to prepare young people for college. East Orange offered a rigorous regimen of classics, math, and languages, which included German as well as the Latin and Greek required for admission to most colleges. In Will's journal at this time are occasional passages written in careful old-style German script, awkward but correct—usually bemoaning difficulties with the girls he was beginning to notice as something more than pals.

The opening page of Will's 1895 journal.

On the last day of the year, Will made a chart comparing his 1894 expectations, noted at the year's beginning, with his performance. He found that, in regard to his lists of species sighted and animals collected, stuffed, and mounted, he had exceeded his expectations eight times, equaled them twice, failed to meet them four times (for number of bird species sighted), and doubled his expectations twice. (12.31.1894)

Sailing North

I will be living the most superb existence of any human being in the world....

William Beebe, *High Jungle*

W HEN WILL'S JANUARY 1895
copy of *Harper's Young People* magazine arrived, he was thrilled to see
a brief article he had submitted. "I wrote it Nov. 13 and had given up all
hope of seeing it in print." The article itself, the first giddy venture into
print of one who was to become a prolific writer, was a short, accurate
description of the appearance and behavior of the brown creeper, a
small lively bird that makes its living picking insects off of trees. The
editors commented at the end that they hoped to hear more from the
"observant youth." He had not told his parents he had sent it, and noted
gleefully that they were "greatly surprised." (1.15.1895)

Will's all-around outstanding performance in school (despite several
days when he skipped classes to go shooting) excused him from mid-
term exams, giving him a long vacation, which he spent setting his collections to rights and cleaning his gun. On a solitary bicycle trip to his
favorite birding area, Llewellyn Park, he noted, "There is now the skeleton of a roadway extending from Springdale Avenue to Bloomfield, and
when opened, of course the woods each side will be quickly destroyed.
So I am glad that I graduate next year and will have a chance of going

into more profitable fields of collecting." (2.16.1895) His dream was to travel to the West Indies after graduation, to see at last the tropics that his reading and imagination had painted so alluringly, and collect everything that moved.

Both Will's writing and his powers of observation were improving exponentially. He was beginning to narrow his exuberant range of interests, and to demand a deeper look into the lives and behavior of the creatures he had so long collected, classified, and admired. As small parasitic wasps hatched out of the large cecropia moth cocoon that had been their nursery, he observed patiently for an hour, describing the process in detail and drawing conclusions from what he saw. "There were only two holes and four wasps came out, so I suppose the wasp-cells must be near together." When he picked up some prized bombardier beetles, he described their startlingly effective defense system with interest. Bombardiers are known for shooting a spray of hot acid at attackers. "When I picked them up, a little piff! could be heard, & a big blue cloud of mist was projected over their back. I thought I could avoid it by grasping them by the head but I seemed to get the full force of it there. The wing coverts were wet after each discharge." (4.28.1895)

That summer, there was no vacation travel. Charles had hoped to take his family to Nova Scotia, and Will took voluminous notes on the birds and other fauna of the Bay of Fundy. But money was short, and the trip never materialized. Will's senior year began on September 11, 1895. Finding his schoolwork not particularly challenging, he launched into a study of birds on his own "in a systematic way—devoting perhaps a week to external and internal characters & then each family in order." He grew pond organisms to observe under his microscope, and found a "regular Dante's Inferno" in a drop from Pierson's Pond. And when fall brought the inevitable football season, he worked hard to make sure he was on a winning team but still let his mind wander. "As I was rushing the ball around the left end, just as I fell I happened to look up and saw a bat flying around. I think this is pretty late for them." (11.18.1895) Always sensitive to the slight variations in sight and smell that signal changing weather, he was quick to pick up a promise of winter in the air: the insects were slow-moving and easy to catch, the hickory nuts had fallen and burst their shells. When the smell of the woods came to him, he described feeling exhilaration and a hint of primitive savagery.

He enumerated the books he owned that would help him in his

studies: the standard reference, Elliott Coues's *Key to North American Birds*; *The Birds of Pennsylvania*; Woods's sumptuously illustrated *Animate Creation*; the *Encyclopedia Britannica*, a massive parental investment that was his constant companion. He had read Alfred Russell Wallace's *Darwinism* that summer, so evolution was very much on his mind. Did his roosters crow because their primitive ancestors, the jungle fowl, would flutter to the tops of trees and sing a morning song? The fear the city-bred horses displayed when a troupe of dancing bears came through—was it inherited from distant feral ancestors? How could these tantalizing questions be answered?

The phenomenon of ongoing evolution had been eagerly embraced by the scientific community, but the mechanisms remained elusive. Darwin's concept of natural selection as the force behind evolution required an understanding of genetics that was not yet in place. Mendel's groundbreaking work on inheritance, which would catalyze the field, had been largely forgotten and was not rediscovered and appreciated until 1900. In the absence of a known mechanism, it seemed of paramount importance to find physical evidence of change in every species, to establish a continuum of evolving forms. Will was armed and eager to join the regiment of avid collectors of irrefutable evidence, convinced that the naturalist alone could show the beauty of natural selection's work.

His reading, of course, had a tremendous influence on the subjects and viewpoint and rhythm of his writing, and that influence spilled over into his perceptions of the world around him. Deep into his lifelong devotion to Thoreau, he tuned his senses to register impressions in evocative detail.

> Last night I woke up about two o'clock and for some time I lay listening to the chirps of migrating birds. How wonderful is nature, whose wheels never stop, whose machinery runs as smoothly in the darkest night as in the daytime. Thoreau says the sounds which affect him the most are the Aeolian Harp & the Wood thrush. I know four which send a thrill over my body—the Wood-thrush and Pewee & Whip-poor-will,—reminding me of dark, deep forests; and the crow in early spring,—of the hot summer, even the thought of which is refreshing after the long winter. While listening to the birds' notes last night, a policeman drummed a signal on the pavement,—what a contrast. (9.20.1895)

He carefully copied quotes that registered with him as particularly important or witty, and tried to save money by copying entire books he would have been hard pressed to buy. The economy was in a severe slump, and Charles's income was keyed very closely to the economic health of the nation at large. Paper use plunged rapidly, and a downturn in construction ate away at the profits of the technique for milling fibrous talc that Charles had invented. Will used his study periods to fill blank books with large sections of zoology tomes, material he believed would stand him in good stead in later life. He also copied the drawings in great detail. In addition to improving his handwriting and drawing technique, which he felt needed work, copying etched the material itself indelibly into his mind. "Money is so scarce now, & I want a cabinet & other zoologies so much that I get satisfaction in knowing that I am really getting $3.- or $4.- worth even with so much labor. I have been going to school now for more than twelve years, and it will be almost a new life to change. It seems as if it brought me nearer to Mama & Papa." (11.10.1895) This early schooling in thrift, reinforced by the positive feelings it brought, influenced the way Will conducted every venture of his later life.

The family's stressed finances, which resulted partly from their determination to put Will through college, made that Christmas a quiet one. Will spent much of his vacation "cleaning house," a yearly project that had taken on something of the sacred or superstitious for him. He caught up on his journal writing, and appended a meticulous list of the state of his collections and his hopes for the year to come. He cataloged and organized and labeled his specimens so that the new year would find them in good order. He listed 288 bird skins of 158 species—an increase of 158 specimens and 48 species. "I have skinned only 65 birds, and this I should really be ashamed of (183 last year.) and also of the fact that I made notes this year on only 113 days as compared with last year's showing of 190 days." He noted with dismay that his insect collection was being eaten by moths as the result of poor preparation in the past. In a gesture every collector will recognize as excruciating, he threw away any specimen that showed signs of infestation. His grandiose new plan was to have 1,500 insects by 1897. His report put his current insect collection at:

450 Coleoptera (beetles)
40 Butterflies, 25 moths; 82 Hymenoptera (bees, wasps, and ants); 15

Diptera (flies of various sorts); 17 Orthoptera; 48 Hemiptera; 47 Neu-
roptera, for a total of 725 specimens.
Most birds seen: 4/27 and 5/9. Nine new kinds on each day. (12.31.1895)

Many students in his class were college bound, and conversations
tended to come round to schools and courses of study. Will's sights had
never slipped from Columbia, whose zoology faculty worked hand in
glove with the American Museum of Natural History staff. In fact,
Columbia's head of zoology, the brilliant Henry Fairfield Osborn, was
also the president of the museum.

Will was again excused from many of his exams for good performance; even better, he learned that his outstanding academic achievements, never envisioned by him to be anything out of the ordinary,
would enable him to enter Columbia with advanced placement. Because
the downturn in the economy had continued, finances were growing
ever tighter in the little household, so this must have been heartening
news. On June 26 he wrote, "My feelings are curious to analyze. I graduate from the East Orange High School & in the fall enter the Junior
year of Columbia in the Dept. of Pure Science. . . . It is like a door
which opens & shuts on a beautiful room near the entrance of my man-

*The graduating class at East Orange High School, 1895. Will is in the back row,
third from left, in dark bow tie.*

sion of life. I would gladly go back a year if <u>everything</u> would recede that length of time."

On July 29, 1896, Will "got up at one minute to five this morning, & for one minute enjoyed the thought that I was eighteen years old. Now I am nineteen." His father had just sold a 100-ton order of talc, so the family would be able to go on a long-anticipated trip to Nova Scotia for the rest of the summer. The steamer *Puritan* took them as far as Boston, where they spent a day sightseeing before boarding the steamer to Yarmouth, Nova Scotia, where they would board a train to Weymouth. Will saw Faneuil Hall, the Bunker Hill Monument, and Boston Common, and secured *The Beetles of New England* for fifty cents at the Old Corner Bookstore.

On the steamer to Yarmouth, Will reveled in the wild motion while others retired to their berths or to the "suggestive" basins placed strategically around the deck. "About half the passengers went below in a rush & it would have been much pleasanter for the rest of us, if some had gone sooner." Will was beginning to discover one of nature's brightest gifts: he was almost entirely immune to seasickness. He passed the time counting the seabirds and watching his first whales spouting and breaching. "I cannot express the peculiar sense of weirdness & freedom which it gave me to see these tremendous primeval monsters absolutely masterless, rioting in this vast expanse. Now, such expressions as the 'boundless deep' . . . have real meanings." (8.4.1896)

Weymouth Bridge (now joined with the town of Weymouth), on St. Mary's Bay in southern Nova Scotia, was a revelation to Will. In the Poconos he had enjoyed a measure of freedom: as long as he didn't stray too far or miss meals, he had been free to explore the woods at will. But here he had the wonders of the coast on one side as well as the river, forests, and several freshwater lakes within hiking distance. Just across tranquil St. Mary's Bay he could visit Digby Neck, a slim peninsula whose far shore was pounded by the Bay of Fundy's gigantic tides, alternately depositing and revealing marine wonders that he had only read about before. A short walk took him to the unpopulated shore, where the river met the bay. "Oh!!! It was grand! Sublime! It was twilight, & I could see far down to the bend in the river. . . . At last I could not stand it longer, so I stripped & plunged in. The water was pretty cold but I felt fine when I got out & rubbed down with my handkerchief." (8.4.1896)

The small village of Weymouth Bridge was populated by hardy seafaring and farming families, augmented in season by a companionable and unpretentious summer crowd, mostly from Boston. Locals and "boarders" quickly made friends, and Will soon became a keystone of the social life of the place. With his love of music, games, and theatrics, no week passed without visits, picnics, and parties. In years to come, the friends he made there would be closer to him than the companions of his childhood.

Gregarious as he was, Will was never unhappy alone, and was perhaps happiest when fishing. He often rose at five-thirty and dug a box of worms, had a breakfast of oatmeal and eggs, and left with gun and rod for Sissiboo Lake, a two-mile walk through pine forests resounding with birdsong. He waded in and caught large trout, each individual fish glowing with its own patterns of iridescent color. His journals record details of shade and morphology as well as the thrill of the chase and his joy in the wild scenery. Nearer shore were univalves and whirligigs with red wing covers, and the cry of the loon broke the silence. In the evening by gaslight he skinned his specimens; made notes on the behavior of a whirligig he captured, quite at home in its basin of water; and wrote up his notes from the day.

Other days, inspired by the fish and vertebrate zoology book he was working through, he went at low tide to the mud flats along the bay shore and dug clams. "Dissected a large one according to my zoology,

Sissiboo Lake, Nova Scotia.

and found everything as represented—foot, siphon, gills, mouth, 'feelers,' etc. Could not find the little white hearing organ in the foot, however." (8.21.1896) He was deep into Alexander Agassiz's *Seaside Studies,* which helped identify the marine and tide-pool creatures he was finding. Between the basins of water beetles, anemones, and sea urchins, the jellyfish carefully preserved in alcohol, and the multitude of bird skins he had prepared, Will's room in the Journeays' capacious boardinghouse in Weymouth Bridge was looking like a biology lab and smelling like low tide at a marine dump.

It was hard for all of the Beebes to leave their summer friends and return to New Jersey. The early fall of 1896 was an idle time for the graduated seniors. Will explored New Jersey's wilderness of forest and reed-bound meadowlands, practiced with the high school football team, and spent long summer evenings with friends, singing, reminiscing, and playing instruments on the cool porches. He alternated collecting in all the old places with trips into the city with friends and studying zoology and geology for the Columbia entrance exams. Watching and listening for migrating warblers from the high school's tower one night, he pondered the wildness of the skies, even over the city. He imagined the skies, like the oceans, to be untouchable by human materialism. "No civilization can mar the heavens at night or can disturb the migrations of birds. If it were possible I suppose some people would put advertizements all over the firmament." (9.24.1896)

CHAPTER 5

The Scientists' Apprentice

*No human has fulfilled his manifest destiny of joy and
awe in this life if his eyes have never looked through a
telescope or microscope.*

William Beebe, *Unseen Life of New York*

AT THE END OF SEPTEMBER
Will went to Columbia for the scheduled exams, and to talk with Professor Henry Fairfield Osborn, the head of zoology, about his planned course of study. Because he would be enrolling as a "special" student, it turned out, he was not required to take the exams he had been studying for. A special student was one who was certain what he wanted to do and was already well on the road toward a particular profession. Having mastered many of the prerequisites, he was not enrolled for matriculation in any particular division of the university, but was free to take courses as his professors determined. Will, who seems not to have been privy to this information earlier, was elated. "I do not know why but for some reason I felt a peculiar sense of freedom when I thought I am not bound by any school. I can do anything I like but one thing is certain, that I am going to study for all I am worth." (9.28.1896)

Columbia, when Will arrived, was just beginning a move from its old midtown campus, girdled tightly by a prospering business district, to Morningside Heights, a newly fashionable area north of Central Park that would allow for major expansion. Will's first year was at the old midtown campus, but in 1897 the impressive Low Library was completed, the centerpiece of an array of stately new buildings—including a grand new building dedicated to the natural sciences—and the university officially moved.

Will quickly showed his mettle at Columbia. Knowing full well that he could have applied himself more to his high school studies, he had never set himself down as particularly talented in any special way, and he attributed his early success at Columbia to his thorough grounding in science: "I find my Natural History knowledge is doing me lots of good in my college recitations. The others never heard of Amoeba, pseudopodia, etc. I illustrated intussusception & accretion before the class." (10.16.1896) His boundless energy and drive, combined with a strong background and powers of concentration rare in one so young, marked him as an exceptional student.

By the end of October, Will had established a routine. Mondays he spent at the American Museum, reading natural history texts and comparing them with the museum's mounted specimens. The museum was like a second home to him, and several of the researchers there were now his professors at Columbia. Tuesdays were spent in South Orange

Professor Henry Fairfield Osborn, president of the American Museum of Natural History and head of the New York Zoological Society.

studying ornithology and taxidermy with Mr. Scott, a naturalist and writer who had trained many prominent scientists, including the great Professor Osborn himself. Mr. Scott was notable for having shot more than 600 of the 768 species of North American birds that were known then. He promised Will that after a few weeks' lessons, he would be able to mount even dried bird skins, and to skin twenty-five to thirty in a day.

Wednesdays, Thursdays, and Fridays Will spent at Columbia reading texts, studying skeletons, and attending lectures and labs. Saturdays he divided between "Reading & Football; Collecting & Skating." (11.28.1896) He still lived at home and joined wholeheartedly in parties with his childhood friends and their parents on weekends. As in many small communities at that time, parents and children in East Orange socialized naturally together, joining theater, instrumental, and choral groups and attending church socials and literary discussions. Will numbered many girls among his friends, and especially enjoyed the company of Eleanor Waite, his friend Winthrop's sister. But the easygoing intergenerational companionship discouraged any overt pairing off among the young people; in Will's constricted social circle, even in the late teens sexuality remained mostly theoretical.

In October Will attended a meeting of the New York Academy of Sciences at which several of his professors were speaking. One lecture that particularly caught his imagination was by William Temple Hornaday, a brilliant zoologist who had been recently chosen to direct the construction of a grand new zoological park to be built in the Bronx, in the northern section of New York City. Mr. Hornaday spoke about his tour of the zoological gardens of Europe, and what it had taught him about zoo design and the care and keeping of captive animals. "He said it had been found out that tropical Mammals & Birds could live far longer by having a yard with trees, to which they could have access in the coldest weather even. A magnificent garden is going to be constructed in upper New York. It would be a fine sight to see, a Chimpanzee, for instance, climbing through a grove of enclosed trees." (10.12.1896) In Will's last entry for 1896, he reflected over his accomplishments of the year, as was his habit, and anticipated improvements.

I think that my collecting during 1897 will be far more profitable than heretofore, as I have learned from Mr. Scott & others the advantages of discarding many useless habits. I can easily skin a dozen birds now in the time which I before spent on half that number. But this page painfully reminds me that I could greatly improve in handwriting.

One hope of being able to do more in '97 is the reduction of my "specialties" from almost everything to Ornithology, (& Coleoptera [beetles] & Lepidoptera [butterflies and moths].)

He added later that, using his new skills, he had successfully mounted his Florida gallinule, his screech owl, and his rose-breasted grosbeak. "Soaked them overnight in a pail of water and dried them in corn meal. I improve on every bird in some way. I feel now that I have really learned a trade." Three days later he bought a dead bald eagle in the market, skinned, and mounted it. "It was very hard work and each operation took about two hours. It was a male, and measured Lgth. 32 in. Ex. [wingspread] 6 ft. 7 in." (1.8.1897)

Will spent much of his free time with Harry Macdonough. They went shooting near Harry's home in Maplewood, attended lectures at the American Museum together, and went to the theater. Many evenings were spent singing and playing with the motley group of artists and performers that gathered at the Macdonoughs' welcoming home. Will loved the theater, and the irrepressible, good-natured folk he met at these informal gatherings gave him a respect for the camaraderie and teamwork that was so evident in their performances. Their cheerful acceptance of each other's talents—as well as their quirks, which were legion—encouraged Will to lose any lingering self-consciousness, and he joined in their playing and singing with abandon.

The instrument of choice for young men at the turn of the century was the mandolin, followed closely by the banjo. Will played both, and the banjo club he had formed with ten of his friends and their parents played for private parties, dances, and concerts. He also practiced the guitar, and he played the piano with more gusto than technique. His tastes were catholic, but he was devoted to popular music, particularly the exciting new rhythms of ragtime, as well as music hall favorites he could sing. Despite his demanding extracurricular interests, however, Will was still able to do well at Columbia. When the spring term was over, he and two other boys were excused from their practical exams because of the excellence of their previous work.

In April Will was finally able to buy a bicycle to replace his old wheel, which had been stolen. With $100 he had raised by selling his mounted birds, he bought a Barnes White Flyer. It meant freedom to come and go from home as he pleased, and to collect the specimens he needed to mount and sell. It also allowed him to ride the eighty miles round trip to visit Harry Mac in Wading River, Long Island, where Harry had bought land and was building a house. And perhaps most important, "everyone admires it."

Will had hoped to accompany his uncle Clarence in July on a coveted cruise to Florida, but money was too tight. Instead, he traveled upstate to visit his Columbia friend Seward Wallace in Nyack. As he boarded the steamer that would take him up the Hudson, he watched the *Trinidad* preparing to head off on a West Indian route. "Don't I wish it was one of those that I was going on," he pined. Instead, he and Seward collected, dissected, played tennis, and worked on their music. From Florida, Uncle Clarence sent a flying fish he had caught, and Will preserved it carefully in alcohol. It would become a sort of siren, calling to him whenever he looked at it, singing to him in a sodden way of the wonders of warm tropic seas rich with animal life.

When school began again in October 1897, Will started an informal work-study program in which he combined course work with classifying and arranging Columbia's collection of mounted birds. His plans for the school year included working hard on his journal, cleaning out his room, arranging his collections, reading through all his biology texts, mounting some birds to "keep his hand in," and then mastering photography. His beautifully detailed and illustrated notes from his biology and embryology lectures are punctuated with hypothetical football lineups alongside lists of birds and bugs.

In November the American Museum of Natural History hosted a meeting of the American Ornithologists' Union. Frank Chapman, whose work, like his pitching, Will had long admired, sponsored him as an associate member. This allowed Will not only to hear the lectures by the country's most noted ornithologists, but also to socialize with them and to exchange ideas. He invited Ferd to go with him to the lectures and hear such luminaries as Charles Merriam, the head of the U.S. Biological Survey; Elliott Coues, the author of many handbooks on birds, including the famous "Coues's *Key*," the closest thing to a field guide the age had produced; and Olive Thorne Miller. Miller, a doughty writer and naturalist, was determinedly bridging the gap between the sentimental and didactic literature that dominated the natural history market, and rigorous scientific treatment of habits and physiology.

Will was invited to give a lecture on "Birds in General" to Uncle Clarence's Bergen Point Culture Club in December. It was the first time he had been asked to speak as a "professional" ornithologist, and he was flattered and excited. Will and Ferd took the train up, and Will spoke for thirty-five minutes without once referring to his typewritten lecture. An hour and a half of questions was followed by music: Ferd played a piano accompaniment to Will's mandolin. Elated by the ease with which words had come to him and the warm reception of the audience, he felt confident in his choice of career and capable of great things.

At Christmastime that year his mother presented him with a study Bible, which marked her increasing interest in Christian Science; Will, who had been raised Presbyterian, attended some lectures on the subject with her (although in his journal he gloats over her gift of Newton's *Dictionary of Birds* and makes no mention of the Bible). He was intrigued by what he saw as the rational idea that each person is a compendium of Spirit, Mind, and Body, "the two latter being the covering of the first. It seems true and is wonderful." (1.16.1898) He was searching for a way to incorporate his deep awe of the wonder and mysteries of Nature with his pragmatic and scientific knowledge of the workings of the world; as an adult, he would describe himself as a Presbyterian-Buddhist. Coming back with Ferd from a concert in the spring, they heard a flock of wild Canada geese overhead.

> We hardly breathed as they passed over us . . . we both felt the same feelings. . . . It was a sound that stirred every emotion however latent— a sound which would well be worth half a night's waiting to hear. How the worry and bustle of our musical, and the thoughts of next week's dances, dwindled beside these few notes from Nature! . . . The sound of their voices seemed to merely descend to the streets & return, not being marred by any sound of civilization at this hour—the first of Easter. (4.10.1898)

That year Columbia closed early to accommodate students eager to enlist for the Spanish-American War. Will managed to interest his professors in underwriting a research trip to Nova Scotia for him, Seward Wallace, and their friend Paul Miller, all students in the special program in biology. As Will presented the trip to his professors, Seward would study "the Fishes," Paul Miller anthropology and botany, while Will was to study and collect birds and insects. Professors Osborn and Dean agreed to provide alcohol and formalin, necessary but expensive adjuncts to any research, as they were the accepted way of preserving specimens.

Professor Gratacap, one of Will's longtime mentors at the American Museum of Natural History, helped with photographic supplies. Photography was fast becoming an obsession with Will, who had bought a Solagraph camera in April, only to replace it with a Premo Long Range in May. The Solagraph focused down to only thirty-two inches; the Premo, with its eighteen and a half inches of bellows opening out front and back, was infinitely more flexible. Will hoped that with it he would be able to take professional-quality photos of birds, which he could then sell for more money than mounted birds.

In the end Will's friend Ferd also went along to Nova Scotia, as did the Irwins, friends from Pocono summers, and Charles and Nettie. As the group steamed past his new place on Long Island, Harry Macdonough sent up fireworks. From the steamer *Yarmouth* Will watched

The woodshed lab in Nova Scotia.

great jellyfish floating by, fragile and intricate but pulsating resiliently through mountainous oceans. It gave him a weird sensation, he wrote, "to see an immense Cyanea floating gracefully along 10 or 15 feet below the surface, & to realize that there was probably over a mile of water below it. At night when I looked out of the port-hole I realized as never before the <u>tremendous</u> waste of waters!" (7.6.1898)

In Weymouth Bridge the boys settled into the Journeays' boarding-house, where the Beebes had stayed two years before. Their accommodating landlady had already cleared out a room to act as a darkroom, and she now let them set up a lab in the woodshed for the "dirty" work, with a doomed hope that the worst of the smell would remain outdoors. Will was ecstatic over the new lab space, and proceeded to set it up in the way that always made him feel in control: shelves on every available inch of wall space, specimens mounted and in jars, and a table or desk for each worker. They made a rule that "if one of us brings in a specimen & does nothing with it for 24 hours, anyone of us three can throw it away or appropriate it for himself." (7.22.1898)

The whole group of young people, locals and boarders, soon coalesced into a noisy, cheerful gang. The social scene revolved around music, dancing, tennis, churchgoing, and the sea. There were often grand parties thrown by the parents, and every night less formal groups gathered on porches and steps to sing and play banjos. The Yarmouth paper published a long article on the boys and their fieldwork, which made them feel very important and professional, and drew hoots of derisive laughter from their friends.

Some days the whole crew would set sail for Sandy Cove, a tiny village across St. Mary's Bay on Digby Neck. They picnicked and collected interesting stones and plants on the wide beach behind the fish-weir, or in the many fascinating tide pools.

In the Bay of Fundy the water was so frigid that today swimming is forbidden on that beach, but Will dared the other boys to dive in. Setting his teeth against the cold, he seized armloads of rocks to take him deeper and deeper below the surface to collect graceful anemones, spiny urchins, and brittle stars. Shooting up from below, he would gasp in the warm air, deposit his treasures, and dive again.

Will took dozens of photos of plants, animals, landscapes, and people, and worked late developing them so he could learn from his mistakes. He spent hours photographing his tide-pool specimens under the microscope, connecting the camera to the eyepiece with a tube of drawing paper and using a candle for illumination. Always he was devising new schemes to photograph difficult or hidden scenes—parent birds feeding their young, a chick just hatching out of its egg, a flower hidden in a secret niche. Preparations for these photo shoots could be trying and lengthy. To photograph a sapsucker nest, Will had to shinny ten feet up a tree before Ferd passed him the camera, which he hooked over a limb while Ferd struggled up with a board to brace it with. Just as Will got settled, the axe fell out of his belt and hit Ferd on the head. Poor Ferd had to clamber down to retrieve the axe, which Will hauled up by a rope. They nailed the board to the tree, tied the camera on, and then found that the angle was wrong and the whole thing had to be moved.

Whenever one of the parent birds arrived with food for the young, the boys had to freeze in whatever awkward position they were in. Once, after they got the camera fixed and the focus adjusted, they spent five minutes propping a page from Will's notebook over the lens to shield it from reflection. At that point they heard crashing below, and saw with considerable dismay that a bull and three cows were nosing Will's coat, which held the other two fragile glass plates.

> Then I attach the line to the lever, Ferd meanwhile swearing continuously & righteously at the bovines below. Ferd takes the line & descends 8 ft. I put in the plate, by using two branches (rotten) for a support & withdraw the slide & set the shutter, & descend. Hardly have I reached Ferd's side when I hear the rustle of wings of the male. He is in beautiful plumage, & jerks continuously down to the hole, hesitating a moment & I pull the string. Then undo everything & go home & find that for some reason the diaphragm 25 & time 1/50 sec was wrong or the focus was wrong, & while I have some fine bark, most of the plate is very indistinct. (7.16.1898)

On July 29 there was a gala twenty-first birthday party for Will. There were fifty or sixty people, "about 35 being girls, & I had a fine time."

> Seward played accompaniment & I played on the mandolin, & everyone thought it was splendid for dancing. We had 5 freezes of ice-cream, cake, and lime-water all the evening. I enjoyed it immensely. . . . After the last dance we all joined hands in an immense ring and sang "God

Save the Queen," & then "The Star-Spangled Banner." And then, led by
Mr. Robinson, the editor, everybody gave three cheers for me . . . &
three for the Queen. Imagine it! (7.29.1898)

Will mastered the art of sailing that summer. Fascinated by the life he
observed beneath the boat in the glassy waters, he fashioned a home-
made dredge and took it out in St. Mary's Bay in a friend's boat. What
he collected made his head spin.

Dredging is the most fascinating work I have ever done. It keeps the
excitement up to the highest pitch all the time, no one knowing what
will come up in the next haul. There was a long, heavy swell, and the
dredge weighed a good deal, but we had a magnificent time. Seward
would be feverishly examining a new specimen with a hand lens, when
a wave would throw him flat into a mass of sponges, etc. in the bottom
of the boat." (8.10.1898)

The thrill he got from the uncertain sport of dredging—bringing up
from the depths creatures that he had known only from books, racing
against time to identify and study them before they expired or dis-
solved into a pool of briny goo—outshone the breathless excitement of
the grab-bag packages from Lattin's. The sheer wonder would never
pall for Will, and would provide the motivation for some of his most
important work.

Will driving friends to a picnic in Nova Scotia.

The boys—Will still referred to himself and his friends as boys—did work hard during long days of observing, collecting, classifying, dissecting, and preserving. Will amassed a book of delicately detailed photographs of birds, flowers, and nests, many of which he was later able to sell. But they played hard, too. Will photographed the research staff picnicking, boating, playing tennis, and giving concerts, as well as in the lab.

Several wonderful photos preserve a "beautiful, grand day" when all the town's summer people, along with their hosts, hired a tug to take them to Sandy Cove for a day of picnicking and games. They played football, ate voluminous baskets of food, and explored the rocky shore of the bay. A number of the more adventurous decided to go swimming on the Fundy shore, but the boys had left their suits on the boat. Never one to hold his dignity sacred, Will would not be balked.

> Ferd borrowed Mr. Silvers bathing suit, & I tried to get into Minnie Burrills but the trousers were the only thing I could get on. They had long flounces and I looked very ridiculous. I put my coat on & when Ferd & I came out from behind the big boulder where I had been dressing, the people nearly died laughing. I walked down as far as possible then took off my coat & dived in. (8.15.1898)

The party came home to a somber house. News had arrived that Aunt Hetty was near death in Glens Falls (Aunt Abby had died four years earlier), and the Beebes said hurried goodbyes and left the next day. As the clan gathered for the funeral, Will found time to explore the area with his numerous Geer, Clark, and Van Zandt cousins and read Herbert Spencer's *Principles of Biology*. With his father he explored some of Lake George's myriad islands, and the idea came to him that an island, with its well-defined boundaries, offered an ideal study site. "It would be great fun to take one of these small islands (or better still a tropical one) and work up the fauna and flora thoroughly," he wrote. (9.8.1898) Will's vision of studying an entire ecosystem top to bottom, instead of focusing on one organism, was to become the driving force in his career. So lightly sketched initially, it remains one of the best ways of understanding ecological interactions that scientists have yet discovered.

Back in New York, Columbia was about to begin its fall 1898 term. With his wealth of background study, Will had already completed most of the course work that Osborn felt he would need to work as a naturalist. Some of his professors encouraged him to continue on the scholarly track to a PhD. While the idea of being a member of the academic elite appealed to him, the life of a professor, tied to an office for most of the year with only the summer to do fieldwork, did not.

Sensitive to finances, he had written a preface, introduction, and first chapter of a (never finished) book, *With Camera, Microscope, and Gun in Nova Scotia,* and had great hopes of making money with his writing as well as his photography. Several of the professors at the American Museum had bought his pictures to make lantern slides to accompany their talks. Frank Chapman paid him a regal $2 apiece for negatives of his photos of a chipmunk, a sleeping nighthawk, a junco nest, and some tern eggs.

On January 1, Will's annual day to assess his character and goals, he wrote that he had "read a good deal and thought more." He had to take time off from classes to nurse Nettie through the flu, and used the time to write an article to go with his junco nest photo. Frank Chapman, now an established ornithologist, had begun a natural history magazine entitled *Bird Lore* — later to become *Audubon* — and asked Will to write several articles. He managed to produce them during intersession, despite the demanding social schedule noted in his journal: Monday to his friend Winthrop Waite's house; Tuesday to the theater with Ida Miller, a high school friend, to see actress Julia Arthur; Wednesday to dancing class with Eleanor Waite, Winthrop's sister; Thursday to the fourth "Season Club" dance with Ida; Friday to play mandolin at a party at Seward Wallace's in New York City. Saturday morning there was a lecture about the St. Lawrence River in which the museum's Professor Albert Bickmore, who had known Will since childhood, used Will's junco picture, which he had made into a lantern slide. That evening he played banjo and mandolin at a soiree at the home of Mrs. Hulitt, a cheerful friend of Nettie's who loved giving parties. Girls were playing an increasingly large part in Will's life. At home Eleanor Waite, Agnes Pelton, and Florence Stevens were constant companions, and in Nova Scotia a dozen young women figure in Will's social calendar. Later in the summer, Eleanor and her parents joined the Beebes in Weymouth Bridge.

Will once again talked his Columbia professors into underwriting his Nova Scotia research that summer, and his friends Wallace ("Weary") Irwin and Winthrop Waite, as well as his cousin, Paul Clark, came along. Will planned to study development in the marine creatures of the Fundy tide pools. The obliging Mrs. Journeay had cleaned out the lab and improved it with a new floor and a large window so the young men would have more light for their microscopes and dissections. Then the boys got down to work fishing for trout in DeLaps Lake, swimming in Sissiboo Falls, and riding their wheels around the bay shore.

For the first time, Will found himself tiring of the group's light-hearted approach to life. To escape the good-natured but constant society of the others, he took to riding his bicycle alone through the woods or around the lake, or to the falls, where he dove in alone and fought the fierce current. He wrote lyrically of these trips, which fed something in his spirit that was increasingly demanding attention.

The tide was coming in and every little wave gained a little in advance on its predecessor . . . the silence was intense, only broken by the musical break of the wavelets on the pebbles, and the "peet sweet" of the Sandpipers. Every sense was satisfied;—my seat was a large rock cushioned with dry elastic sea-weed; the air carried faint suggestions of salt and sweet-smelling sea-weed, and the ear and eye received noth-

Inside the woodshed lab, Weymouth Bridge, Nova Scotia.

ing to disturb the wild serenity of all nature. . . . For half an hour I was
contented to sit still and drink it all in, and it seemed impossible to
think of anything but the highest and best things. (7.9.1899)

Will's prose, when not tinctured with sentiment or self-conscious
"fine writing," was growing sharper and sparer. He was increasingly
able to convey the beauty or grandeur of a scene without the fulsome
descriptions that had come to characterize his style since high school.
On July 16, for example, he walked with Wallace, Paul, and Winthrop to
the point and watched the sun setting in a bank of purple and orange
clouds.

There was a mackerel sky. Later it grew darker and the groups of spruces
and occasional bare spots of hill were sharply silhouetted against the
bright sky. The river was dark and the Polyphemus eye of Weymouth
light gleamed brightly, continually gaining power. 50 yards from shore
a little schooner reflected faintly the dull sunset, and when a Great
Heron slowly flapped in near us, the effect was perfect . . . the picture
full of contrasts, light and shade. (7.16.1899)

With a young Radcliffe student he had met collecting starfish on the
beach, he fertilized some of the eggs and was fascinated to see, over the
course of a week, the development of the cells into swimming gastru-
lae. Will and Wallace also grappled with some of the less appealing
obstacles to research when they went to Lily Lake to photograph water
lilies. They would wade in and focus for a few minutes, then lift their
legs in turn and tear off dozens of leeches. Both were covered with
blood by the time they finished the job.

When the fall 1899 term began again at Columbia, Will went to see
Osborn to plan his course of study for the year. "How, one day, we
human beings rest in blissful ignorance of what another day will bring!"
he wrote in his journal afterward. Professor Osborn had given the budding ornithologist a choice: since Will had completed all the required
courses for the three-year degree in pure science except the math,
which he had been postponing,[1] he could repeat Course Five at Columbia, with an additional course in birds, or go with Osborn to the new
zoological park in the Bronx to interview for a job actually working
with birds.

Two trains and then a horse-drawn cab took them to the grounds:

They are <u>immense</u>! 261 acres, and the buildings will be very fine. We saw Mr. Hornaday, who made me this proposition. I would act as Mr. Loring's (who now has charge of both birds and mammals) assistant, until I had learned perfectly about caring for birds, at a salary of $60.—per month, & when I was able to take all the birds in charge I would get $75.—or $100.—a month. (10.3.1899)

This was the "magnificent" zoo that Hornaday had prophesied to an enraptured Beebe three years before. The prospect of being part of such a grand scheme was irresistible. Quite apart from his increasing restiveness with his college studies, Will had felt for a long time that he was taxing his family's restricted finances. This opportunity would not only ease that strain but remedy it.

The next morning he escorted a stunned Nettie proudly over the grounds, describing his duties and showing her where the many new buildings would be. The smaller of the bird houses was nearly finished,

The flying cage at the Bronx Zoological Park.

and Will was to have two rooms in it for study, as well as a camp bed-room to use when animals needed round-the-clock care. The great main room held tropical birds, with appropriate landscapes, on one side, and Northern birds on the other. In the center was a large wire cage, lofty enough to allow the birds inside to fly. Outside, the building was sur-rounded by cages for hawks, eagles, and owls. A large bird house was still in the planning stages, but a huge outdoor flying cage was nearly complete—"an immense structure of steel girders, wire-covered, & even enclosing several full-grown trees."

A new era was dawning in the Beebe household. Although Will blithely spoke of commuting, Nettie soon realized that the idea was impractical: he needed to be closer to work, and he needed looking after. Even if rent had not been an issue, she wanted to be sure that her son ate properly. Besides, the house in East Orange, with Will gone and Charles frequently away, would be both lonely and expensive. The whole family would have to move to the city.

CHAPTER 6

Bronx Zoological Park

*A free zoological park containing collections of North American
and exotic animals, for the benefit and enjoyment of the general
public, the zoologist, the sportsman and every lover of nature.*

William Temple Hornaday,
Annual Report of the New York Zoological Society, 1896

O CTOBER 16, 1899, DAWNED
misty and cool. Walking through the great iron gates for the first time
as an employee, the new assistant curator of birds was overjoyed to be
part of the great Bronx Zoological Park. Every detail stood out in sharp
relief: the chiseled edges of the imposing new brick and stone buildings, the infant plantings just taking hold as the autumn came on, the
muted feral sounds of the zoo animals, the undercurrent of wild birdsong that constant exposure never drove from his consciousness.

The zoological park that Will Beebe walked into was very much a
work in progress. The New York legislature had purchased lands for
public parks in 1884, and although a zoological garden was planned, the
New York Zoological Society itself had not come into being until April
1895. Politics and the attendant red tape had devoured years and tried
the patience and the purse strings of the determined founders.[1]

In 1887, Theodore Roosevelt and ten friends had resolved at a dinner
party to found an activist group, the Boone and Crockett Club, to stop
the wholesale destruction of game animals in the United States.
Granted, these men were interested in big game because they were

inveterate hunters, but their concerns were critically valid. Market hunting was decimating deer herds, while "game hogs" took vast numbers of prey by underhanded, unsportsmanlike methods for profit or éclat. The Boone and Crockett members—whose numbers built rapidly to 100 wealthy, socially prominent men—were determined to use their power and influence to bring the indiscriminate slaughter to an end before whole species went extinct in their native environments.

One of these men, Madison Grant, was a Columbia and law school graduate full of zeal for this cause. Not only were he and his brother DeForest passionate about big game hunting, but they had also dreamed of creating a zoological park that would allow North American animals to roam through naturalistic habitats. It would provide sanctuary for endangered species, building their numbers while the problems of restoring their dwindling habitats could be addressed. The establishment of such a park in New York City would enable the huddled masses to see and experience a wilderness they were unlikely to encounter any other way.

Grant asked Roosevelt to establish a committee that would have the dual purpose of putting an end to the "hounding" of deer in the Adirondacks, where the animals were run to exhaustion by packs of dogs and slaughtered for market, and establishing a zoological park in or near New York. The committee, which was composed of Grant, lawyer and statesman Elihu Root, and Grant La Farge, a young MIT-trained architect, joined forces with William White Niles, assemblyman for the areas in northern New York City that had been slated for parkland. The newly elected board of managers of the nascent society was top-heavy with Boone and Crockett men. To Roosevelt, Grant, and La Farge were added nearly a dozen more, including Niles and Osborn. On April 26, 1895, the New York Zoological Society was voted into law.

The director for this daring new enterprise had to be not just an animal keeper or veterinarian but an established, active conservationist with creative vision, business acumen, and scientific interests. The enterprise was one of considerable daring and scope: in their 1895 manifesto, Osborn and La Farge wrote that they envisioned a zoological park at least twice the size of the largest then known (the National Zoological Park in Washington, D.C., was 168 acres; Europe's largest, the Berlin Zoo, weighed in at a meager 63 acres), with "both native and foreign animals of the tropical, temperate and colder regions as far as possible in the natural surroundings. Thus the larger wild animals of North America—Deer, Elk, Caribou, Moose, Bison, Antelope, Sheep— would be shown not in paddocks but in the free range of large enclo- sures, in which the forests, rocks, and natural features of the landscape will give the people an impression of the life, habits and native sur- roundings of these different types." [2]

The society's board drafted the brilliant but irascible William Temple Hornaday to direct the design and management of the new park. Hornaday had been the impetus behind the founding of the new National Zoo but had left in disgust when his authority did not match his vision. The ideas of Hornaday, a master taxidermist, were revolu- tionizing the display of animals in museums. Instead of presenting them in stiff, unnatural poses, he believed in using groupings that seemed caught in motion, with painted backdrops depicting their native turf, and real rocks and plants. It's an approach that we take for granted today, but that museums were just then beginning to implement.

The new zoo promised to be Hornaday's in a way the National Zoo had never been. His vision of free-ranging animals in a natural setting jibed exactly with that of the board. The scope of the undertaking was

William Temple Hornaday, director, at his desk at the Bronx Zoological Park.

vast, in terms of both area and financing. With private subscription as their springboard, and many of the wealthiest, most philanthropic people in New York eager to help—plus significant pressure from equally well-heeled sportsmen eager for conservation—the sky seemed the limit.

Hornaday spent several weeks scoping out the various areas slated for parkland, finally settling enthusiastically on Bronx Park, whose 261 acres of unspoiled wilderness in the growing sprawl of the city sent him into raptures. "I shall never cease to enjoy my discovery of South Bronx Park! Nor will I ever forget my unbounded astonishment at finding . . . that there nature has made a marvelously beautiful and perfect combination of ridge and hollow, glade and meadow, rock, river, lake and virgin forest, and that man has mercifully preserved it all from defacement and destruction." [3]

Naturally politicians weighed in with objections, not the least of which was that this particular tract of land—adjacent to the site of the projected New York Botanical Garden—was a mile from the nearest public transportation. Hornaday was unfazed, certain that New Yorkers would cheerfully walk a mile for his grand creation. Other objections were to the private nature of the New York Zoological Society itself. In a widely reported speech, Mayor W. L. Strong said that if New Yorkers wanted a grand zoological garden, "let us get one of our own." Why hand over to a private corporation land that taxpayers had purchased for their own pleasure grounds?

But the press rallied to the society's aid. William Randolph Hearst's incendiary *New York Journal* quoted the mayor's inflammatory speech and then commented, "In other words, if we want a great zoological garden, let us tax the people several hundred thousand dollars to pay for it, instead of accepting the offer of a number of public spirited citizens to establish it at their own expense. Especially let us put it in the charge of politicians, instead of allowing it to be created and managed by scientific experts." [4] In the end the city acquiesced, but it drove a hard bargain. The parkland grant, for instance, was contingent on the society's ability to raise $100,000 in one year. After that, it had three years to establish a zoological park and raise another $150,000. These stipulations ensured that the zoo's early years would be driven by grim financial necessity and constant fundraising efforts.

With his suitcase and mandolin, twenty-one-year-old Will Beebe boarded the train in East Orange and moved in with his uncle Clarence, his wife, and his daughter, Cornelia. In their solid brownstone at 227 West 121st Street, he could be near the zoo while his parents searched for an apartment. When his father was not upstate, his office was in Brooklyn, so it made both economical and practical sense for them all to live together in New York.

To reach the zoo from Manhattan, Beebe had to take the New York and Hudson River Railroad to the Fordham station and then walk Hornaday's mile. On the first day of this new routine the cool gray mist did its best to dampen his spirits, but they would not be vanquished. The walk invigorated him, and when he reached his quarters in the bird house he changed into his Nova Scotia exploring suit of "old clothes, leggins, and black and orange jersey." Then he inaugurated his reign by setting out to move the flamingoes and cranes from the dark cellar of the bird house, where they had been held, to a temporary animal yard.

Even for experienced handlers, moving the flamingoes and cranes would have been a grueling task. Will had an extensive background, but it was mostly in theory, so he had a great deal to learn. Although their wings had been clipped to prevent flight, the great birds still had their large beaks, strong legs, and powerful wings, and they were not interested in relocating. In the damp and chill of the October morning, Will and J. Alden Loring, the new assistant curator of mammals, were thoroughly wet and exhausted by the time the birds were settled.

Loring had worked for Merriam at the U.S. Biological Survey for eight years as a field naturalist, and had had two years of experience working at the Zoological Garden in London. At twenty-eight, he was well qualified for his position. But Loring had been doing at least double duty with not only his mammals but also the rapidly expanding population of birds, and he was glad to have Beebe to help out.

Loring introduced Will to twenty-three-year-old Raymond Ditmars, the new assistant curator of reptiles. Ditmars, a reptile fancier, had made his living as a court reporter for the *New York Times* before joining the zoo. When Ditmars arrived, he brought with him his own impressive collection of reptiles "numbering about 40 specimens, comprising a representative collection of the reptiles found in the vicinity of this city, and a number of the larger poisonous snakes of the United States among which are a large diamond rattlesnake with four young, and several of the largest cottonmouth moccasins in captivity."[5] Despite vast differences in their educational backgrounds, the three young assistant curators got along well. They shared powerful common interests, and each had idiosyncratic skills and areas of special knowledge to share.

After two weeks Will felt he had a good grasp of the feeding and cleaning of the birds. He enjoyed the responsibility of caring for the birds himself, and Loring must have been relieved to have the new man catch on so quickly. "I cannot tell how glad I am to be working for myself," Will wrote. "It will be hard work for a week or two, but the thought of my $2-a day (& incidentally grape nuts at dinner) helps to make the work which I thoroughly enjoy, still more enjoyable, and easier." (10.16.1899) A week later he wrote that "my work is easy, & so pleasant that it will seem queer to take pay for it." (10.28.1899)

By late October Beebe had 172 birds of 38 species under his care. His list includes four species of ibis, five of herons, four of owls, and two of swans, plus various storks, cranes, hawks, brown pelicans, and a vulture. In addition, he had three bald eagles in an outside cage. The end of October, however, brought torrential downpours, and the temperature plummeted. Will had to take the five flamingoes he had brought outside on his first day on the job, plus three Demoiselle cranes and four of the brown pelicans, into the bird house's large central cage. They seemed to Will to enjoy the warmth, and played and splashed in the bird house pool.

In November Charles and Nettie left the East Orange house Will had grown up in and settled in a light and airy apartment in the Sans Souci building on West 124th Street, facing Mount Morris Park. The rooms were "finely finished," Will exulted, with both gas and electricity, and inlaid hardwood floors. Will had a bedroom and a "lab room" with a wash basin between, which he speculated would be ideal for filling his numerous aquaria. A sixty-trip train ticket from 125th Street to Fordham, the closest station to the zoo, cost him $3.85, and by leaving at eight in the morning he could easily make it to the zoo by nine, including the longish walk.

Everything about his new job, even the hard dirty work, delighted him. In addition to the mantle of responsibility he had donned so eagerly, he enjoyed the sense of being a contributing member of his family. He lists "Board: nominal, $10" in his careful accounts of his first salary check. "Drew my first salary yesterday afternoon. The six five-dollar bills looked very large to me. This morning I bought Mama a fine picture of the Prophets for $3.75 & had it sent to the house." Charles received a genteel smoking jacket from the new wage-earner. (11.4.1899)

When Will walked from the station to the zoo every day, the air was fresh and "country-like," and the white-throated sparrows greeted him at the gate. He relished his independence, delighting in a Spartan lunch and, when he had to work late, a dinner of Grape-Nuts (which he considered the ultimate in scientific nutrition) and doughnuts—a meal that Nettie would certainly never have served at her carefully regulated table.

It had become obvious early on that the zoo would not be ready to open on schedule in July 1899; by the time Will signed on in October, it looked very much as if even the new November date would be impossible. Construction was slow, and no single attraction was complete. Crowds of people came anyway, disrupting the workmen and adding to the chaos, only to discover that there was nothing to see but buffalo and elk in their outdoor ranges. One Sunday in October, 20,000 visitors filtered through the ineffectual closed gates.

The Bronx Zoological Park finally opened on November 8. (Hornaday fought a vocal battle his entire life to make the public refer to his creation as "the park," but "the zoo" was unstoppable.) It was still woe-

The bird house, Beebe's domain.

fully incomplete, but the staff had good reason to be proud of what was there. Twenty-two exhibits were open, linked by neatly raked gravel paths. There were 843 animals of 157 species. A special train brought visitors to the Fordham station for the opening, where carriages were in waiting for the most honored guests. The Honorable Levi P. Morton, the New York Zoological Society's second president, was a former U.S. vice president and governor of New York. A contingent of Osborn's colleagues from Columbia were there, as well as the distinguished staff of the American Museum of Natural History. Some of the zoo's biggest donors, such as Percy Pyne, J. Pierpont Morgan, William Dodge, and William Whitney, were in attendance.

Outside the elegant aquatic bird house, Osborn, the acknowledged architect of the Zoological Society itself, gave a brief address and sent the crowd, which had reached nearly 2,000, on a tour of the grounds. Curators and keepers had been instructed to remain in their areas to guide the guests, but pride in their domains would have kept them there without orders. Will's aquatic bird area was the tour's first stop. In the spacious outdoor cages were several species of hawks and owls as well as specimens of the local bird population. Inside was a central pond area where storks, geese, ducks, penguins, and other waterfowl swam or preened, and around the walls were cages of various species of

The aquatic bird house at the zoo.

exotic birds. Many visitors lingered in the warm building with its lush foliage, and Beebe was in his element identifying and explaining what they were seeing.

Will's first Christmas as a wage earner was marked by lavish gifts to all his friends. To the girls in Nova Scotia he sent pearl pins, shirtwaists, and books. Nettie gave him books by Thoreau and the *Riverside Natural History*. As a Christmas treat he took his school friends Agnes Pelton and Winthrop and Eleanor Waite to a skating rink, then to his house for supper and to see *Ben-Hur* at a theater. "It was very realistic and we all enjoyed it. I paid $7.50 for the seats and enjoyed it all the more as it was my earned money." (12.28.1899) On the eve of the new year Will pondered the changes in his life and what the 1900s might hold. Nettie had just bought a booklet, *Right Living as a Fine Art*, based on the great Unitarian preacher William Ellery Channing's essay "My Symphony," and Will was so taken by its premise that he copied it into his journal:

> To live content with small means; to seek elegance rather than luxury, & refinement rather than fashion; to be worthy, not respectable; & wealthy, not rich; to listen to stars and birds, babes and sages with open heart; to study hard; to think quietly, act frankly, talk gently, await occasions, hurry never; in a word,—to let the spiritual, unbidden and unconscious, grow up through the common—this is my symphony.

For all his love of play, Will Beebe was earnest in his determination to add meaning to his life, and to sharpen his awareness of the spiritual. "To-morrow—Monday—a new week;—January 1st—a new month,—1900 a new year and <u>Century</u>. What a day for new resolves to be made on, and kept. . . .

—December 31st 1899—
I have spoken!"

CHAPTER 7

Widening Horizons

Keep your eyes on the stars, but remember to keep your feet on the ground.

Theodore Roosevelt;
Prize Day address to the Groton School, 1904

ONE OF WILL'S FIRST ACTIONS at the zoo had been to install his natural history pals in jobs. Seward Wallace came first as Will's unpaid gofer and eventually, after a series of random positions, rose to the rank of head keeper. Wallace Irwin filled in sporadically but never settled into a steady position. Will also hired his cousin Bob Van Zandt, whose impecunious father and alcoholic uncle were forever in debt, giving him $20 a month out of his own pocket until the zoo was willing to take over his salary.

Gradually Will's frequent visits to East Orange became fewer, and more of his friends made the trip to New York to see him. He loved showing them the zoo and treating them to shows or lectures. In February Eleanor Waite came for a visit, and they spent the afternoon in the American Museum of Natural History. "We talked—nay, we settled the fate of Universes!" he wrote. For Will, as for all his companions, lack of opportunity had joined with a thorough schooling in repression and self-discipline by wary parents to thwart intimacy. Despite circumspect behavior, though, East Orange gossip—unbeknownst to Will—had long paired him with Eleanor. But Will was not ready to settle down, even if his salary had permitted it; he vowed to himself to avoid the fleshly temptation that had nagged at him since puberty and that now, in the form of Eleanor, was "nearer than ever before." Strengthened by his mother's quiet encouragement—and less subtle vigilance—he resolved to "try" to resist. (2.18–19.1900)

In fact, Will was having great fun—and making a profit—with his work, and particularly with his photography, which may have been a more dangerous temptation than Eleanor. When he first took the job at the zoo, he rejoiced in the opportunities it would give him to take great pictures of animals. The famous naturalist C. Hart Merriam had bought several of Beebe's photos and had encouraged him to keep it up. Mr. Hornaday, however, was less enchanted by his young employee's hobby. "It is my opinion," Hornaday wrote to Beebe,

> that you are not sufficiently interested in your work as a whole. I fear that you are interested in outside matters, which should not be allowed to divert your attention from your legitimate work. You have no time for photography during working hours, and no matter what the object, I wish you would suspend it entirely during working hours.
>
> You still come to the Park late every morning, and leave shortly after four o'clock . . . and if you cannot hereafter report for duty promptly at 8 o'clock every week day, and remain at the Bird House until 5 o'clock, I will be obliged to reduce your salary to $80 per month. . . . I see no reason why you should not render the same hours of service as the other officers of the Park.[1]

Will presumably reformed: no more warnings are on record at the zoo, and his salary remained intact. But his energy could not be contained by captive creatures in a bird house, however spacious and elegant. Regretting his abandoned degree, he pored over his notes from his biology classes and formed an ambitious reading plan. He read Thoreau to nourish his spirit and collected muck from Lake Agassiz and Beaver Pond, two of the zoo's natural lakes, to study under his new, dearly bought microscope.

As a curator, part of Will's job was to breed and rear his birds to provide a natural succession of species. His youthful chicken-raising experience served him well here, and he reported many successes—as well as unexpected disasters. His smaragd duck killed her own nestling, which Will had been coddling and feeding by hand. Someone bumped an incubator, unwittingly turning the heat up and effectively cooking a large number of eggs of rare and precious species. And one day the water supply to the aquatic bird house abruptly ceased, and Gannon, the keeper Will had been allotted, discovered a two-foot eel plugging the intake pipe. But the East India ducks were laying by March, and Will put the eggs under a hen to brood. By April, one of the wild turkeys was tending a nest with twenty-three eggs.

By summer Will had trained Bob Van Zandt to work under him, and although the zoo had not yet taken over Bob's salary, Will felt confident leaving the birds in his care while he visited friends or took long weekends at the Peltons' constant house parties at the New Jersey shore. On these extended weekends, his parents and other old East Orange friends would join new friends from the city to sail, play tennis, and perform music, skits, and magic tricks for each other, as well as discuss subjects from New Thought to Will's old love, musical comedy. On May 15, Will and his cousin Paul Clark, Eleanor Waite, Agnes Pelton, and the girls' college friend Mary Blair Rice went up the Statue of Liberty, "and had a fine time. The view was beautiful from the crown." This is the first ref-

Will, right, and Harry Macdonough, ca. 1899.

erence in Will's journal to the woman who would become his wife; soon
Mary Blair would have recurrent cameos there.

Life at the zoo, though becoming routine, still held its moments. On
a blisteringly hot day in June, Will recorded "the most exciting experience of my life." Whistles shrilled from several keepers at once, and he
ran with them across Baird Court after a large bear that had escaped.
The bear took to the water and swam back and forth across the Bronx
River, Loring and his men swimming after him with a lasso. The bear
scrambled into a thicket, and Will and Hornaday began to work slowly
toward him, trying to get a noose over his head.

> Three or four men on our right with boat-hooks & several with sticks
> & pitch-forks on our left protected us against his rushes, or tried to. He
> rushed 40 or 50 times and again & again I thought we were both goners.

The men with the boat hooks panicked and dropped their weapons
and the bear escaped, only to be cornered in the woods. When Beebe
and Hornaday tried to lasso him, he turned and bit Hornaday's hand
clear through. They managed to noose him at last, but the assistants
pulled so hard on the ropes that the bear was strangled.

When the panic had subsided, Will shamefacedly admitted to
Hornaday that he had been so frightened that he felt sick. The older
man reassured him that "no coward could have stood the bear's rushes
as I did." Although they were sorry to lose the valiant—and valuable—
animal, Hornaday's savaged hand and Will's trousers were the only
other casualties.[2]

That summer's too-brief trip to Nova Scotia included his friend Joe
Fobes as well as Florence Stevens and her mother, a close friend of
Nettie's. Florence was a good sport who played and sang and entered
into all the boys' work with zeal and good humor. She and Will sneaked
off to the woods a few times to play "Truth or Dare"—the era's version
of strip poker—in a secluded "pine-needly" glen, often returning after
dark with hair and clothing awry.

Will spent New Year's Eve at Mrs. Pelton's in Brooklyn with Eleanor,
Winthrop, and Agnes, who was studying painting at the Pratt Institute.
He greeted 1901 on their roof, watching fireworks and meditating on
resolve. January also brought raises for both Will and his father. The
senior Beebes were planning a much-anticipated trip to Europe, and

Will celebrated by taking Mary Blair Rice to *Floradora,* a well-known musical revue that was breaking theatrical ground on the New York stage with its proto–chorus line of six perfectly matched "Floradora girls" and their accompanying swains. One of its songs, "Tell Me Pretty Maiden," had become wildly popular. This was the first time Will and Mary Blair had been out alone together, and they apparently liked the experience: in the next month they saw *Floradora* three times, as well as the vaudeville team Weber and Fields, and Julia Marlowe in *When Knighthood Was in Flower,* which had "the finest duel I ever saw on the stage."

At this point, pages have been torn from Beebe's journal—a regrettable practice he was to follow all his life when he had written of personal matters. Since he often shared his journals with friends, it is not surprising that he censored his softer side. He never flinched from recording his mistakes, or even sentimental spiritual effusions, but whenever his thoughts turned to romance, he evidently wrote about it as usual but edited himself ruthlessly. So when Mary Blair Rice entered his life, the first of several jagged hiatuses appears in his narration of events.

Mary Blair Rice, ca. 1900.

On August 15, 1901, Will set off for his first expedition with zoo funding. With Charles and Nettie in Europe, Will headed for Nova Scotia with Weary Irwin, who was attending Penn, and Ferd Knolhoff, who was working for a bank in New York after having failed to get into Harvard. They collected anemones and urchins from the rocky tide pools, and dredged for animals and plants from the depths. Will tried to capture some live fish for the aquarium but had better luck finding interesting specimens to dissect. His fascination with embryology and development, spawned by the previous year's starfish, incited him to seek out eggs and larvae; from a female dogfish he salvaged some fine embryos that kept the collectors occupied for hours.

With the responsibility of collecting for the zoo on his shoulders, Will worked more and socialized less than on previous trips. To eliminate the chancy commute by boat from Weymouth Bridge to their favorite collecting areas, they moved from Mrs. Journeay's comfortable lodgings to a boardinghouse out on isolated Digby Neck. This departure from his accustomed social routine in order to maximize work time marked a sea change in Will's attitude. He began to find the escapades of Weary and Ferd troublesome, and in letters to his parents told of their exploits in humorous but rather impatient terms. Eventually Ferd, having committed the indiscretion of kissing Kitty Macdonald—twice—went home early, his life having been made a living hell by the teasing of the other two. Will lent him money to go, and had to lend Weary funds to stay.

When the two collectors headed back to New York, Weary remained with Will and was both a good companion and a trial, smoking and drinking and complaining about bedbugs—a complaint Will scoffed at, not denying their existence but denying that they would bother anyone who had the least sense about avoiding them. The cockroaches in his office at the zoo, however, were a real nuisance: they were eating his books. For them he called an exterminator.

Nettie wrote that Charles was working on an order for 6,000 tons of talc, and that he would get it if they had to stay abroad till Christmas. She described the royal palaces and gardens they had visited, but reminded Will sententiously that he belonged to a different sort of royalty, the intellectual and moral elite. After all, Will was home alone—or, worse, with Weary, who smoked and drank. And perhaps consorting with girls whose parents and upbringing were unknown. "You have the divine power within you to resist temptation & do right no matter what others do," she wrote. "I am so thankful and proud of you."[3]

Ten days later, the city was in an uproar over the shooting of President William McKinley in Buffalo. New Yorkers felt that an attack had been made on them directly. "I believe he is still living, but they have had to call out the entire Buffalo police to keep the mob away from his assassin's jail," Will wrote his parents. "They are trying to lynch him."[4]

On September 14, Theodore Roosevelt, an object of Will's admiration because of his intrepid explorations, took the oath of office. "Everyone says Roosevelt will be the next to be assassinated," Will wrote, "but he carries a revolver in each hip-pocket."[5]

CHAPTER 8

Nestbuilding

*[T]he woods abound with full-sized but awkward young birds,
blundering through their first month of insect-hunting and fly-
catching, tumbling into the pools from which they try to drink,
and shrieking with the very joy of life. . . .*

William Beebe, *The Log of the Sun*

IN APRIL 1902, NETTIE RECEIVED
a cryptic letter of condolence from her cousin Anna in Amsterdam,
New York. Just back from a visit to the Beebes, Anna told her dear
Nettie how she has "wished I could put my arms around you and tell you
how sorry I am for you . . . my hope is that something may yet inter-
vene to make a change in the conditions, as they were when I left. . . .
Try to keep up your courage, my dear, hoping for the best, at least til the
worst comes." Her wisest counsel she saved for last: "If come it must
don't make up your mind too firmly what you will do in such an event."[1]
The event in question was Will's engagement to Mary Blair Rice.

It would have been difficult for a mother as devoted as Nettie to rec-
oncile herself to any daughter-in-law, but her objections to this particular young lady must have been great. Habitually mild and used to
approving whatever Will did, she felt so strongly that she took the risky
step of writing to her son in strong terms. Though the letters do not
survive, she seems to have felt that there were many problems with the
match. For one, the Rices were dyed-in-the-wool Southerners and had
owned slaves, many of whom still lived on the family plantation as ser-

vants. Given that Mary Blair had grown up in privilege and ease, Nettie feared that she would be unwilling or unable to care for Will and minister to his interests as Nettie's idea of a proper wife should do. In addition, she seems to have had qualms about Mary Blair's moral character that put Will on the defensive. In March, he wrote—which was odd in itself, as they lived together—that he was going to Ithaca to visit his friend, the artist Louis Agassiz Fuertes, because he was feeling tired and more than a bit browbeaten.

> Please remember that I love you with all my heart, and believe I am trying to do right. . . . I love you and Papa lots and lots. (kiss, kiss). Trust me, Mama, that I will never say anything to Mary about the letters, for I know you believe it was right. I will see if we can't both think such grand, true thoughts, & not all suggested by me, either, that you will live to see that she is good & lovable in some ways. I will think of you as happy no matter what you say, for if we both do our best to do right we must be happy.[2]

A month later Will was writing his parents from The Oaks, the Rice family plantation on the Staunton River in Virginia.

The house, which still stands largely unchanged, was imposing and square rather than graceful, with a grand double stairway and two large wings and an attached chapel complete with stained-glass windows. It stood on twenty-three acres of tobacco fields. City-bred Will was

The Oaks, the Rice plantation at Coles Ferry, Virginia.

impressed and a bit awed, particularly by the black ex-slave servants, who were feudally cheerful and solicitous of the family's comfort. The hospitable Rices made him welcome: with Mr. Rice he went quail hunting, and Mary Blair's irrepressible brothers, Harry, Theo, and Roger, showed him the property and every creature that inhabited it. They hunted fossils and shot opossums and went fishing in the smooth, muddy Staunton.

Henry Crenshaw Rice, Mary Blair's father, was a gentleman tobacco planter and farmer who served in the Virginia legislature. He had fought in the Civil War with his college regiment, the "Hampden-Sydney Boys," and then in the Fourteenth Virginia Cavalry, and had lost a brother in the conflict. He had married his cousin, Marie Gordon Pryor—the daughter of General Roger Atkinson Pryor, "the loudest mouth in the South," and his equally spirited and opinionated wife, Sara Agnes Rice Pryor.

After the war, completely impoverished, with his Virginia property destroyed and six children to feed, General Pryor had begun a law practice back East. After several years of poverty Pryor began to make a name for himself, eventually becoming a justice of New York's Supreme Court. Sara Agnes wrote articles for various magazines and eventually two fascinating books, *Reminiscences of Peace and War* and *My Day: The Story of a Long Life.*

Marie Gordon Pryor, their daughter and Blair's mother, had been educated side by side with her brothers by her intellectually formidable mother. Known as Gordon, she read Latin and Greek as well as French, and her father expected her to be the most intellectual woman in the South. In New York, Gordon graduated from Brooklyn's prestigious Packer Institute. When she surprised her transplanted family by marrying Henry Rice and moving back to Virginia, she varied the duties of plantation life with frequent trips to her parents' gracious home in Brooklyn. By the time Mary Blair Rice was engaged to Will, her mother, Gordon, had published several articles and book reviews in *Macmillan's* and the *New York Times* and, like her own mother, was well known in New York intellectual circles.

Gordon made sure that her country-bred daughter received a thorough education, though by Mary Blair's account she was more grounded in the life of the plantation than her peripatetic mother and grandmother had been. Sensitive to the insularity of life in the postwar South, her parents decided that she needed the finishing touches of a New York college. She was a student at the new Pratt Institute in Brooklyn when their mutual friend Agnes Pelton introduced her to Will Beebe.

The Pratt Institute had been founded in 1887 by Charles Pratt, who had amassed a fortune making paint and wanted to endow an institution to teach the practical arts and sciences. The course of study for women at that time centered around art, teaching, and domestic science, as well as more technical careers. The "normal" track, in which Mary Blair was enrolled, prepared teachers for secondary schools and was exceptionally rigorous: requirements for admission included examinations in algebra, geometry, physics, and physiology. The two-year course itself required botany and bacteriology, zoology, physiology, chemistry, physics, dietetics, drawing, psychology, emergency medicine, cookery, and sewing.

Will found Mary Blair fascinating. She was a vivacious, intelligent, well-read, and well-spoken young woman with strong opinions, yet with all the then-proper feminine graces. Her Southern accent would have lent charm to whatever she said. She was poised and accustomed to command, yet growing up with boys in the country had made her easygoing and keen for adventure.

In Will, Mary Blair would have seen a tall, wiry, earnest but humorous and self-deprecating young man, intellectual but fun loving and adored by his friends. He was a respected scientist and a published author who loved vaudeville and music. Perhaps most important, his respect for his mother had made him think of marriage as an equal partnership, and he wanted a mate who would be a best friend as well as a lover.

Despite his citified air and Yankee accent, the Rices adored Will from the start. His good-humored, gregarious nature and love of the outdoors and sports endeared him to Mary Blair's extensive family, and his knowledge and responsible position appealed to her parents.

The wedding took place at The Oaks at dawn on August 6, 1902—according to the *Richmond Times* of that date, "because of the freshness and beauty of the day at its beginning." The oppressive heat of a Virginia summer may also have figured in the early hour. The elder Beebes were notably absent. Will wrote to his mother the day before:

> When you get this I will be "all done," but please always remember that
> my love for you is very different from that for Blair; if they could be

compared yours would never be less. You are so good to me that when I think of any of my blessings, it always works around to you in the end (except my stomach at present).

I will be just as good as I possibly can and will always try to keep up to my highest ideals.

Don't worry about me. I will take good care of myself and not get sick.

My very best love to Papa and you. Will.[3]

The *Richmond Times* played the story of the local belle and the Yankee ornithologist to the hilt. The front page accompanied a rather hard-featured sketch of the bride with an article, "The Rising Sun to Witness Novel and Romantic Wedding." The next day the story made the front page as well, under the headline "Sweet Birds Sang Wedding Hymn."

The party passed around the box-bordered flower beds to a point facing the east. There they divided, leaving a space beneath two arching pink crepe myrtle trees for the bridal pair, who soon appeared and took their stations facing the rising sun. Its rays smiled upon them during

Will and Blair, August 6, 1902.

the reading of the marriage service of the Episcopal Church, and soon
disappeared behind a gray curtain of clouds. During the ceremony the
garden was vocal with the singing of birds, and fragrant with the fresh
odors of the dawn.

The paper goes on to list a rich assortment of wedding gifts from
prominent friends such as Cyrus McCormick, of reaper fame, whose
daughter also gave the pair a pianola and a library of music. The couple,
it said, would live in New York, "where Mr. Beebe is well known in
scientific circles as a writer and lecturer. He holds the position of
ornithologist of the New York Zoological Park, and is engaged upon a
book on ornithology, to be published by Henry Holt and Company."[4]
For their honeymoon Will and Blair traveled to Weymouth Bridge
and lodged in the familiar boardinghouse, leading Mrs. Journeay to
facetiously suggest that Will's room, formerly known as Gehenna,
should be renamed Paradise. Almost immediately, the newlyweds began
what would be the pattern of their life together: back in his native element, Will set out early on their first morning with Mary Blair (whom
he, at her insistence, called Blair) in a boat to begin collecting. They
took photos of periwinkles and flickers and collected barnacles, which
Blair, in a rare moment of humor, referred to as "carbuncles on Digby
Neck." They went rowing through dense clouds of Aurelia jellyfish and
collected the tiny parasitic shrimp that live in them. Will showed Blair
eagerly how, white and transparent when they first came from their
hosts, the shrimp soon begin to turn reddish brown, the pigment growing almost perceptibly.
For the honeymooning couple there was a busy social schedule, similar in many ways to the Weymouth of old. There were tennis, ping-
pong, and whist, calls to pay on old friends, and myriad dances to attend.
Will wrote home that Blair was a great success: "[E]veryone likes her
here & she entertains our numerous callers finely." They went on mara-
thon fishing trips: one day they caught thirty-four flounders ("Blair
caught the most"), then took in a cricket match, then sailed the boat up
to Gates Falls, where they basked in the moonlit loveliness of the scene
and watched a barred owl soar silently through the tops of the dense
spruces, black in the moonlight.
In addition to being blessed with Southern social skills, Blair proved
a great help with Will's ever-present writing. He wrote his parents that

she was "a marvel in criticizing." He had dictated to her part of a book he was planning on bird anatomy, and boasted that "the way she corrected it will make it very easy for me to get through the whole thing in less than a year." His well-intentioned praise may have aroused more jealousy than admiration, though, in Nettie's lonely breast.

In fact, in each of his letters home, Will seemed to be trying to present a different facet of Blair's excellence: she was well liked by the family's oldest friends, entertained politely, had good aim with the shotgun, edited beautifully and so would help with his all-important work—and she could even play the mandolin. Hoping that even obdurate Nettie would be unable to withstand such a barrage of virtues, Will beseeched his parents to come up and join the fun. "I love you with all my heart, am trying to do right, & Blair sends her best love. <u>Please</u> come." [5]

At last Nettie's devotion to her son overcame her scruples, and she relented. She and Blair had a week to get to know each other better before Will had to return to the zoo to resume his position. It was early to leave, he wrote, "but I have so much happiness ahead of me that I did not mind it as much as other years." (9.1.1902)

In the city the newlyweds moved into a top-floor apartment in the Sans Souci, where Will had lived with Charles and Nettie. Will set about making their marriage and their home match their imagined ideal:

> Last night Blair and I slept for the first time in the top apartment. We have fixed it up very nicely, and my books & collections make it the home of a Naturalist. My work room is light, & we can see the N.J. Shore from our windows. The pianola is perfect and we have some 15 or 20 rolls of music. My newspaper articles are coming out regularly and I hope to be able to do a lot of work this winter. (9.23.1902)

For a few months the couple continued taking their meals downstairs with Nettie, who would otherwise have been very much alone. A sort of uneasy camaraderie arose between the two women, who really had little in common. Charles was still away frequently, finding new markets for the fibrous talc he had developed, which was making his company a force in the trade. As a result, the inhabitants of the little honeymoon suite upstairs, and the bereft mother downstairs, were at once allies and competitors. Not surprisingly, the catering arrangements soon became a source of mild friction.

Will's life of journal-keeping and correspondence was paying off in the astonishing facility he had with words. He was becoming a sought-after writer for newspapers as well as for the burgeoning magazine market that was bringing knowledge and culture into the homes of legions of readers. Always on the lookout for animal stories, which never failed to generate human interest, editors quickly found that the young ornithologist who wrote so easily of his romantic outdoor life and the creatures that populated it was a favorite of readers everywhere.

As his journalistic career blossomed, however, a new career as a book author was also demanding attention. In November he finished his first book, which Henry Holt had contracted to publish, and began to discover how much time was involved in the preparation of a manuscript.

> Yesterday I took my book to Mr. Holt. There are 110 illustrations. Blair and I have rewritten and typewritten the 43,000 words and printed and mounted the photographs. . . . This first one is dedicated to Mama, and the next one will be to Blair. . . . We have been eating in our own apartment for some time and we are both trying to be all that our ideals yearn for. (11.21.1902)

This volume, tentatively titled *Notes of a Bird-Keeper,* proved not to be the comprehensive tome Holt had in mind. So Will forged ahead, undaunted, on a major work on birds that incorporated and expanded the earlier manuscript. Meanwhile, he was contributing articles on creepers and nuthatches, juncos and nighthawks, and the wild turkey to Chapman's journal, *Bird Lore,* and to the zoo's bulletin. One of his most popular articles, "Enterprising Eagles," described two bald eagles at the Bronx Zoo who were hard at work incubating a stone. Complete with photos, the article was widely reprinted and brought many visitors to the zoo to see for themselves.

In 1901 alone, he contributed two articles to the *New York Evening Mail,* seven to the *Evening Post,* and two to the *Bulletin of the New York Zoological Society,* bringing his total journalistic output for the year to seventeen articles. "Feathered Ocean Waifs" was about migrating birds, blown off course, that had been sent to the zoo. "Strange Friendships" detailed life in the great flying cage; "Evolution Study in the Park" illustrated the relationship between reptiles and birds; "Some Snake-Killing Birds" described the feeding habits of roadrunners. All of these intriguing, well-crafted pieces generated interest in the zoo and helped bring Beebe, already under the wing of the great Henry Fairfield Osborn, to the attention of the society's leadership.

Fortunately for Beebe and the animals under his care, he was so adept at his duties that he had a great deal of spare time in which to write. His experience in the field, his emerging leadership qualities, and the ease with which he picked up new techniques and ideas made the job a snap. And for the first time, it began to dawn on him that the rest of the world was not necessarily as quick, as energetic, nor even as lucky as he was in that respect. He could not help noticing that Ditmars, for instance, would work three weeks on one agonizing lecture that Will could have prepared in an hour and a half, or that, although Loring had worked doggedly to get and keep his position, he had lost it in the end. It gave him pause to think "how easy my position is for me & how little real study I put into my college work. . . . I suppose the very realization of these things at present shows that I am waking up, & I am very glad." (3.5.1901)

PART II

Ornithologist

Flying South

While in my own estimation my chief profession is
ignorance, yet I sign my passport applications and
jury evasions as Ornithologist.

William Beebe, *Edge of the Jungle*

THE VERY EASE OF HIS JOB WAS
an irritant to Will's restive nature. Rereading in January some of his
high school journal, he realized that the zoo routine was robbing him of
his youthful joy in nature. The icy winter kept most of the birds
indoors, and Will's job primarily consisted of doctoring his birds, corresponding with other ornithologists in an effort to increase the zoo's
collection, and writing up his experiments on feather coloration.

> Sometimes fears of having lost a little of my . . . interest come upon
> me, but I think it over and realize it is only the sobering over of all my
> indoor work and reading. When I remember the delight of seeing
> mockingbirds and Ajax papilias in Virginia last Spring, I know there is
> underneath the feeling of love for all Nature as strong as ever. Ah, . . .
> the love of the Universe will ever be strongest in me, overbearing in the
> aggregate all technical science. (1.26.1903)

He pined for the warm, fecund South, in his imagination teeming with
life and interest, with theoretical puzzles to solve and mysterious creatures to discover.

This brief but intense testimonial turned out to be prescient. Despite the couple's attempts to live up to their ideals and to think "right thoughts," Will continued to fall victim to the respiratory ills that had plagued him since childhood. A serious bout of abscessed tonsils and what doctors termed "false diphtheria" prompted a decision on the part of zoo officials—acting under a mandate from Osborn—to send Will away for the worst part of the New York winter. At that time a septic throat could be deadly, and Osborn was taking no chances with his protégé.

However miserable the throat may have made him, Will was ecstatic at the result: he was at last going—under orders—to the warm climes he had dreamed about since infancy. Although the true Spanish-speaking tropics still eluded him, he and Blair were "sentenced" to Florida, and they prepared eagerly for another honeymoon. Florida, particularly the Keys, was a paradise for biologists. The American Museum's Frank Chapman and the popular naturalist John Burroughs had written extensively about it as a biological Eden. Both had stayed with the legendary Mrs. Latham near St. Augustine, who ran a sort of scientific boarding-house. A self-taught naturalist, she knew every plant and creature along the Indian River, and acted as guide, counselor, and nurse for her "children," as she called the continual flux of researchers who landed at her door.

Will and Blair set off on February 12, 1903, on the steamer *Comanche* from a New York dock veiled in swirling snow, and sailed for two days through a sea so rough even the captain was sick. "Blair was not sick," Will recorded proudly, "and of course nothing could affect me." As the ocean calmed and the air warmed, Will spent more and more time on deck, watching for birds and studying the water. He was enchanted by the sight of his first live flying fish, a tiny creature no more than two inches long, which skimmed the surface beside the great steamer as it labored through the swells.

At the railroad station in St. Augustine they met the beloved nature writer and philosopher John Burroughs himself, white-bearded and imposing, sitting on his luggage and writing intently. A boat took them all up the Indian River to Mrs. Latham's. "She is a thin, weather-beaten, hearty, hospitable, keen, intelligent, sympathetic woman—one in a thousand," Will wrote. "Her experiences are wonderful, & I am making many notes." (2.16.1903) The palms alone—"and not in pots!"—made Beebe ecstatic, and the birds and butterflies and lizards that inhabited them seemed miraculous. In a letter to Charles and Nettie, he tried to describe the scene at Mrs. Latham's: he was writing with his back against a cabbage palm, Blair was knotting a butterfly net, and Burroughs wrote on the porch. A visiting collector had just returned with two birds he had shot. Carefully folded within his letter are several carefully pressed flowers and a bit of coarse palm fiber.[1]

As Will's strength gradually returned, he and Blair began to make forays alone to the shore and swamps. Burroughs took an avuncular interest in Beebe, and they had long discussions about the duty of the naturalist and writer to present nature accurately, and in ways that would educate and generate interest in conserving threatened areas and species. When Burroughs left, his rooms were taken by the nature illustrator Charles Livingston Bull and his wife, Fanny. The two young couples got on well together; they frequented the shore, fishing and looking for sea birds and their nests, and took boats upriver to find new species of marsh birds and other animals. Will captured live creatures for the zoo, which Bull sketched; and Bull, a skilled taxidermist, shot and skinned specimens to be mounted and painted back home in the action scenes he was becoming known for.

Harpy eagle, by Charles Bull.

Blair wrote her own description of a trip to a small offshore island to observe a pelican colony. Her account paints a portrait of the Beebes' life together in the field, and provides an insight into the gradual metamorphosis of the Southern belle into a dauntless explorer and travel writer.

> We took some pictures — of nests and eggs — of baby birds — of half grown birds — of birds on their nests and also several Kodak snaps and some pictures of the birds flying. I carried the bag of plate holders and handed them to Will when he was ready. I was so excited that my hands trembled so I could scarcely open the bag. Will took two pictures of a bird on her nest that he thinks will be fine and he deserves to have them good for he lay on the ground a long time waiting a favorable moment to pull the shoe thread and thus take the picture. . . .
>
> Friday the twenty-seventh was a memorable day to me because it was the day of my first plunge into the Atlantic ocean — my first plunge indeed in anything larger than a bathtub. Oh! how I did enjoy it! I did not realize it was such fun. (2.27.1903)

Though Blair enjoyed escaping the confines of her domesticated existence, not all of her forays into the wilderness were so idyllic. The combination of the natural discomforts of their rugged life and her husband's goal-driven personality made setbacks difficult to overcome with equanimity. When she fell off their sailboat head first, her rubber boots filled with water and her encumbering skirts were soaked; Will had to drag her back in, laughing, "a forlorn but funny sight." When rain robbed them of a day's collecting, or the savage mosquitoes made exploration of the woods impossible, he feared for the expedition's success. "This morning was rainy and dismal," she writes of a day when Beebe was obsessed by his unmet aims. "We were plunged in the depths of despair. It was doubly hard for me for I felt so bitterly Will's disappointment." (3.1.1903) Oil of citronella helped ward off the mosquitoes somewhat, but the threat was no mere annoyance: yellow fever and malaria were endemic in the area, and even seasoned explorers were wary of being bitten.

They managed a brief, eye-opening excursion to the Keys. Nothing in Will's long acquaintance with Nova Scotian tide pools had prepared him for the riches of Florida's warm tropical coast. The endless variety of chitons, anemones, holothurians, and the tiny fish that coexisted with them opened dreamy vistas of endless, blissful research to the enchanted Will. His first trip out to a coral reef was an epiphany. Through water transparent as glass he watched, breathless, as the unimaginably varied life of the reef community went on undisturbed beneath him. A school of black angelfish swept into view, careering slowly on their sides and looking up at Will and Blair, their graceful streamer fins curving in folds as they swam. They passed over a Portuguese man-of-war drifting along, its poisonous tentacles trailing far beneath, its crystal float glistening in the sunlight. "Its tiny bow was a deep blue, its stern green, while a purple stain ran along the sides, and a glint of pink was reflected from the sail. There is something about these little animals which reminds me of old-fashioned galleons." (3.18.1903)

Here he discovered for the first time the wonders of the water glass, a bucket with a disk of glass inserted in the bottom, which could be sunk to its rim to act as a window into the world below. As they neared the reef the transparent surface grew disturbed, but through the water glass they could still view the incredible creatures of the coral reef: mountainous brain corals, surrounded by waving fronds of sea-feathers and sponges, fish without number in all the colors of the rainbow—sometimes all on one fish. To call the water glass a window, Will exclaimed, did not do it justice: "It is magical, the nearest approach to an optical miracle I have ever seen." Brilliant blue tangs swept majestically by schools of black and yellow sergeant majors that darted back and forth with one mind. One fish was so unlikely in its colors that he sketched it: blue tail, then a wide band of green, then narrow bands of black, pale blue, and black, and a blue head. "What could seem more artificial than such a gorgeous poster fish!" (3.18.1903)

The time allotted for the Keys seemed wretchedly truncated; Will had to be back by the first of April. But he was healthy again and looking forward to summer. The next months sped by in a whirl of social engagements. There were dances and parties with the Columbia set, and picnics with their new friends the Bulls. Will's position as curator of birds at the fledgling zoological park, and his growing reputation as an engaging writer for several newspapers and magazines, made him a sought-after lecturer and guest at gatherings of the literati, as well as the charity events of the "glitterati."

Osborn's dream of using Will as a magnet for publicity was coming to fruition even faster than that great visionary had foreseen, and he became increasingly protective of his young lion. To Hornaday's considerable annoyance, Osborn insisted that Will be moved—for reasons of health—from his tiny, dark office in the bird house to other accommodations. Hornaday, whose pedestrian attempts at writing cost him great labor but were barely noticed by the critics, had little patience for the too-fortunate youth who dashed off polished and engaging articles and books in what must have seemed an effortless stream to the hardworking director. He grudgingly agreed to let Will move into a small "portable house" that had been set up in front of the bird house to accommodate work during construction, which the great labor strikes of the period were delaying indefinitely.

One of the little prefab's first visitors was a young Wall Street lawyer with a consuming passion for birds and—most important—a sailboat and a piece of property on one of Virginia's barrier islands. Louis Whealton proposed a week's tour through the islands near Cape Charles, where gulls, terns, skimmers, ospreys, plover, and oystercatchers breed. Beebe was enchanted with the scheme but had given up his summer vacation in compensation for the extended winter leave in Florida. Whealton's potential as a donor to the zoo, however, coupled with the possibility that the area might be a valuable source of photographs and specimens, turned even Hornaday's heart. Will was given permission to go under the aegis of an official expedition for the zoo.

Expedition or no, "this is our third honeymoon in a year," Will wrote excitedly. "We are to have several tents, cook, etc, a guide and sailboat, and spend a day or two near each colony. . . . I have a new camera, plates, etc and hope to learn a lot." (7.11.1903) Whealton's "sailboat" was actually a sixty-foot sloop with two masts and a jib, complete with crates of live chickens and two tons of ice. This would, Will wrote presciently, be their first trip "travelling like millionaires." It would not be the last.

When they reached the barrier islands they set up camp on Cobb Island, once a popular resort of the wealthy until erosion ate half of it away, and with it the hotels and houses. The lifesaving station was the only building remaining, but nature had repopulated the island thickly with nesting seabirds. Colonies of gulls filled the air with cries. Herons, sandpipers, and curlews waded in the marshy shore, and secretive clapper rails exploded from the undergrowth. A black skimmer performed a pathetic broken-wing display to lure them away from its nest and young. On tiny Morcou Island, a sudden rainstorm caught them off guard as they photographed a lively heron rookery. They scrambled to erect a makeshift hut of seaweed and an old umbrella to protect the camera, but finally had to make a run for the boat, the precious camera clutched awkwardly under Blair's wet and sandy skirts.

Each day Will got up at five-thirty, swam, skinned the previous day's birds, and wrote up his notes before breakfast. Birds were everywhere, and remarkably tame. They photographed and made exhaustive records of their breeding habits and feeding patterns for use back at the zoo. Blair learned to swim her first tentative strokes, and when they left to board the sleeper back to the city, Will deemed it a week of "splendid experience, and enjoyment, which red-legs [the local term for chiggers] and fleas and mosquitoes & rain could not spoil." (7.20.1903)

Sorting through his booty from the trip, Beebe discovered that despite rough treatment and cold temperatures, some of the eggs he had brought back to preserve had living embryos. He put them in an incubator and hatched two terns, then a skimmer, a laughing gull, and a little green heron, "pot-bellied, and with long yellow feet." He began feeding the nestlings killifish and tiny shrimps, carefully boiled and mixed with ant eggs. At first he had to tease the tiny bills open with forceps and use a medicine dropper. As they grew, he had to work quickly to appease their voracious appetites.

Beebe devoted himself to the small creatures, whose progress he described with the passion of a proud parent. He named them, made careful notes on their behavior, and worried constantly about their health. After two weeks Samson, one of the terns, died. "I am more sorry than I can write, I had become so attached to these tiny fluffs of life," Will lamented. True to his calling, the ornithologist dissected Samson carefully. Finding that the tern had had a digestive problem, Will castigated himself for not having figured this out in time to help, "but the others are in good health & I have learned a lesson." (8.1.1903)

He wrote up his experiences as a surrogate parent for the nestlings in an issue of the New York Zoological Society's *Bulletin,* and the article's appealing candor, coupled with scientific detail, won him accolades among the society's membership. He also wrote an article on the Cobb Island expedition for the zoo's annual report, which dwelt on the pressing need for conservation of the fragile seabird colonies along the

Eastern seaboard. Citing records of their presence in vast numbers in the time of Henry Hudson and Captain John Smith, Will was quick to assign blame for their decline. "The reduction of these beautiful creatures to a pitiful remnant, has for its cause the robbing of untold thousands for their eggs for food, and the worse craze for adornment which has sacrificed cart-loads of breeding birds, to gratify an instinct . . . harking back to savagery."[2]

In September, Will and Blair moved out of the Sans Souci into a smaller apartment at 2307 Loring Place, University Heights, within bicycling distance of the zoo. With only five rooms to care for and a rent of $20 a month, they felt they could travel and still save money; Blair must have been able to breathe more freely, too, out from under the watchful eyes of her well-meaning in-laws. She was working diligently on her drawing, which she hoped to turn into a career. In the evenings, after long hours at the zoo, Will worked steadily on his bird book. *Ladies' Home Journal* had published "Night in a Zoological Park" and had already paid him $100 for "City of the Birds," another short article on his work at the zoo.

By November, Will had completed the hardest part of his bird book—26,000 words on anatomy. He wanted to make the subject as clear and approachable as possible while still being accurate, building on such classics as Elliott Coues's famous but formidable *Key*. He hoped that his book would lay to rest any challenge to his academic status as a serious ornithologist, which his detractors—including Hornaday—accused him of sacrificing to the gods of more lucrative writing.

With the arrival of winter, however, came the sore and swollen throat that was becoming Will's yearly bane. This time Hornaday bowed to the inevitable, and the board voted to give Beebe another leave, from mid-December to April, with one month at full pay and the rest at half pay. Seward was given all Will's work with the birds, at $75 a month. Will wrote his annual report—a duty exacted from all the curators—to be circulated to the board and all of the zoo's contributing members. He also polished off several magazine articles he had promised, along with a number of newspaper releases. All told, twenty-six-year-old William Beebe had published more than thirty-four articles and photos that year. More important to Will, his achievement in science brought him election as a fellow in the American Association for the Advancement of Science.

Migration

A summons has pulsed through the finer arteries of
Nature, intangible to us, omnipotent to the birds.

William Beebe, *Two Bird Lovers in Mexico*

Hıs APPLICATION FOR LEAVE
granted, Will was determined to press farther south. Frank Chapman
recommended Mexico for its variety of topography and bird life. On
December 18, 1904, Will and Blair set off on the SS *Esperanza*. Accompanying them were several trunks, clothing bags, and suitcases, two
folding cots, and the massive impedimenta of a collector and photographer. Will had packed his Kodak and his 4 x 5 Premo, twelve dozen
films, and twenty-four dozen glass plates. There were the scores of
tubes and vials he never traveled without, along with secure cases to
carry them in, as well as assorted nets, dredges, and chemicals. He had
treated himself to a "beautiful" 28-gauge double-barreled shotgun and
300 shells, besides reloading tools and extra powder and shot. He also
brought his .22-caliber rifle and shot cartridges and his .38 revolver. For
Blair he had purchased a small but effective revolver. Madison Grant
had obtained a letter of introduction for the Beebes signed by Secretary of State John Milton Hay, an impressive document addressed to
"All Consular and Diplomatic Officials of the U.S." whose golden seals
were guaranteed to facilitate their passage through customs.

The winter-weary couple brightened with the slow progress of the steamer as they left New York, with its coal-choked air and dreary December skies. They felt the air gradually warm and watched the water change from slate gray to emerald. The boat docked in Veracruz on Christmas Day, and a train took them 600 miles west to Guadalajara, where they were joined by the Bulls.

The Bulls arrived with a Christmas box from Will's parents that held a little silver purse for Blair and a $10 gold piece for Will. In the evenings, Blair wrote, they went down to the Plaza to hear the music.

> The town at night is <u>fascinating</u>—so weird and so picturesque. . . . I <u>love</u> it all—the strange figures in their big sombreros and gay serapes, the lights—the fragrance of orange blossoms and [gardenias]—the cloudless sky bright with more stars than we ever see in the north—the balmy air and the music.
>
> We went into the Crystal Palace and had strawberry ice which was delicious and only quince centavos a plate. (1.12.1904)

Beebe hired a young Mexican to go with them as translator, guide, and chef, and in mid-January the ragged little line of horses and mules left for Colima, the great volcano that was their goal. Will and Blair

The pack train.

rode horses, and the baggage—cameras, photographic plates, chests of medicines and collecting vials and chemicals, guns, baking oven and its accompanying pots and pans, plates and cups, tents, folding cots, suitcases and steamer trunks of clothes and books and blankets—were stowed on the mules' backs. Although she had never learned to ride, Blair was a natural horsewoman. "Will had a fine spirited black horse and mine was <u>very</u> nice," Blair wrote. "We enjoyed the ride immensely and Will and I had several gallops. I just love horseback and I have not an atom of fear." (1.16.1904) The Southern belle was relishing the rough life of a naturalist's companion in arms.

Their first campsite was at the bend of a rushing river, at the bottom of one of the barrancas, the deep cool gorges that wind through the Mexican Plateau. To negotiate the treacherous hillside, the sure-footed mules had to feel their way at each step, and the riders leaned far back in their saddles to help the horses labor down the slopes. At the bottom, the horses and pack mules had to wade, and in some places even swim, to reach the tiny sandbar.

Solitary as it was in the barranca, the place teemed with animal life. Two large black hawks were nesting in a tree on the hillside above the camp, keeping the resident sandpipers and killdeer wary; insects whizzed by, and lizards rustled among the dry leaves. Diminutive canyon wrens filled the gorge with silvery notes, and a pair of green Texas kingfishers went fearlessly about their business. Meals at the camp

Dinner in camp;
Will and Blair on left;
Charles and Fanny
Bull opposite.

tended to reflect whatever had been shot that day, whether it had been collected to be skinned and mounted, in which case its flesh was available for the cook, or shot for meat. One night they had fried iguana for supper, "with the 40 or 50 large yellow eggs which were in it, and bird pie, consisting of jay, parrakeet [sic], & two species of doves. They were all delicious." (1.22.1904) Another meal was "a stew made of something none of us had ever expected to eat—a cormorant! And it really was not bad."

Except for food, most of the party's hunting was done with camera and glasses, a sport Will found as exciting as the leveling of a gun, and requiring infinitely more skill. "To my mind," he wrote, "a bird in the bush is worth a whole flock in the skin drawer, but the characters of modern identification often require more than the eye and the opera-glass can reveal."[1] When he needed a specimen to dissect, to fill a gap in the species record or to establish its species, he would shoot it, but merely to add to a collection was no longer a valid reason to take a life.

Will's increasing respect for the animate world conflicted not only with his drive to amass scientific knowledge, but with Charles Bull's desire to acquire material for his sketches. Back in his studio, he would need to have stuffed models to create the naturalistic scenes he was

Will and Blair at a campsite in Mexico.

known for. But to Will, it seemed as though he was collecting wantonly. To make matters worse, Fanny Bull was not used to roughing it. From her first brief burro ride in Guadalajara she had disliked riding, and she particularly hated the steep paths with their sheer drops that Blair found thrilling. Fanny feared bandits, illness, and particularly the simmering volcano that lured Will and Blair ever closer. When the Bulls decided to abandon the expedition and head for the coast, Will lent them money to go.

The unanticipated loan to the Bulls put the rest of their trip in peril, and Will wrote anxiously to his father for additional funds. At all costs, they were determined to come as close as they could to the volcano, whose eerie flickering tongues of flame held them spellbound.

Their next camp was still nearer. Will wrote ecstatically that it was "in the tropics themselves . . . in the dry bed of a stream and not two hundred yards from two clear, fresh, pure streams which spring from the very rocks. . . . There are great palms and green trees draped with lianas, tall fig trees, and scores of others not named to me." (2.25.1904) In a swampy area where the stream pooled were dense eight-foot ferns and huge elephant ears, the whole area redolent with an almost overpoweringly heavy, spicy scent. At dusk every night a vast exodus of tiny bats flew through their camp and disappeared as quickly as they had come. Then the bass duet of a pair of great horned owls serenaded for an hour or more. Vampire bats flew through the tents, squeaking and making eerie shadows against the tent walls but never bothering the human inhabitants.

One of the springs was warm from the volcano's underground heat, and Will and Blair swam and washed early every morning—Will often naked, reveling in the freedom from Northern restraint. Then they jumped in the icy waters of the stream lower down and emerged ready for a day's work. They explored, photographed, shot and skinned birds Will needed to study, and watched the behavior of animals whose habits were unknown in any book. At dusk and in the early afternoons, when the entire country shut down for the hot hours, villagers poured in wanting cures for all sorts of maladies, from parasites and malnutrition to tuberculosis and blindness. Blair found her small store of medicines running out after a very few days, but they were powerless against most of the ailments anyway.

One afternoon about four a tremendous blast and a great rumbling signaled the eruption they had been hoping for, and with opera glasses and camera they hurried to get a good view. The mountain's outline, usually so distinct against the sky, was dark as night and mantled with swirling black clouds. The air was heavy with ash, and after a few hours a terrific rush of wind blew toward the mountain as the heated air rose, drawing in new, cooler air. Small earthquakes continued for several days, and explosions sent down intermittent showers of small stones.

The book Will wrote about this expedition was the first of what would be a long line of natural history books that combined travel, excitement, humor, and keen scientific insight to produce a work that had wide appeal. *Two Bird Lovers in Mexico* is a vastly readable account of the trip written almost directly from his daily notes.

Blair contributed the cover design and a chapter on "How We Did It," in which she tells the "camping woman" of 1904 how to plan and execute a perfect wilderness vacation, from clothes, food, and medicine to riding, demeanor, and the hiring of a cook. For clothing, she recommended a khaki hunting coat, knickerbockers, and a soft felt hat for men, divided skirts for women, and sturdy shoes for both. "A pair of canvas hunting leggings, like my husband's, were the joy and comfort of

Volcano at Colima, Mexico.

my life; for whatever Mexico may lack, it is not thorns!" As for riding, "my theory is that all one has to do is get on and *ride*. . . . I had never ridden before but I simply got on and rode off. Of course for the first few times one cannot ride long distances, but that soon comes with a little practice." "No one," the nouveau expert asserts, "should attempt to ride side-saddle over these steep mountain trails; indeed the woman who does not intend to ride cross-saddle should never undertake a camping trip in the wilds of Mexico." She assures the woman who travels in the right spirit, however, of a marvelous trip: "[B]e interested in everything and . . . have one's mind firmly made up to ignore small discomforts. . . . Pluck and a philosophic spirit will soon make a good rider, and a good camper, and a very happy person indeed." On the advisability of carrying a gun, she is adamant. Every Mexican male in the wilder parts wore one, and "it is but the part of safety to do likewise," although it was considered most unfeminine.

Reviewers then began their long love affair with Beebe's writing: dozens of reviews praise the spirit and accuracy of the style, the author's contagious enthusiasm, and the quality of the photographs. The reviewer of *The Nation* applauds Beebe's advocacy of the Mexican "revolver law," a vigilante approach that was proving significantly more effective in protecting the diminishing egret population from feather pirates than American laws had yet managed to be. The *Tribune* of Cambridge, Massachusetts, called it "the very latest word in hunting with a camera." Beebe's career as a popular author was launched, and Blair received kudos for her travel writing that would influence her future as well.

CHAPTER II

The Naturalist as Author

To have discovered [these facts] is sheer joy, but to write of them is impertinence, so exciting and unreal are they in reality, and so tame and humdrum are any combinations of our twenty-six letters.

William Beebe, *Jungle Days*

Oₙ THE LONG VOYAGE BACK to New York Will kept writing, but the ocean's mesmerizing power drew him to the deck rail more and more. In his journal, minute details bring to life each tiny crab and colorful worm; questions of survival in the perilous depths and about the interactions of the varied life forms erupt through carefully noted observations. The dolphins that frolicked around the ungainly steamer attracted everyone's attention, but Will was even more fascinated by the tiny grasshopper-like flying fish that skimmed along the surface, southern cousins of his Uncle Clarence's consolation prize, which still held a place of honor in his collection in its jar of formalin. "A few tiny pale-grey forms shot ahead just below the surface and suddenly emerged, the great fins spreading taut instantly. The smaller rear pair occasionally close, but the pectoral ones remain stretched. . . . Some appear as tiny as a half-inch, others as large as 6–8 inches." (4.4–7.1904) Beebe's attention was also arrested by the large floating mats of sargassum weed, each strand an all-sufficient home for a complex community of organisms, floating maddeningly out of reach. Important questions formed in his mind, but his occupation as ornithologist was with airborne creatures rather than those that flew through the oceans.

It was mid-April 1904 when Will returned to the zoo, and breeding season was beginning. Seward, who had taken his place during the winter, had left to set up the National Zoo's exhibit at the St. Louis Exposition, and Will had his hands full. There were nests to be located, eggs to be watched, and incubators to be tended as he strove to keep the bird populations productive and in balance. Some of the rarer birds had to be coaxed into laying their eggs and tending their nests, while the more common ones were apt to be too prolific. In the large waterbird cage the mallards had bunched their young, one female tending forty-two eggs and the other females setting again. The extensive data Beebe had collected on the diets and habits of birds in the wild enabled him to rear many species successfully that had failed at other zoos.

The comprehensive bird book Beebe had been working on for Holt was well along and even had a title: *The Bird*. With the manuscript more or less in order, Beebe could devote himself to taking photographs that would illustrate his points precisely. His fascination with photography compelled him to keep up with the rapidly growing field, and he always found reasons to buy the latest equipment. For *The Bird* he invested in a reflex camera with a telephoto attachment that could get good photos in a thousandth of a second. Almost unmanageably heavy and awkward, in 1905 it was the very latest in sophisticated equipment. He experimented with a colleague's X-ray machine, and produced a photo of a pigeon skeleton and an ostrich embryo for the book. Blair stepped behind the machine and Beebe was amazed: "Saw her vertebrae and ribs plainly, as if no clothes or flesh intervened. Saw the marvelous phosphorescent light which the ultraviolet rays arouse in brown crystalline minerals. The Doctor sent a current through four of us which lighted an electric light bulb which Blair held. It was all wonderfully interesting." (2.25.1906)

A throat specialist was treating Will's throat with hydrogen peroxide and iodine, and Beebe got through the winter with only minor irritations—a good thing, since he had three books in the works. With only a few jaunts to Virginia to visit Blair's family, they stayed home for the next four years.

By January 1906 *The Bird* had been sent to Holt, although a few photos remained to be taken, and he was already in the throes of another book. *The Log of the Sun* would be a "yearbook" of the sort that was then popular for gift giving. It was to include fifty-two short essays, illustrated with his own photos and full-page drawings by the nature artist Walter King Stone, who had done the color frontispiece for *The Bird. Two Bird Lovers in Mexico* was selling energetically, helped along by rave reviews in papers across the country. Nature writing, like travel writing, was at its zenith, as audiences hungry for entertainment experienced vicariously the thrill of exotic places and creatures they could hardly imagine. *Two Bird Lovers* was both travelogue and nature study. President Roosevelt, who had recently published his own take on nature, *Outdoor Pastimes of an American Hunter,* told John Burroughs that he had read and enjoyed it.

Burroughs himself had just printed *Ways of Nature,* in which he blasted the growing coterie of nature writers who, trading on the public's ignorance of the natural world, passed off a world of imagination as fact. This battle royal hit the press under Roosevelt's aegis as the "nature-faker" controversy. In a tradition that had flourished since at least the time of Aesop, writers such as Ernest Thompson Seton had been telling an eager public tales that romanticized the "antics" of animals. But the nature fakers crossed the line when they wrote stories of

Blair on horseback in Mexico, 1903.

imaginary animal accomplishments under the guise of science. There are people alive today who remember and believe William Long's convincing tale of a courageous woodcock that set its own broken leg.

The controversy had far-reaching effects on the future of nature writing, as scientists labored to make an absolute demarcation between "true" or scientific accounts and "false" or imaginary stories about nature—a line that was sometimes set by the number of descriptive adjectives in the prose. With *Two Bird Lovers* Beebe had set his course determinedly along the contested border of this dispute, adhering rigorously to the path of scientific truth but clothing it in language that would engage readers at all levels, not just scientists.

Burroughs gave *Two Bird Lovers* his enthusiastic imprimatur, as did countless other writers and critics. Beebe's scrapbook, labeled carefully in Blair's hand, contains ninety-two reviews—all positive—clipped from journals and books as diverse as the *New York Times, The Churchman,* and *Our Dumb Animals.* Burroughs called Beebe's book "so keen, so scientific, and yet with the atmosphere of poetry and romance over all. You tell me just what I want to know about the wild life in Mexico, and you tell it with rare purity and charm of style." The *Chicago Herald* said Beebe's book was written with "a refreshing enthusiasm that is contagious. . . . There is always something about which Mr. Beebe is as enthusiastic as an unspoiled boy."[1]

Blair was studying Spanish, becoming fluent in preparation for what they both hoped would be an ongoing series of expeditions to the tropics. For Christmas 1905, Will had given Blair a "new Smith Premium" typewriter for writing her own articles. Encouraged by Will and her grandmother, she had already published several articles in the *New York Post* and *Harper's* about the Mexico trip and the Aztecs, and planned to write more. Her sympathy with the native peoples of the regions they visited and the novelty of her adventurous life appealed to a wide audience.

To clear some space in the small apartment, which was rapidly filling with the manuscripts of two prolific writers, Will presented his precious collection of 990 bird skins, nests, and eggs to the zoo, to be used in exhibits and educational presentations, and as a research bank for scholars. This donation, along with his big lantern slide camera, entitled Beebe to the honor of life member of the New York Zoological Society. The same year he was elected a fellow of the New York Academy of

Sciences, an unexpected honor that pleased him greatly, defensive as he was about his professional status.

In October 1906 Holt released *The Bird* to enthusiastic reviews: "a new departure in the literature of bird study" *(Review of Reviews)*, "a book of rare authority and interest" *(New York Herald)*, "absorbing as well as scientific" *(Hartford Courant)*. A technically precise volume with exquisite photography, it is still vastly readable. Beebe was able to convey the intricate morphology of a bird's wing with exactness and clarity, and to make deciphering its structure as compelling as a mystery. *The Bird* broke new ground in combining scientific expertise and even technical biology with graceful and attractive prose, and in this it accomplished Will's stated goal: "To take a few dead facts and clothe them with the living interest which will make them memorable to any lover of birds, and at the same time to keep them acceptable in tenor and truth to the most critical scientist—this has been my aim."[2]

Will dedicated the book to his friend and mentor, Henry Fairfield Osborn, who had guided him from untutored enthusiast to methodical scientist. Osborn had also influenced Beebe's response to the living world, turning him from the rhapsodies of his early writing to a more coherent view. For from the youthful collector who had loved nothing more than adding skins to his collection, Will was metamorphosing into one of his era's strongest voices for the nascent conservation movement. In the first chapter of *The Bird,* he makes a statement that has become a touchstone for conservationists everywhere: "The beauty and genius of a work of art may be reconceived," he wrote, echoing the biblical book of Revelations, "though its first material expression be destroyed; a vanished harmony may yet again inspire the composer; but when the last individual of a race of living beings breathes no more, another heaven and another earth must pass before such a one can be again."[3]

> And the next time you raise your gun to needlessly take a feathered life, think of the marvelous little engine which your lead will stifle forever; lower your weapon and look into the clear bright eyes of the bird whose body equals yours in physical perfection, and whose tiny brain can generate a sympathy, a love for its mate, which in sincerity and unselfishness suffers little when compared with human affection.[4]

The zoo's board was pleased with Beebe's growing reputation, as well as his skill with the birds, and raised his annual salary from $2,000 to $2,400. It was fast becoming a wonder how he managed to write so voluminously and still keep up with the work, but he handled the birds with a dexterity born of love and long experience, and the writing with the relentless efficiency his driving energy impelled.

The Log of the Sun, his calendar book of essays, received glowing reviews. It contains brief essays for each week of the year, some of which had been published previously in various magazines. Like the work of John Burroughs, the essays were rich with detail, forging images that readers would not easily forget while staying true to scientific fact. The preponderance of the essays involve birds, providing a bird's-eye view of the world in a particular situation in a particular season. But other chapters touch on some of Beebe's keener interests—the ongoing debate over protective coloration, the mystery of insect sound production, the "poetry and romance of evolution" as evinced by the development of a polliwog, the tiny hairs that keep water-boatmen afloat, the thrill of seeing the nighthawks that populated the roofs of New York City—and several intriguing comments on deep-sea life, which foreshadow the shift in his interests that would later take place. But his own prognostications stand like challenges to himself:

> The time is not far distant when the bottom of the sea will be the only place where primeval wildness will not have been defiled or destroyed by man. He may sail his ships above, peer downward, even dare to descend a few feet in a suit of rubber or a marine boat, or he may scratch a tiny furrow for a few yards with a dredge: but that is all.[5]

Christmas that year (1906) was festive. *The Log* was selling briskly, and Blair had sold several new articles. Will had been made science critic for *The Nation* and was receiving books to review for them at a great rate. He cataloged each review carefully, noting the value of the book, which he received gratis, as well as the amount he was paid for each article. He received G. W. James's *Wonders of the Colorado Desert,* for instance ($5), and was paid a penny per word. Thus he considered himself to have made a princely $15.50 for the 1,050-word review. As art critic for Holt's series of nature books, he would get $100 for the first volume and a penny per word for a lengthy encyclopedia article. In addition, Will wrote optimistically, Blair's dowry of copper stocks had netted about $500 "so far." (2.24.1907) Her brothers Harry and Roger came for Christmas and stayed for three months, ensuring a continual round of parties, theater, opera, and family gatherings—in addition to reading T. H. Huxley's *Letters* aloud by the fire in the evenings.

Huxley, "Darwin's bulldog," was a pivotal figure in the ongoing metamorphosis in the field of science. Not only had he shaken a sluggish religious establishment into action with his strident, no-holds-barred advocacy of Darwin's concept of evolution by natural selection, he was also a scout for the changing of the old guard in academia—a change crucial for young scientists like Beebe. Previous generations of scientists had come perforce from the leisure classes—clergymen, country gentlemen, tenured civil servants with assured incomes that allowed for a modest amount of scholarly tinkering on the side. Darwin himself was a well-bred gentleman with a healthy respect for entrenched morality, who had never looked forward to stirring up religious controversy.

But as the son of an impoverished schoolmaster, Huxley, with his rough manner and uncompromising language, had a living to earn; like the members of the working class whose increasing education was propelling them into places previously reserved for the independently wealthy, he had to battle for recognition. The social revolution that was empowering a new class of academic entrepreneurs was also changing the face of academe: institutions dedicated to scientific progress had to come up with money to fund these hungry young lions and their groundbreaking studies—funds that would previously have been provided by the researchers themselves, or their wealthy families and friends.

Determined to be independent, Will was trying to support his research and a lifestyle Blair could enjoy with a small but respectable zoo salary and the money he earned from writing. While Osborn, born to unassailable wealth, applauded his three books, he feared that Will's growing reputation as a "literary" naturalist could interfere with his reputation as a scientist. He counseled his young protégé to stop writing and spend the rest of the winter in scientific reading and work. Will took his advice to heart and submerged himself in the experiments on feather coloration that had been receiving favorable attention from the scientific community. In particular, he had proven that the plumage of many birds changed color when subjected to intense humidity, a phenomenon that had never been studied. This discovery held great significance for ornithologists and taxonomists, who depended heavily on coloration in classifying subgroups, and Osborn judged wisely that it would make considerable stir among evolutionists. (4.12.1907) Hornaday, pleased to see his curator of birds actually in harness, gave Will $300 to hire a helper for his experiments.

In August, Will and Blair accompanied Professor and Mrs. Osborn to Boston for a meeting of the International Zoölogical Congress. The big guns of the zoological scene turned out in force, and the Osborns made certain that the Beebes were introduced all around. Will read his paper on the effects of humidity on the feather color of doves, which intrigued Cambridge professor William Bateson, who with Reginald Punnett was working to establish the new field of genetics. The conference finished with a week of meetings in New York, and the Beebes were invited for the first of many visits to Castle Rock, the Osborns' villa on the Hudson, fifty miles north of New York City. As an occasional guest of the Osborns, Will made the acquaintance of many wealthy and powerful people who valued science—contacts that would pay off royally in the future.

Will's New Year's journal entry for 1908 reported great progress. Mr. Holt, the publisher, had appointed him editor of the entire American Nature Series, with full responsibility and a good salary. Thanks to the contacts he had made at the zoological conference in Boston, he had been elected a corresponding fellow of the London Zoological Society, which appealed to his desire to be well connected abroad. His research on variation in plumage was to be published in the flagship issue of a new journal the New York Zoological Society was launching. *Zoologica* was to be a scientifically rigorous companion to the member-oriented, newsy *Bulletin of the New York Zoological Society.*

But most important was what he considered the "victory of my life": Hornaday announced—with what reluctance we can only guess—that he and Osborn had decided to put Will on the same footing as the research scientists at the American Museum of Natural History. This post was in essence a primitive sort of research grant: two full months off each year, one on full pay, the other to be paid for by a newly formed scientific fund from the New York Zoological Society. The shorter periods the zoo normally allowed were intended for collecting purposes, not for research; only with time for lengthy field study could any real scientific work be accomplished. Beebe excitedly set plans in motion for a spring expedition to the tropics.

CHAPTER 12

Rain Forest at Last

*Some day, if we do not delay until the destroying hand of man is
laid over this whole region, we may hope partially to disentangle
the web. Then, instead of a seeming tangle of unconnected events,
all will be seen in their real perspective: the flower adapted to the
insect; the insect hiding from this or that enemy; the bird showing
off its beauties to its mate, or searching for its particular food.*

William Beebe, *Our Search for a Wilderness*

O~N~ FEBRUARY 22, 1908, THE
Beebes sailed out of the harbor on the steamer *Trent* of the Royal Mail
Line. The elder Beebes, the Bulls, and many other friends had sent them
off in a mad swirl of candy, gifts, and good wishes. They were bound for
Trinidad and British Guiana, well-known naturalists' paradises. Not
only were these lands endowed with natural riches, but they were
English-speaking and British-ruled. In Mexico, Will had found that his
schoolbook Spanish left him at the mercy of translators. Worse, the
notorious instability of Mexico and many Central and South American
countries—Mexico had been in the grip of revolution and lawless bandits when the Beebes visited—made travel chancy at best. In countries
ruled by the Empire, some sort of order would reign. Frank Chapman
had visited Trinidad in 1893 and found it refreshingly civilized.

When the *Trent* docked at Colon, "the waist of the world," the Americans aboard opted to charter a train to travel along the route of the
Panama Canal, which was in the first stages of construction. The train
stopped often as it made its way through the tropical forest, allowing
Will to get off and collect fossil shells from the canal cut and catch

some of the magnificent insects he spotted from the windows. Great electric blue morpho butterflies wafted along the edges of the jungle, and giant mantids looking like anything from twigs to orchids minced delicately across the path or waited, motionless, to test their powers of camouflage against Will's practiced eye. Scattered along the canal cut were rusted cars and engines, covered with vines—relics of the disastrous French attempt to dig a canal. "There is a saying," Will wrote, "that every sleeper [tie] on this R.R. is the tombstone of a man's life." (3.1.1908)

The saga of the making of the Panama Canal makes depressing reading. The failed French attempt alone—engineered by the great Ferdinand de Lesseps, who had built the Suez Canal—is a testament to human inadequacy in the face of heat, disease, rock, and the sticky, slippery red mud that oozed or slid into every cut. But by 1908, Teddy Roosevelt's determination was making the canal happen. Workers' houses were neatly swathed in netting to keep the deadly mosquitoes at bay, sanitation projects eliminated typhus and cholera, and realistic engineering replaced the French plan of a sea-level canal with one of locks. Aggressive funding had procured modern earthmoving equipment that seemed capable of leveling the mountains that had defeated every previous attempt. As a result, the cut stood as a triumph of the American "can-do" spirit that Roosevelt represented, and that Will admired.

The train took them back to the ship, and they steamed past Aruba and Curaçao, stopped briefly at Guiara on the Venezuelan coast, and anchored at last outside Port of Spain on the island of Trinidad. At that time Trinidad, along with its neighbor island, Tobago, was a British colony. Its native population of Carib and Arawak Indians had been superceded first by Spanish and French settlers, then by the British. Until slavery was abolished in 1834, planters had brought in large numbers of African slaves to work the sugar plantations. Indentured workers from India replaced the slaves, bringing Hinduism to the already rich mix. The diverse population that resulted made the island an exotic, immeasurably strange place to the Beebes, however cosmopolitan they had considered themselves.

With their impressive letters of introduction, they were welcomed effusively by the acting governor, who issued Beebe an all-important permit to shoot two of every species of bird on the island, which would enable him to make a collection for the zoo that could be studied, and also provide information to naturalists on Trinidad that would help their efforts to regulate hunting.

The diversity of nature and culture on Trinidad could have kept the Beebes occupied and content for a lifetime. The insects in their multitudinous forms were especially compelling, particularly in light of the debate on the purposes and effects of protective coloration. A devout Darwinian, Will was attracted to the idea that every morph or body type served an adaptive purpose: moths that rest on tree trunks have patterns that camouflage them for protection; waterbirds have webbed feet for swimming; woodpeckers' bills are long and hard, and their tongues have back-curved barbs for fishing insect larvae out of holes.

But to be a Darwinian in 1908 was to be in the minority among biologists. Although *Origin of Species* had provided compelling evidence for evolution—that is, change within species, as well as the creation of new species—the mechanism that would allow this change was hotly debated. Darwin proposed that natural selection was the engine of change; but genetics, as it was then understood, seemed unlikely to be able to generate such change. The understanding of inheritance at the time came largely from breeders, who saw evidence of blending, in which the characteristics of offspring seemed to represent a compromise between the features of the parents. Such a system would tend to lead to greater similarities among individuals, rather than to the varied forms that selection required.

Random chance and inbreeding were other possible explanations

Tropical jungle in Trinidad.

for the clear evidence of evolutionary change that was being docu-
mented. Mendelian genetics had resurfaced only eight years earlier, and
Bateson and his Cambridge group were working to apply his findings to
animals. Later in 1908, Godfrey Hardy and Wilhelm Weinberg would
build on Mendel's model to provide the first explanation of how variation could be maintained in populations. It would be fifteen years
before Ernst Mayr and others would construct the "modern synthesis,"
reconciling variation and selection with genetics in its modern form.

As often happened, Beebe's intuition was ahead of the textbooks.
Although agreeing that some examples of adaptive camouflage and
behavior had been strained beyond the point of belief by overeager
Darwinians, Will still found justification on every side. Every minute
he was halted by a curled leaf that resembled a caterpillar, a partly
decayed fruit that turned out to be a curiously marked beetle. Many of
the fantastically varied butterflies displayed habits that took advantage
of their coloration: one species with black wing margins alighted only
on the undersides of palm fronds, where their markings duplicated the
venation. A giant owl butterfly flew with audible snaps of its great wings,
camouflaged from above but flashing its startling eye spots below as
each snap propelled it distractingly in unpredictable directions, making it almost impossible to catch.

Reluctant to leave Trinidad's riches but eager to press on to the Venezuelan streams that were their destination, they hired the *Josefa Jacinta,*
a twenty-one-ton schooner with a courtly captain, an ethnically diverse
crew, and a Malay cook, for $6 a day. They stocked up on tinned food
(spending $17 on supplies) and 150 pounds of ice, blew up their air mat-
tresses, and settled in to the boat that would be their home.

Blair, whose appetite for adventure seemed to grow keener with
each new experience, described the practical side of the cruise in a
chapter of their book about the expedition, *Our Search for a Wilderness,*
which she called "A Woman's Experiences in Venezuela." Her description of that first night aboard the schooner adds a dimension of domestic realism to Will's paeans to sunsets and science. After depicting his
vain efforts to inflate the new air mattresses with the elegant gilded
foot pump, she adds a plaintive subtheme to his previous accounts.

I doubt if either of us will forget that first night. Beneath the flooring
and behind the planked sides of the vessel was a mysterious under-

world, densely populated by rats of most sportive disposition. . . .
There seemed to be some kind of a running track extending around the
hidden depths of the sloop. A race would start near the stern, the con-
testants tearing around W— —'s bunk; then the footfalls would die out
toward the bow to become audible almost at once on my side—a medley of sounds indicating a mob of invisible rushing creatures, galloping
down a mysterious homestretch. . . . [1]

The *Josefa Jacinta* jolted them queasily across the Gulf of Paria to the
mouth of Venezuela's Caño San Juan. As they progressed slowly up the
caño, which wound through a dense mangrove swamp, the dream of a
wilderness became reality. At midday there was no sign of life, but as
soon as the sun got low they were surrounded by thick clouds of raven-
ous mosquitoes. Bird sounds began tentatively, then increased to a
cacophony of shrieks, screams, and whistles, produced by vast numbers
of waterbirds of a variety that beggared imagination. Their eager eyes
picked out snowy egrets, brown pelicans, frigate birds, turkey vultures,
spotted sandpipers, a yellow-crowned night heron. "On all the snags
were perched either terns with great yellow bills, or beautiful blue and
white swallows. A skimmer passed, skimming as his near relations do
in Virginia." Muscovy ducks the color of the mud itself and a rufous-
bellied kingfisher appeared together. "A living blaze of color shot past
against the dark green foliage and we had seen our first Scarlet Ibises

The sloop Josefa Jacinta.

within 5 minutes of our first monkeys." (3.26.1908) Soon clouds of hundreds of ibises soared past, swooping low over the tiny boat as it idled slowly along the channel among the dense trees.

Several species of land crabs also made their homes in the grotesque tangle of mangrove roots, each root's population colored to match that particular tree's vagaries of tone and texture so exactly that they defied vision. Savage biting flies of an astounding variety feasted on the bare feet and legs of the crew. Since Trinidad was under a yellow fever quarantine, Will and Blair had been dosing themselves religiously with arsenic "tonic," which was thought to kill the disease-causing organisms in the blood. They had taken so much arsenic that they were hardly bothered by flies or mosquitoes. In fact, at two tablespoonfuls a day they had by this time ingested a half pint each, and Will had developed the first signs of poisoning—a large bronze-colored swelling that made it difficult for him to sit down. He decided to back off his daily dose for a few days. "But we are immune and need have little fear of any fever, yellow or green!" he assured anxious parents back home.[2]

For eleven days they drifted up the Guarapiche with the tides, raising anchor as the water level in the *caño* rose and moving upstream with it, tying up when the tide went out, and either exploring the surrounding jungle or rowing upstream in the small dugout canoe they kept onboard. In this way they were able to access small streams and backwaters where birds and insects lived undisturbed.

Moving their sleeping quarters to the rough benches on the deck and surrounding them with shrouds of white mosquito netting turned out to be a blessing. In triumphant contrast to those first two hellish nights, "never were nights more beautiful than those which we spent on the deck of that little sloop, and never was sleep more dreamless and peaceful." The crew of three worked well with the captain and were chivalrous toward Blair. But the cook, Maestro, was of uncertain mood and certain uncleanliness, and was eventually left marooned in a small jungle village after threatening the captain with his machete. "I often shut my eyes and see him with streaming eyes stirring some fearful concoction over the little stove," Blair wrote in *Our Search for a Wilderness*. "Or again on his knees mixing dough for the leaden dumplings to be boiled in the pig-tail stew which appeared at every meal. We so often wished we had brought graham flour. White flour does show the dirt so!"[3]

Will's journals make no mention of these very human aspects of the trip. They are filled with excited notes on rare species glimpsed for the first time, rhapsodic depictions of glorious tangled banks or mischievous monkeys or thrilling night sounds whose mysteries resolve themselves in the light of day. A peculiar gurgling, sighing sound was particularly bone-chilling, seeming to come from the banks near the boat at certain times of day. After days of nervous wonder they spotted a giant anaconda slipping noiselessly from its burrow at water level; when it had gone, the tide lapping the empty hole gurgled and sighed as it met the empty space.

After nearly two weeks afloat they reached Guanoco, a small town that housed the workers and management of La Brea, a vast lake of pitch that supported an astounding array of wildlife. Guanoco was a picturesque little town of thatched huts built on a hillside, with a midget railroad that carried workers back and forth to the lake running along the foot. The lake itself was a wonder of nature. A small sea of slowly heaving asphalt that had been used for centuries to caulk the ships of pirates, explorers, and conquistadores, it provided most of the asphalt used to pave the streets of the United States' burgeoning East Coast cities and towns. The pitch itself was solid enough to walk on in most places, if you kept moving. Small ponds of fresh water filled the

The train on the pitch lake near Guanoco.

places where pitch had been dug, supporting populations of strange and idiosyncratic species. Weedy flowers grew on the crackling surface, and an oasis of trees in the center was the roost for hundreds of Amazon parrots.

The little train traveled back and forth seven or eight times a day, so Will and Blair could come and go at will, sharing the open cars with the men who worked the pitch. The railroad had been in place for twenty years, so the jungle animals were accustomed to it and had no fear. Monkeys swung across the narrow cleared space along the line, anteaters shambled slowly across the rails, and alligators sunned in the ditches alongside. Once a puma sauntered across, snarling. Only the sloths, heavy and slow and incapable of navigating on the ground, were unable to bridge the gap; they swung up to the edge of the forest and gazed uncomprehendingly at the impassable space before turning slowly back.

After two weeks in this collector's paradise, the Beebes headed back to Trinidad to pack up their menagerie. In addition to three parrots and three macaws, which the Indians had caught for him in Caño Colorado, Eugene André, a prominent Trinidadian naturalist, sent five crates containing forty birds he had trapped for the zoo's collection: ten silver-beaks, three blue tanagers, two palmistes, five black tanagers, two bare-eyed thrushes, two kiskadees, twelve euphonias, and four cowbirds. Will had collected two white-faced teal, a scarlet ibis, two green macaws, a sun bittern, two porcupines, and a tree snake. All of these creatures had to be carefully crated and tended during the long voyage to their new home at the zoo.

On the long voyage from Trinidad to New York Will wrote and read, and competed with the polyglot, multinational passengers—Dutch, African, Chinese, Venezuelan—in deck sports to keep fit. But his favorite shipboard sports were photographing flying fish from the bow and snatching sargassum weed as the ship sped along. "The first is equal to any quail shooting for excitement, and requires a steady hand and a quick eye. A small net tied on a boat hook or pole and dashed into the water at the moment a wave brings a bit of weed close & high enough requires more practice than would seem to be required." (4.25.1908) In the weed mats he discovered a myriad of creatures: minute crabs, snails, and fish, all clinging fiercely to life on a frail, tenuous, ultimately doomed raft.

The sudden shift to the cool, brisk air of a northern April as they steamed up the East Coast was invigorating after the velvety tropics, and Beebe shifted into New York gear. In his journal he made a list for their next voyage, already taking shape in his optimistic mind:

Bring pajamas for use on boat; also bath robe & slippers.
Linen suit. Less reading matter.
Folding boat for mangrove swamp.
Dress tuxedo suit; white and black.
Small stiff cards for bird lists!
A folding canoe.
Long board to land over mud when tide is down.
Brush for mud on clothes.[4]

In Search of Wildness

When you look for things . . . year after year, and train your senses
to continue their concentration after the less important parts of
you are sleeping or eating or playing or merely talking, then sooner
or later, very special things happen within sight or hearing.

William Beebe, *Nonsuch: Land of Water*

WILL SPENT THE SUMMER OF 1908 increasing the zoo's collection of birds, writing, and playing tennis. Blair had become an avid horsewoman and rode out often for exercise. In the fall she spent a month in Virginia while Will worked and wrote with a fury. An expedition to British Guiana the following spring was in the works, and he wanted to complete some experiments on feather color before he left. Having been allowed to increase his staff by two, he hired May Howard, a family friend from Brooklyn who needed work, to help with the typing, and Lee Crandall, a newly fledged ornithologist, to help with experiments.

In early December, Will made some contacts that would help shape his future career. On a trip to Washington, D.C., he and Blair called on President Theodore Roosevelt, whose admiration for the young naturalist's writing had resulted in an invitation. As a rancher, hunter, and explorer, Roosevelt had a keen awareness of the fragile ecological balance. As an empire builder, he felt strongly about the mandate to protect and preserve nature. The president warmed to Will's eager manner and ingrained respect for the natural world; his friendship would be a passe-partout for Beebe, opening safe harbors and vistas of opportunity beyond even his eager imagination.

In a cold February drizzle, Will and Blair and Lee Crandall left New York on the Dutch *Coppename,* bound for British Guiana. They planned to go up the Demerara River from Georgetown, and explore the wild Essequibo and Potaro rivers. Beebe was determined to amass a noteworthy collection of South American birds for the zoo, particularly the strange hoatzin, and was obsessed by the idea of finding a perfect site for a field station. Every time a research expedition went to the tropics it returned with a wealth of data, but the difficulties of personnel and logistics ate away at limited time and funds. If there could be an outpost with scientific facilities, already set up and established—perhaps something like Mrs. Latham's in Florida—researchers could use the surrounding jungle as a study site and collect reams of valuable information. Although there was as yet no funding for such an enterprise, Will knew that Roosevelt was interested, and T.R.'s support might turn

*Will with his friend
Teddy Roosevelt,
Georgetown, British
Guiana, 1916.*

dream into reality. British Guiana seemed a likely place for such a station, and Will planned to look about him while he was there.

At the captain's table they made friends with Henry Gaylord Wilshire, an eccentric Socialist millionaire whose spectacular land deals in
the late 1800s (notably the purchase of a desolate tract in Los Angeles
that would become Wilshire Boulevard) had brought him tremendous
wealth. His true vocation, though, was Socialism, and much of his fortune went to the cause—particularly to *Wilshire's Magazine,* a political
and philosophical publication that, having been evicted by the United
States, was forced to publish from a Canadian base. Politically naive, as
he would remain to the end of his life, Will described Wilshire's Socialism as "of the sanest description and very interesting—no anarchy. It is
really a sort of brotherly love idea, with every man earning just according to the real value of his work. No division of wealth or anything like
that." (2.21.1909)

The Wilshires owned some gold mines in Guiana and had traveled
extensively in the interior. The hospitable couple invited the Beebes to
accompany them in their launch, first to their mine at Hoorie in the
northwest, and then to Aremu in central Guiana. The Beebes would see
much more wildlife this way, they promised, more untouched wilderness, than on their proposed route along the more frequented and
"developed" Demerara. They would have seasoned guides and good
company, and be spared the trouble of making their own arrangements.
Besides, it would save a good deal of the expedition's scanty budget,
which Will was constantly trying to stretch to fit his hopes for more
birds for the zoo.

In Georgetown Blair bought bloomers like Mrs. Wilshire's, which
she highly recommended for wading through mud and creeks and
climbing over thorny terrain and fallen trees. An immediate convert,
Blair wrote enthusiastically, "One never felt the burden of a single super-
fluous ounce of weight, and when thus freed from the drag of heavy
clothing one would come in unfatigued from tramps which would have
been impossible for a woman in orthodox dress, no matter how short
the skirt."[1]

For four days, the little expedition prepared for their trek through
the wilderness. They purchased cigarettes, beads, and candy to trade
with the Carib Indians, tinned meats and fruit, straw hats and mosquito
netting. They knotted butterfly nets and fishing seines, and bought

hooks and line. In the evenings they watched bats swoop low for insects, and the great toads, which spent the day in ditches, crawl out to sit in solemn groups under the street lamps, waiting for insects to fall.

Will and Blair would learn quickly that free travel had its drawbacks. Because of either Wilshire's expansive hospitality or Will's personal magnetism, several of their fellow passengers decided to join the expedition as well, which they saw as a great adventure. The river steamer *Mazaruni* took them to Morawhanna, where they discovered that the mining company's little launch had not arrived. It was four interminable days before they could leave the tiny backwater, where some of their number slept in hammocks in the sad, elegant house of the one lonely government agent, and some at the local police station. They took small field trips into the jungle, but their numbers scared off the birds and sent all of the other wildlife packing. And the plague of chiggers took its toll on everyone's spirits.

Will was becoming increasingly sensitive to chiggers, or "bêtes rouges," which are common to warm wooded ground. Like minute ticks, these tiny mites live in the bush, waiting for unwary mammals, and then tunnel into the skin, setting off an immune reaction. The fiercely itch-

Blair in her new knickers.

ing bites made the sand flies and more dangerous mosquitoes seem immaterial. The persistent societal taboo against scratching in public made any social gathering a torment. "Like sea-sickness or an earthquake, bête rouge is a great leveller of mankind," Will wrote, "like a common disaster doing more to make men 'free and equal' than all the constitutions and doctrines ever signed." [2]

They tried every local remedy that came to hand—dry soap, ammonia, wet salt, wet soda, camphor, chloroform, resinol ointment—even the local favorite, the totally ineffective "Tango," which is still used today in the tropics by the faithful. In Guiana in 1908, the preparation *du choix* was a repellent made with the oil of the crab-wood nut, a foulsmelling substance that was mixed with dry soap. Greasy, sticky, and with a remorselessly rancid odor, it was moderately effective for Will.

But his spirits were beginning to chafe under the delay, and the presence of so many idle people, however cheerful and companionable, interfered with his usual strict schedule of collecting, study, and analysis. To pass the time, he and Crandall trapped some of the lovely tanagers that flocked to cleared areas; the gorgeous silverbeaks, resplendent in wine and maroon, with silver-gray bills; the muted olive palm tanagers; and the bright black-and-white magpie tanagers with their long, graduated tails and startling white irises.

At last the thirty-foot gasoline-powered launch that was to propel

The Wilshires' tent boat on the Barama River.

the tent-boat upriver arrived and the sizable party headed off, the launch alongside, several wooden dugout canoes of Indians carrying the cargo behind.

Down Mora Passage and up the Waini River, Will saw more birds than he had ever thought possible. As if determined to make up for lost time, blue and yellow macaws flew past, always two by two; clouds of blue-headed parrots swirled over and back; gigantic flocks of anis, their black feathers blotting out the sun, swarmed past, more than 4,000 strong. Scarlet ibises rose at dusk, whirling in a cloud of brilliant red to their communal feeding grounds.

Night brought a steady rain, and the eleven people slept in the tent-boat on boards, half in, half out of the steady downpour. Will fretted about his cameras, which despite all his care were suffering. Worse still, the following morning revealed that in all the chaos the tin trunk that held all three guns and their ammunition, most of the photographic plates, and all the collecting apparatus had been left behind on the *Mazaruni*, bound now for Georgetown. Depression overwhelmed him. "I shall do the best I can, but am badly crippled in my work. . . . I hope never to travel in a crowd again." (3.7.1909)

Once they were well under way, however, his resilient spirits bounced back. Scarlet ibises took wing around every bend, curious herons accompanied the strange flotilla for miles before taking their leave, interesting pipefish and piranhas took their bait—and a piece of Crandall's finger—as the hull glided through the languid water of the Waini and then the Barama, and eventually, after three days, into Hoorie Creek, a small tributary thirty-five feet across.

Suddenly the mine's tiny twenty-acre clearing came into view, and the disjunction between primal forest and the handiwork of man was startling. A dam had been thrown across the narrow valley, and on the rim of the jungle lake it made, just beside the thatched huts of the Indian workers, was the powerful electric engine, a "great thing of vibrating wheels and pistons" that seemed strangely out of place in the wilderness.[3]

Beebe spent his days at Hoorie trekking through the jungle with Abel, an Arawak guide who astonished him with his surefooted stealth, keen senses, and knowledge of this alien world—a knowledge that seemed not acquired but innate. This was the noble savage of Will's youthful imaginings, man still in tune with the natural world, still responsive to stimuli invisible, inaudible to city-bred senses. Away from the well-meaning but inept horde of humans, Will and Abel could approach creatures that would have fled at the first footsteps.

As hunter for the camp, Abel provided armadillo, peccary, and tinamou, a wild jungle relative of the chicken. He knew the great red-crested woodpecker that nested in a tall rotted stump of a tree, informing Will that it was mateless and had one offspring. He told Will that the trumpeter he sought nested in trees, always laid two or three white eggs, and slept on one leg.

It rained, but Will found that once he was soaked through, the rain was not noticeable. Walking softly on deep mold or moss he found the silence absolute, broken only by the soft whirr of wings or a humming-bird's shrill cry. Brilliant metallic blue morpho butterflies sailed lazily through clearings, and on the trunk of a fallen tree a tiny movement revealed a "ghost of a butterfly"—transparent except for a trio of tiny azure spots near the edge of the hind wings. When it alighted on a scarlet flower, the red of the flower and green of the leaf showed through, the faint gray haze of the wings marked only by faint venation.

As they were leaving camp, Indians materialized from all directions, bringing strange specimens of all sorts, from five enormous whip scorpions to tree frogs and snakes, all destined for the zoo. Enchanted by the place and its friendly people and exotic fauna, Beebe noted that Hoorie

The gold mine at Hoorie.

would make a splendid research station: "Hoorie is a perfect health resort; temperature good; no mosquitoes; food splendid; place for fine laboratory work; insect life superabundant; lizards & snakes also; perai, electric eels and manatees in the creek; peccary, deer, red howlers and armadillos within short distance, also sloths and anteaters. . . ." In a fit of optimism he imagined that the capitalistic mining operations of his patron Wilshire might conceivably have a good effect: "Another year it will probably be better still as more logs are left to rot and larger clearings are made here and there in the jungle." (3.14.1909)

The Wilshires' second mine, at Aremu, was even better. The crowd of hangers-on, most of whom had never walked farther than the corner to hail a cab, abandoned the expedition and headed back to the cocktails and cushions of the city. The small tent-boat navigated a portage at Perseverance Falls and on into the Little Aremu River, which Wilshire had dredged to make way for his mining equipment, hacking through huge old first-growth trees, their wood so dense that they sank rather than floated, making a barrier more impassable still. They had to be dynamited at last to make a navigable channel. Now, six months later, what had been a coarse cut was softened by vines and new growth, as understory plants strained to reach new sources of light.

The mine at Little Aremu was an oasis in a dense jungle of mighty mora trees, some much more than 100 feet tall. The descent into the mine was made in the mining bucket, "two and two—each clinging to

A ghost butterfly printed over text in Beebe's Two Bird Lovers in Mexico, *to illustrate its transparency.*

until a slanting beam of sunlight struck it before us, was the ghost of a butterfly! It s inches but was wholly transparent save fo of azure near the edge of the hind wings As we looked, it drifted to a double-headed and when it alighted, the scarlet of the flov of the leaf were as distinct as if seen th while the faint gray haze of the insect's wi only by the indistinct veination. The ap ghostly butterfly amid the silence and awe of the reeking jungle was most impressive the wire cable and balancing the opposite person," Will wrote. "Down and down went the swaying bucket, slowly revolving—the heat and sunshine of the upper air replaced by the cool darkness—damp and chilly with rich, earthen, clayey smells."[4] Candles shed a flickering light on the dripping, slimy walls, and it was hard to imagine that eighty-five feet above was a tropical world of sunshine, butterflies, and parrots.

Will was amazed at the variety of birds. In the ten days of their stay at Aremu they identified 80 species and observed at least 200 more, without being able to make an exact identification. His aim was to study the birds alive, to learn about their behavior in their native habitat, in hopes of adding "something of real value to our knowledge of the ecology of these most interesting forms of tropical life. We have the results of the collector, par excellence, in our museum cases of thousands of tropical bird-skins. Now let us learn something of the environment and life history of the living birds themselves."[5]

More eager than ever to establish a research station in the wilderness of British Guiana, Will strove in his writing to emphasize the region's selling points for scientists. "I was able to rise at five o'clock in the morning and with intervals only for meals, keep up steady work—exploring, photographing and skinning until ten o'clock at night, when usually the last skin would be rolled up or the last note written. I would

Blair and Margaret Wilshire descending in the mine bucket at Hoorie, British Guiana.

then tumble, happy and dead tired, into bed and know nothing until the low signal of our Indian hunter summoned me in the dusk of the following morning."[6]

The woodcraft Beebe learned from the Arawaks was leagues beyond the simple lore he had mastered in childhood. His Indian guide heard every sound, ignoring the frequent loud crashes of falling branches but catching the smallest chuck of a bird or rustle of a leaf. He located every bird or beast Will wanted to see, often with only a clumsy sketch or call to identify it.

> He walked like a cat and *never for a moment* relaxed his vigilance, and there he differed from a white man, who would unconsciously relax when he thought game was still some distance away. His figure slipped silently ahead of me, flowing under trunks, passing around the densest clumps of underbrush, while I followed and imitated as best I could, learning every minute more than I had ever known of the art of effacing oneself in the wilderness.[7]

They still wanted to find the elusive hoatzins, which was one of the main goals of the expedition. Evolutionists theorized that the awkward hoatzin, which bears a thumblike claw on the leading edge of each wing much like that of the primitive archaeopteryx, would prove to be an important link in the gradual progression from reptile to bird. To

A hoatzin nest.

secure some of these birds, both for study and for the zoo, had become
almost an obsession with Will, and here the goal was in sight. They left
the Wilshires at Aremu and took a boat to a small island in the Abary
River, where there was a large and thriving hoatzin colony in a brushy
scrub marsh. Every evening twenty-five or thirty of the birds would
come to the edge of the clearing on the island to feed. Will used his field
glasses to study one pair in particular, making careful observations
of their habits, their nest and eggs, and their idiosyncratic crawling-
swimming-flying mode of getting around, which made awkward use of
the clawlike appendages on their forewings.

In mid-April Crandall arrived from Georgetown for a grand hoatzin-
trapping, the crown of all their ambitions. But on the evening of his
arrival, without warning, one knot of Blair's hammock loosed its hold
and she was thrown seven feet down off the veranda, breaking her wrist
badly. "Game little lady," Will wrote, "her first words were 'Oh! We can't
get the Hoatzins'!"[8]

Packing with desperate haste, they set out for the nearest station
and telegraphed for a special train. With Blair woozy from morphine,
Crandall and Will climbed up to the roof to watch dogs, cows, donkeys,
and people jump from the tracks as the unscheduled two-car train ran
through. "The yellow and scarlet blackbirds blew up like chaff on either
hand. Egrets, ibises and jabirus watched in amazement from afar, or
flew hurriedly off at the long drawn-out siren whistle, which hardly
ceased across the whole country." Ten hours after the accident Blair's
wrist was splinted and set, with the assistance of a half pint of chloroform, and "the worst day in our lives came to an end." (4.15.1909)

It was close enough to the end of their time that they had to give up
the rest of the trip and head home. Despite an attack of fever that
floored Will for several days, he managed to gather and prepare all the
animals he had captured or been given during the trip. Each had to be
boxed, caged, or made safe in sturdy aquaria. He and Crandall had to
make decisions about which creatures had requirements similar enough
and temperaments docile enough that they could live together, and
they had to make special provisions for an infinity of dietary needs. The
ship sailed for New York with fifteen boxes of snakes, lizards, turtles,
vultures, ducks, and opossums, a green tree boa that had lived in the
Georgetown Museum's cellar for ten years, and a beautiful sloth from
Aremu. (4.22.1909)

They kept the animals clean and fed, but Will's journal reads like a police blotter. On April 27, "all the birds were doing well, but an opossum escaped." On April 30 they lost a thrush and a flycatcher, then a blackbird, a cacique, and an ani. Two snakes died, and the second opossum escaped. The sloth died of diarrhea—Will saved its skin and skull—and the barn owl escaped. Both of the precious flying fish were found on their sides in their jars, but when put next to a radiator they recovered. At the dock it took all day to clear the animals through customs, after which they were transported in two big moving vans to their new home in the Bronx.

CHAPTER 14

City Lights

The toil of the explorer and field naturalist [brings]
extremes of exaltation and physical pain which no
dweller in cities can ever realize.

William Beebe, *A Monograph of the Pheasants*, Vol. I

T HE BOOK THAT EMERGED FROM
the expedition to British Guiana was *Our Search for a Wilderness.* Dedicated to Blair's grandparents, Judge and Mrs. Roger A. Pryor, it was sub-
titled "An Account of Two Ornithological Expeditions to Venezuela and
to British Guiana." Will made Blair the first author of this work, as she
contributed much of the "human interest." At the close is a statement
of what was to be Will's manifesto, his blueprint for the future he
planned for himself and Blair.

> Some day, if we do not delay until the destroying hand of man is laid
> over this whole region, we may hope partially to disentangle the web.
> Then, instead of a seeming tangle of unconnected events, all will be
> seen in their real perspective: The flower adapted to the insect; the
> insect hiding from this or that enemy; the bird showing off its beauties
> to its mate, or searching for its particular food. These things can never
> be learned in a museum or zoölogical park, or by naming a million more
> species of organisms. We must ourselves live among the creatures of
> the jungle, and watch them day after day, hoping for the clue as to the
> *why* — the everlasting *why* of form and color, action and life.[1]

The book is a pleasure to read, an easy blend of science, natural history, and adventure illustrated with crisp photos of people, places, and creatures. At the end are several appendices which, as in most of Will's books, list all the birds seen, with their common, scientific, and local names, making the volume useful as a reference text as well as entertainment.

Reviewers were universally enthusiastic, citing the novelty and interest of both matter and manner. *The Independent* opined that whether writing of the search for birds or of gold, "or of the life of the jungle, the beauty of its vegetation or the virtues of its silent Indians, masters of woodcraft, the authors never fail to be picturesquely and adventurously interesting. One of the most satisfactory trips one can ever make by proxy in that armchair." Of the forty reviews in Will's scrapbook for this work, from journals as diverse as *The Nation* and *Bird Lore,* almost every one relates a different favorite episode or insight.

Back at the zoo, Will's position as curator of birds was changing. With Crandall to help with the management of the aviaries on a day-to-day basis, he had freedom to write and to plan his next expedition. Under Osborn's and Roosevelt's patronage he had edged into the upper echelons of the New York Zoological Society and was developing into the public relations bonanza his astute professor had predicted. Despite the showy conservation battles Hornaday fought with fanatical zeal, it was Beebe who garnered the most press coverage. He could tell a good story and illustrate it with his own photographs. He had an attractive, well-spoken wife who not only was the granddaughter of a Supreme Court judge and a society author, but accompanied her intrepid husband on his daring exploits. On top of it all, accurate science made his chatty nature writing worthy of perusal by serious readers, and the exotic locales and adventures made the science palatable to a wide audience. At the same time, his technical papers were making real contributions to his field. Articles such as "Ecology of the Hoatzin,"[2] "An Ornithological Reconnaissance of Northeastern Venezuela,"[3] and "Racket Formation in Tail Feathers of the Motmots"[4] shed light on subjects that had never before been researched.

None of this coverage was lost on the zoo's governing board, which, in addition to establishing an internationally acclaimed zoological park, also had to make it a paying proposition for the city. William White Niles, the prominent New York lawyer who had handled the transactions that resulted in the Bronx River Parkway, was one of the founders of the zoo, and its treasurer. He too, along with Osborn and Madison Grant, realized Beebe's worth to the Zoological Society, and saw to it that the Beebes were well connected socially. His nephew Robin, a friend of Will's from his earliest days at the zoo, introduced them to some of the rising generation of entrepreneurs who, in addition to being wealthy, were closer to their own age.

Henry Fairfield Osborn continued to bring Will and Blair forward into the best scientific society. A scion of a wealthy and well-connected family and a Princeton alumnus, Osborn was able to induce many of his friends to underwrite this up-and-coming young scientist's ambitions. The Roosevelts were frequent visitors to the Osborns at Castle Rock, as were the Kusers, who lived across the Hudson in New Jersey. Anthony Kuser adored all birds—he had a small wildlife park at his estate, Faircourt, and many aviaries, and more planned—but his ruling passion was pheasants. When he met Will, a massive project that had long been dear to his heart began to take shape: the creation of an authoritative, comprehensive guide to the pheasants of the world.

Kuser was the personification of the American dream—a self-made man who had started his career as a pants-presser, moving up to be a streetcar conductor in Trenton, New Jersey. He cannily plowed his savings into the New Jersey Light and Traction Company, which became Public Service of New Jersey, just as kerosene power was giving way to Edison's electricity. He increased his power and social standing by marrying Susan Dryden, whose father had founded the Prudential Insurance Company. (Their son, Dryden, would become a U.S. senator from New Jersey and Brooke Astor's first husband.) Kuser's great passion, however, through all his commercial dealings, had remained wildlife—especially birds.

In Beebe, Kuser saw the answer to his prayers: he was the very man to research and write the much-needed comprehensive tome on the pheasant family. Despite the importance of the genus to humans—the domestic chicken being a direct descendant of the red jungle fowl—relatively little was known about these reclusive, fantastically varied creatures. Kuser, a sporting man himself, was determined to end that ignorance.

After a summer of working up the material from his Guiana trip, however, Will was eager to launch what he dreamed would be a lifelong study of South American birds. He faced a challenge familiar to researchers everywhere: without his curator's salary and the income from his books there would be no tropical research at all; yet with the fragmented sort of expeditions his writing and zoo job permitted, there could be little real research and no comprehensive experimental program. The conundrum unresolved, December brought a radical shift in his plans.

> Col. Kuser asked me to write a monograph of the pheasants and Blair & I will sail on Dec. 29th on the "Lusitania" for a year in central Asia and the East Indies. . . . Crandall will take my place at the Park while I am away. It is a most wonderful dream, and we can hardly realize it is true. (12.13.1909)

Beebe's mandate was to find every species of pheasant he could, observe its behavior, and document its habitat in detail, providing data for the American Museum's dioramas and Kuser and Beebe's own rearing projects. He would collect live specimens of any that he could for both Kuser and the zoo, mounting and preserving any dead ones en route. He would be accompanied by Blair as well as an artist who would make sketches from life.

This expedition would be a radical departure from traditional natural

Golden pheasant, by C. R. Knight.

history trips. Instead of traveling for a different purpose and studying whatever happened to be in the way, as Darwin had done on the *Beagle,* or setting up camp to collect specimens for museums and zoos, as Alfred Russell Wallace had done, Beebe would circle the globe, studying in detail a single family of closely related species. Kuser was funding a study that would result in a comprehensive guide to the pheasants, elaborately bound and a monument to his passion. Will's goal was more elusive: to understand the ecology and evolution of the birds in order to establish an evolutionary continuum of this one group in the most accurate and exhaustive way possible at the time.

The zoo's board had debated long and hard over Kuser's proposal. Beebe would have to be away from his duties for an unprecedented time, and Hornaday, never eager to encourage his restlessness, objected strenuously to the plan. The scientific papers that had come from the British Guiana trip, however, had increased the zoo's status as a research institution, and Beebe's popular articles and press coverage brought the zoo the all-important name recognition. In the end, Kuser had his way and was allowed to donate $60,000 for the expenses of the trip and the preparation and publication of the monograph. After all was said and done, he would contribute nearly $200,000 to the project—more than $2 million in today's dollars.

The pheasants that were so dear to Kuser's heart are indeed a fascinating group, varied in color and form, some so outrageously beautiful that they were in danger of being hunted to death for their plumage. They are also so uniformly good to eat that many were nearly extinct for that reason. As Beebe wrote in the introduction to the monograph,

> Handicapped as the pheasants are by long tails, decorated wings, ruffs, and the most brilliantly colored feathers, covering flesh beloved by every carnivore from man to marten, these wonderful birds have found a place for themselves on mountain, plain, and island, and by exercise of the keenest of senses, have outwitted their foes and overcome physical characters which long ago would have doomed less virile groups of birds to extinction.[5]

These survival tactics made them also notoriously difficult to study.

The pheasants also presented an intriguing evolutionary puzzle. The species we know best, such as the ring-neck and the peafowl, are those in which the males have elaborate markings and long tails. But in other members of the group the sexes are very similar. Darwin had formulated the concept of sexual selection, a special case of natural selection, to account for the differences between the sexes. He divided sexual selection into two types: contest and choice. In contest-based selection one sex, almost always the males, fights for social dominance or to control resources that females need. In choice-based selection, females would be expected to choose between males on the basis of their decorations and display behavior, thus perpetuating those characteristics that they favor. Almost no one believed that a system of female-choice selection could evolve, since it put males at such a disadvantage in survival. Appendages such as the peacock's heavy, dragging tail make the birds' lives much riskier.

Pheasants had been Darwin's key example of this sort of sexual selection, and Beebe would be the first to study their mating systems systematically in the wild. He hoped to determine whether the males' exquisitely awkward adornments had any advantage, either in male contests or in female choice, and whether they served any other subtle

Reeves's pheasant, by C. R. Knight.

purpose. He would weigh and measure, watch and wait. He would com-pare the social systems of pheasants that all looked the same, and those that were wildly sexually dimorphic—different in size, shape, or col-oration from one sex to the other.

The pheasants are in the same family as the common domestic chicken, the Galliformes. Beebe grouped them into four subfamilies, the Perdicinae, Phasianinae, Argusianinae, and Pavoninae. All of these groups, after a brief, failed incursion into Europe in the Miocene, make their homes in Asia and certain islands of the East Indies.[6] The familiar ring-necked pheasant was imported into Europe as a game bird, and then into the New World. For the monograph, Beebe would treat the twenty-two genera included in these groups. He would have to look in Ceylon for the jungle fowl, which are the closest relatives of the domes-tic fowl we know today, and in the eastern Himalayas for the tragopans and the Koklass pheasants. The Cheer pheasants and the so-called true pheasants inhabit the high peaks of the western Himalayas; the gor-geous long-tailed pheasants live in China, Japan, and Burma; and the golden and Amherst pheasants are found in the mountains of western and central China. To find the Argus he would have to search Borneo and the Malay States; the peafowl were most plentiful in Java.

The pheasant expedition was a mammoth undertaking, and marked Beebe's induction into the ranks of explorer-naturalists. He carried let-ters of introduction to all the big guns in Europe and Asia—in the safari as well as the scientific sense, for, as in the conservation movement as a whole, some of the bird world's greatest benefactors were themselves sportsmen. In addition, he had to visit some of Europe's most impor-tant zoos to collect information and ideas, and natural history muse-ums to amass data about the birds he would be searching for. Some specimens were so rare that examples of their plumage were displayed in only one or two places worldwide, and to identify the birds he was seeing in the wild, he needed explicit descriptors.

Colonel Kuser threw a festive bon voyage party for the Beebes as they set sail. After five days of the high life aboard the *Lusitania,* whose immense bulk and unparalleled luxury made it a floating palace, Will and Blair landed at Fishguard and boarded a train for London. The British capital delighted them—the quaint streets and ancient build-ings, the helmeted bobbies and the Horse Guards all looked right out of Dickens. At the famous British Army and Navy Stores, they amassed the gear they would need for their extended expedition. They bought pith helmets, botanical collecting supplies, chemicals, medicines, batteries, and food for the first twelve weeks.

Blair added, in a letter to her mother, that she had bought a low-necked evening dress of white net over white satin, and that Will had bought her a marvelous present, a green morocco dressing case "marked M. B. B. in gold and lined with green watered silk." It was fitted up with everything a lady could need on the road:

> 2 hair brushes, two clothes brushes, 1 comb, 1 mirror, 1 case of manicuring instruments, 1 smelling salts, 2 cologne bottles, 2 perfume bottles, 1 brandy flask, 1 powder box, 1 cold cream jar, 1 extra jar, 1 glass tooth brush case, 1 tiny match box, 1 little pin case. The whole thing,—I mean every thing gold topped. The little manicure set consists of knife, scissors, Etc.[7]

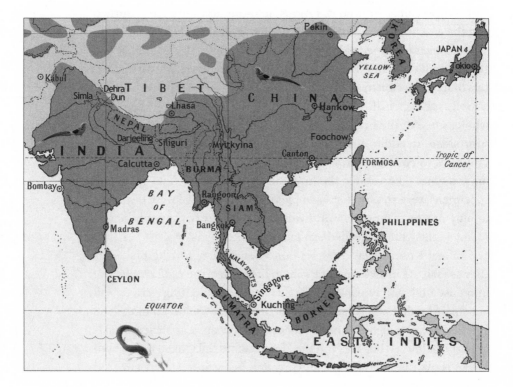

The Army and Navy Stores also furnished an entire artist's kit for Robert Bruce Horsfall, the experienced wildlife artist who was to accompany them. The forty-year-old Horsfall had a good reputation as a museum painter, and with a wife and young child to support, he needed the work desperately. In addition to making paintings for the book, Horsfall was to sketch rare specimens in the wild, including the coloration and behaviors of living birds, and enough of the landscape to allow taxidermists and designers at the zoo and the American Museum of Natural History to create realistic settings to display the specimens Beebe would bring back.

They left London for the Channel steamer with regret. Will's first taste of the culture that had spawned the Empire he so admired had been heady, and the éclat with which he had been received was gratifying. The prospect of the adventure to come was thrilling, but his responsibilities weighed heavily. He was acutely aware that many people depended on his success—not only the small expeditionary force, soon to be increased by as yet unknown servants and bearers and guides, but the vast expectant audience back home. Colonel Kuser, whose dream of a pheasant monograph could be realized only by Will's efforts; Osborn and the zoo's half-skeptical executive board; the American Museum, which confidently expected rafts of skins for new Asian exhibits—not to mention the zoogoers and newspaper readers who would receive periodic updates on their progress—all waited, trusting in his energy, determination, and resourcefulness. And then there was the added specter of Hornaday, who was sure to experience a perverse pleasure if Beebe should fail on any front.

A nightmarish trip across the storm-tossed Mediterranean took them to Egypt. At Port Said they boarded the SS *Mongolia,* which would take them through the Suez Canal and the Red Sea and across the Indian Ocean to Ceylon, where the search for the world's pheasants would begin in earnest.

CHAPTER 15

Pheasant Jungles

Every hue, every pattern, every habit, every character in these birds is often a two-edged sword, cutting toward both life and death.

William Beebe, *A Monograph of the Pheasants*, Vol. I

T HE CEYLON OF 1910 WAS, LIKE so much of the world at that time, a mind-boggling blend of the exotic East and the sedate, supra-British ambience of the Empire. Civil servants in derbies and high starched collars strode to work through dusty bazaars redolent of ripe fruit and buffalo dung, rubbing shoulders with Sinhalese, Tamils, and Malays draped in deep-dyed, loosely flowing robes. Colonial tea planters maintained rigidly British households and customs even in the isolation of small jungle holdings, while their servants went home to large and noisy Hindu, Buddhist, and Muslim families. "Great Afghan money-lenders charging 18%" sat cross-legged on the turf, waiting for customers, while "in the distance, always the patient little zebu bullocks tugging at the big wagons." (2.20.1910)

Here Will established what would become the pattern of this expedition and of future travels. After contacting the local Cook's travel agent, who was generally waiting at the dock, the group would book into a pleasant British hotel in the capital city, check the mail and other business matters, and then make contact with government and scientific officials of the district. There was usually someone Beebe knew from scientific meetings, the zoo network, or correspondence, and he carried impressive letters of introduction that helped ease the party and their luggage through the quagmire of officialdom. Then he would learn what he could of local natural history from museums and scientists, plan a route, hire upper and lower help, and provision his forces.

In Colombo, the largest port in Ceylon, Will and Blair registered at the Grand Oriental Hotel and then went to the Natural History Museum, where Arthur Willey, a prominent marine biologist and friend of Osborn's, presided. The Willeys invited Will and Blair to stay with them, see the sights, and meet their friends. Party followed party as the Beebes were introduced to displaced Westerners who, though they lived in and off the country, were never *of* it. Starved for news and gossip from the West, they vied with each other to entertain the visitors. "I never was such a fashion plate before now!" Blair wrote to Mrs. Beebe. The ladies were all copying her clothes, her shoes—even her hairpins. Mrs. Willey was entranced by her nightgowns, never having seen one that went on over the head.[1]

At the museum they photographed ancient Sinhalese relics that bore emblems of peacocks and jungle-fowl cocks, for an article Blair planned to write on the ritual significance of the pheasant. She tried her hand at developing the films herself, but in the intense heat they wilted hopelessly and "ran off the celluloid like thin gravy," despite the forty pounds of ice used to cool the water bath. From that point, they used the little native Kodak shops that graced the marketplace of practically every city.

Blair's courses in practical medicine at Pratt came in handy, as she also acted as the party's doctor and prided herself on keeping them "bloomingly well." Malaria was raging, and Horsfall, the artist, was particularly nervous about contagion. "I started my patients and myself on 2 gr. Quinine pills 3 times a day for a week," Blair wrote her mother, "then iron and arsenic ditto—leaving off quinine. Then nothing for a week. Then repeat treatment. It is a splendid anti-malarial tonic. By omitting every 3rd week you give the system a rest and the medicine has more effect."[2] Poor Horsfall, left as often as not to fend for himself, suffered a "slight attack of ptomaine poisoning" but recovered quickly. Self-conscious and alternately too diffident and too pretentious, Horsfall did not fit comfortably into any society. Although he had been trained abroad and was well respected for his wildlife paintings for museums and universities, this was his first encounter with the tropics; nor was he gifted with any genius for social engagements.

With the information they had collected in Colombo, the party headed inland to the ancient hillside city of Kandy. At the several taxidermists' shops in the colorful bazaar, Will secured many good animal skins for the American Museum's planned Ceylon group, and a fine collection of local butterflies for himself. Working with local naturalists and zoo officials, he laid out a route that would take him as near as possible to known and established territories of the Ceylon species he sought. He drew information about tracking and hunting methods from British civil servants who, because they were inveterate hunters, often provided the most reliable information.

Lovely as Kandy was, the party was happy to escape civilization. The *Lady McCallum* took them south from Colombo to the Bay of Galle and Hambantota, on the southern coast. With Horsfall, a cook, a taxidermist supplied by Dr. Willey, a Sinhalese manservant who could also skin animals beautifully, a Tamil tracker, three bullock carts, three drivers, six oxen, a game license, a boar's skull, a jungle-fowl egg, five peacock feathers, and two dozen bottles of soda pop generously piled on by a young government agent, they were off to the wilderness.

The trip by oxcart was memorable, and Will was to write of it years later in *Pheasant Jungles* as a precarious, noisy, unsavory journey. The

The infamous bullock cart, fording a stream in Ceylon.

carts bumped slowly and inexorably on, the bullocks' wooden bells setting up a constant clamor to drive away evil spirits—which also drove
away any interesting wildlife. The drivers flogged their beasts stolidly
over every obstacle and through streams that bubbled up through the
flooring of the carts and over the passengers' feet. When they stopped
to eat, the drivers freely intermingled the dish-drying and bullock-
wiping cloths. "This," Beebe commented dryly, "accounted for many
things."[3]

They arrived at length at the first of a succession of remote government *daks*, small bungalows maintained for the benefit of traveling officials and officers, and used with permission by hunters and tourists as
well. They were kept clean and minimally maintained by caretakers but
contained no furniture or supplies, on the assumption that travelers
would necessarily have all that was needful with them. While Will, Blair,
and Horsfall slung their hammocks inside, the support staff prepared
their own shelter outside and set up camp. Soon the ground was covered with mats and rugs, fires were lit, and tables and chairs were being
fashioned from bamboo with the deftness and ease of long experience.

Although arid, this place—called Welligata—was teeming with
birds, and Will wrote Kuser that he was experiencing hot sun, untold
millions of thorns, stinging ants, ticks, and midnight noises, combined
with wonderful animals, splendid health, enormous appetites, and joy
and delight at being in the jungle again. After ten days, Will was able to
report that he not only had located the jungle fowl Kuser coveted, but
had been able to creep close and observe them "feeding, crowing, even
sleeping." He collected specimens of both sexes of jungle fowl and peafowl, and packed boxes with their food and shelter plants, along with
nests, birds, and insects to make a display for the monograph artists as
well as for the museum Kuser was hoping to establish himself. Triumphantly, Will announced that he had secured for Kuser rare living
hybrids of the domestic fowl and the Ceylon jungle fowl.[4]

After the first week, Blair sent Nettie a description of a "typical"
day's work in the jungle:

Six or 6.30 A.M. coffee, cocoa, crackers, jam, bananas & mangoes.
Working in the jungle until 10.30 A.M.
Breakfast 11.30 A.M. Jungle Fowl croquettes, okra, potatoes, rice,
limeade. Fried banana, bread, butter, Peters chocolate.

Write up notes, photograph, draw, skin until 3.30 P.M.

At 3.30 o'c, cup of bouillon & crackers.

Work in jungle again until 6 o'c P.M. At 6 o'c I take a luxurious hot bath in a basin of hot water. Don my pajamas plus a skirt for modesty. Braid my hair. Dine at 6.30.

Dinner. Soup (flour dried soup), roast or stewed chicken, or eggs on toast, vegetables (various), potatoes, okra, stewed fruit (pears, apples, prunes, apricots or peaches made from dried fruit—a French preparation), cold boiled rice & custard, cocoa.

After dinner we load [photographic] plates. W. & Horsfall take their baths. In hammocks by 8 or 8.30 P.M. and asleep at once after our long day in the jungle.

The cook is a treasure and I greatly enjoy my task of housekeeper, for I get a lot of fun out of my coolies.

I am as you see quartermaster, assistant photographer, scribe, etc., etc. Also Doctor. I wish you could have seen me this morning, with half a dozen forlorn natives around me and my medicine kit.[5]

Both Will and Blair were becoming disenchanted with Horsfall. Perhaps because of his inexperience and unease, he tended to speak patronizingly to the native people, exasperating both Will and Blair, who were accustomed to dealing with servants. Worse, when socializing with others Horsfall tended to exaggerate the importance of the expedition, his position in it, and even its findings. This concerned Will

Blair in Borneo, 1910.

greatly because if reported, it could cast doubt on the scientific reliability of the whole endeavor—a possibility that haunted him constantly. A scientist without a bona fide degree on a rich man's expedition, his writing often walking the fine line between strict science and popularization, Beebe would spend his life waging a battle to be recognized as a scientist first, a writer second. As Will wrote Kuser, Horsfall was "afraid of the dark, afraid to eat strange food, and yet is foolhardy enough to pick up a stunned snake by the tail."[6] As long as his work was good, however, Beebe determined to make the best of it.

Five days on steamers and trains took the travelers to Calcutta. Several rare species of pheasant could be found only at 15,000 feet on certain mountains in the Himalayas. Will had promised Kuser that he would study and capture specimens of every one, and he was determined to make good. As rapidly as he could, he established dealings with various animal traders in the city and arranged an animal exchange agreement with Nelson Annandale, head of the Calcutta Museum, as he had done with the London Zoo. Ditmars was eager for cobras and other exotic reptiles, which were primarily available only at prohibitively high prices—and often near death—from less-than-honest dealers. Annandale immediately shipped a dozen fine cobras to New York, in exchange

Beebe preparing the camera, while Horsfall paints.

for some wood ducks that bred easily in the Bronx but were unavailable in India.

Will and Blair spent days in the museum, taking notes on every species of creature they were likely to encounter. "Time is my trouble," Will wrote. "Try as I will I have only succeeded in finishing my studies of one genus at the museum. Mrs. B and I stop not a moment from morning to night except in the worst heat."[7] Evenings were devoted to paying the necessary social calls and attending the endless festivities that were planned for the guests. Will put the lost hours to work by grilling longtime residents and sportsmen about routes and strategies. The unsettled nature of some of the country they most wanted to survey was a recurrent theme. The hill tribes were restive, and the Chinese were making forays into Indian territory. British troops and native Gurkhas were preparing to mount punitive missions through the prime pheasant areas, and locals warned the visitors to avoid the hills.

Undaunted, Will and Blair continued their extensive preparations. The vast Himalayas beckoned, where blood pheasants and Impeyans pecked and scratched amid the snows of the peaks, regardless of men's petty strife. Horsfall, however, became increasingly nervous and demanded peevishly that they stay close to civilization—which to him meant well-defended British compounds. Instead of convincing Will, his feeble protests only made him an object of greater pity and scorn. In the oppressive heat, the party packed the thick woolens and sweaters they would need in the cold nights ahead and blithely mapped a route that satisfied their optimistic goals.

A hot and dusty train ride took them to Siliguri, where they changed to a tiny narrow-gauge "toy" railway that still weaves a tortuous route up to Darjeeling, one minute puffing under high bridges, then doubling back to cross over the same bridges as it zigzags toward its lofty destination. As the little train forged ahead, its passengers passed from the baking plains of the Terai, with its brush and stunted timber, to humid forests of orchids and tropical trees, through the zone of towering rhododendrons and tree ferns, and on to the cool and fragrant pine forests, where waterfalls dashed over sheer cliffs to nourish the luxuriant growth far below.

The lovely mountaintop aerie of Darjeeling was cloaked in cloud, giving Will the uneasy feeling of being isolated in the universe. The next day, the vast reaches of the snowcapped Himalayas were spread out

before them. To this hill outpost British civil servants were sent to recuperate from overdoses of Indian heat, and wives and children vaca-
tioned in its brisk, pure air. Here Will laid in provisions for a major trip
into the mountains, and hired the servants and porters that would form
the backbone of the expedition.

To this point, their journey paralleled that of countless British and
American tourists, who delighted in following comfortable rail routes
and well-established trading roads to exotic destinations. The ubiqui-
tous Cook's travel agency had blossomed with the tourist trade, thanks
to the spread of the railroads from England throughout the Empire.
From its beginnings as a business ferrying teetotaling rural churchgoers
to town meetings, Cook's expanded its services to include arranging
entire itineraries, reservations, baggage handling, meals, mail, and trav-
eler's checks for globetrotting Britons, helping to make trips to exotic
lands as simple as a jaunt to the Lake District. Guides and agents at foreign stations and ports solved every problem and smoothed every path.
Many of Cook's peripatetic tribe kept copious journals of their journeys, including details of manners and dress at the myriad dinner
parties and picnics they attended, and the names of neighbors from the
Midlands encountered in the Swiss Alps, the Himalayas, or Bombay.

As a young woman, even the dauntless Gertrude Bell chronicled a
claustrophobic 1903 trek from British outpost to outpost, attending

*The Siliguri
train winding
up to Darjeeling.*

dances and dinner parties at every stop. Despite this lackluster introduction, she would build on those experiences to become a specialist in
Eastern customs and politics, a spy for the British in World War I, the
learned and daring mentor of T. E. Lawrence—the intrepid "Queen of
the Desert."

The scientist, explorer, or hunter followed a vastly different agenda.
The Beebes took full advantage of Cook's services, but left the boardinghouses and station cafes of the Cook's crowd behind to follow the
pheasants, which refused to adhere to well-beaten tourist routes. The
famous botanist Joseph Hooker had published a journal of his 1848–50
expedition for London's Kew Gardens, a journal that covered much of
the route Will and Blair now followed. The lure of exotic plants yet
undiscovered, and the government's urgent need of maps and surveys
of the area, drew Hooker to Sikkim and the vast northwest area of the
Himalayan chain. Sixty years before Beebe's expedition, Hooker had
beaten the trails that Will would follow with much the same cumbersome equipment and goals; there had been little improvement since
Hooker's time in the route's amenities.

Tandook.

Tibetan woman.

Tandook, a versatile servant of Chinese extraction, was Beebe's head servant or "coolie" on this leg and hired a motley crew of twenty-seven hill men to carry the vast impedimenta. To Will's surprise, eight of the porters were women. They carried the same heavy loads as their male counterparts and contributed much of the good-humored jesting and complaining that made the day's march pass quickly. "'Our loads are all too heavy,' the porters shouted in unison. 'Your loads are all too light,' screamed Tandook stormily, first in Hindi and then in a medley of hill tongues."[8] The bearers laughed at Tandook and his comical hat and pig-tail and then, still giggling, shouldered the heavy packs.

The expedition crew also included a venerable bearded "sweeper," whose job was to precede the group and prepare the *dak* for occupancy, sweeping it with the broom that was his symbol of office and lighting fires both inside for the sahibs and outside for the servants. Casteless, he had to eat by himself and be careful not to touch the food of the others. Will wrote that he was in this perhaps the most to be envied, "for being unable to sink lower, for him there was no book of etiquette."[9] Will was also accompanied by his "man," a Tibetan named Satan, who could wait table, prepare shaving water, and track animals with equal dexterity. The party included four horses, with their attendant *syces,* or grooms, and specially delegated fodder-gatherers. Das, an animal skinner from Baluchistan, was indispensable, and Masson, a Scotch hunter down on his luck who knew the country, accompanied them, along with his own man, for five rupees a day.

The human train snaked its way from Darjeeling along the steep trail up Tonglo, a mountain where the black-backed Kalij pheasant—"a beautiful steel-blue, white-breasted bird with the hen clad in red-browns and russets"—was said to dwell. In the valleys it was still warm, but they left the bamboo behind and entered a new jungle zone, where bell cicadas sent their tremulous, sweet metallic vibrations through crystalline air. They passed through Ghoom, with its beautiful and ancient monastery, to the bungalow at Jorepokhri that would be their camp, set high on a lovely, windswept cliff. Temperatures dropped below freezing as the trail continued zigzagging up to a ridge that led to the pheasant-friendly peaks of Tonglo, Sandals, and Phalut.

The wind blew hard, and now swirling snow bit into eyes and faces. Will wore sacks over his hands to take photographs and collect moss, barberries, and other everlastings for the pheasant displays that Kuser and the museum planned. The altitude made even Will short-winded, Blair developed sores from the cold, and they all suffered respiratory troubles. Much of each day was spent in silent immobility as Will waited, motionless, for the forest creatures to forget his presence. Much of his frustration was the result of feeble human corporality getting the better of him. Time after time a tortured limb would demand just the tiniest of stretches, and a startled insect or bird would destroy the peace that hours of discomfort had bought.

One hunting technique that gave him a bit more comfort and freedom was to erect a small, inconspicuous tent or blind, well hidden beneath a towering spruce or deodar, whose long shaggy branches nearly swept the ground. He would leave the blind in place for several days before daring to use it, so the forest creatures would become accustomed to it. Inside, although he spent hours lying quiet, his eyes or field glasses glued to the open tent flap, he could occasionally rise and massage his angry muscles. When a tent blind was inconvenient, he sometimes used a large umbrella, staying hidden beneath it until the wildlife resumed its accustomed rhythm. Other times, he simply found a low, concealed spot and lay at full length, motionless for hours at a stretch. Most forest animals live in a dim world of shadow and light, and though keen of ear are generally better at spotting movement than in discriminating shapes.

One day he lay prone on a flat boulder covered with damp moss, waiting for the Kalij whose nest he had spotted to amble back from feeding. So silent and motionless was he that a laughing thrush landed near his shoulder, startling him so that he turned to see what it was. In a frenzy of fear the thrush let loose a volley of cries, alarming every creature within earshot.

> The terrible silence of fear closed down upon the jungle. Myriads of living things breathed quietly, panted or held their breath, while my thrush continued to shriek its terror to heaven as it fled headlong. For several minutes the moss-hung forest gave forth not a whisper of life. Only the slow flapping of great deaf butterflies showed that anything still lived within the shadows.[10]

Will was pleased with the success of his observations of the Kalijs, but the blood pheasant, the most valuable prize they sought in that area, still eluded him. They heard its distinctive call from the pine forest as they climbed, but could not get near enough to see it. They made camp on the high reaches of Phalut, and the next day Will and Masson set out with two of the Tibetans on a difficult twelve-mile trek through heavy snow that turned out to be fraught with peril.

> We pushed on—Masson, myself, our 2 best Coolies & some chocolate. We had to climb a hill 1200 feet high knee-deep in snow; icy wind blowing, at an altitude of 13000 feet. We found a snow leopard's track; then where it had killed a cat-bear, then where it had flushed a covey of Blood Pheasants—a feather of one of the birds clinging to the edge of the snow crust where they had burst out.

Will led the way, but suddenly, near the top, he slipped off the trail and began to slide down toward a precipitous drop. In desperation, he tried to slow his progress by digging the edge of his precious Graflex into the snow, to no avail.

> As I slid I revolved—and thought. Below me was a sheer drop of 2500 feet. They say that people think of their past life & sins, but strangely enough, each time I revolved, as I faced the snow peaks I pondered only on the resemblance of the reddish clouds & green pines far below to the colors of Blood Pheasants.

He remembered having a terrible desire to see what was over the edge of the precipice, of being almost in a hurry to get to the edge.

> Of course in reality I must have been almost paralyzed with fear, & these strange dreamlike ideas were only momentary. But they were what remain clear to me. I went around faster & at last was about to let go my camera in a last effort to stop when my foot caught in an inequality, slipped out, the other heel went in—and held. This left me upside down on the slope, & you can imagine with what care I reached around & got my hand in—the leopard's track. He had walked along here when the snow was soft & sunk in deeply, & his tracks had now been frozen. Didn't I bless that animal!

Characteristically, Beebe short-circuits the tension with a wry anticlimax:

When . . . my knees had lost some of their shakiness I crept along the leopard's track & looked over the edge. Now I had a second shock— one of mortification for there, about 6 feet lower down was a nice broad rock covered with snow, on which I would have landed! The sheer drop began from that rock. . . . But while revolving down that slope I knew nothing of the rock & its safeguard.[11]

Will continued to search for tragopans and Impeyans, and to document their ranges with written descriptions and photographs. Blair, impeded by her skirts in the snowdrifts, pursued her own interests closer to camp. In a letter to her parents, Blair told of one day when Will had been away all day "on a terrible hard tramp, too much of a trip for me. So I with 10 coolies have been excavating wild mouse tunnels, photoing, making notes, etc., incidentally finding out interesting things about the coolies, as well as the mice. . . . We are all working like steam engines, and the time flies. But are we not having a really wonderful time?"[12]

By the end of September Will felt he had done as much as he could in the eastern part of the range, but Horsfall was unwilling to venture farther into the unsettled west. At his wits' end, Will decided to leave him in Jorepokhri to paint native flowers and any specimens the locals brought in. There had been a continuous flow of pheasants, nests, eggs, and various beasts and insects since their arrival, so Horsfall would earn his keep of a rupee a day at the bungalow.

CHAPTER 16

Western Himalayas

An ornithologist in the field knows no dignity. . . . And if the
approach is made well, with the patience of a creeping feline,
the reward is sure to be overgenerous, out of all proportion to
the bodily discomfort.

William Beebe, *Pheasants: Their Lives and Homes,* Vol. II

FROM THE CITY OF DEHRA DUN
they trekked on ponies 5,000 feet up to the breathtaking hill resort of
Mussoorie, where the forest agent lent them a bungalow deep in the
woods, surprisingly covered with fragrant white climbing roses and
looking every inch an English cottage. Here Will hired a headman and
a *chowkidar,* or housetender, the *syces* or grooms he needed for his horses,
local forest rangers as guides, and wild hill men who were whipped by
the headman to work.

The forest above Mussoorie, lush with silver spruce and deodar,
roses, and flowering raspberries, was also rife with pheasants. Will and
Blair discovered a Kalij within 100 yards of the bungalow, and Will
heard the calls of Cheer pheasants, tragopans, partridges, and Koklass
pheasants on his walks. And he knew the rainbow-hued Impeyan lived
here too, the most beautiful of pheasants, and he was mad with desire
for an opportunity to study its habits. The only thing that made the setting less than paradisiacal were the malicious biting flies that left blis-
ters of blood beneath the skin that itched wildly and often festered into
ugly sores.

One day in Garhwal Will lay, unmoving, alongside a long mossy mound that had once been a mighty tree, hoping to see one of the splendid Impeyans that he knew frequented the glade. Attempting to ignore the flies, he concentrated instead on a gargantuan slug that slid its way along the trunk on its trail of slime.

> The great mollusk crept along the damp bark, leaving a broad shining wake of mucus, then tacked slowly and made its way back. In the meantime various creatures, several flies and spiders and two wood-roaches, had sought to cross or alight on the sticky trail and had been caught. Down upon them bore the giant slug and, inevitable as fate, reached and crushed them, sucking down the unfortunates beneath its leaden sides, its four, eyed tentacles playing horribly all the while. The whole performance was so slow and certain, the slug so hideous, and my close view so lacking in perspective, that the sensation was of creatures of much larger size being slaughtered.[1]

Unknown to Will, the slug was also unknown to science, and when he secured it in one of the preserving vials he always had about him and sent it home for identification, it became the first organism to be named after him: *Anadenus beebei.*[2]

His wait was rewarded with a slowly assembling group of splendid cocks who dug industriously near him, picking for grubs in the grass with their strong beaks. "When the sun's rays reached the glade, the

Pheasant haunts of the eastern Himalayas.

scene was unforgettable: fourteen moving, shifting mirrors of blue, emerald, violet, purple, and now and then a flash of white, set in the background of green turf and black, newly upturned loam."[3]

An unanticipated bonus of their time in Garhwal was the advent of Halley's comet. Will and Blair had plenty of opportunity to take in the landscape made eerie by its "sickly green light" and to ponder its awesome timelessness. Will thought particularly of the changes the world had seen since the comet's last visit—changes that he cast in the positive light of progress: railroads, the telegraph, the *Origin of Species.* The comet prompted him to think, too, of the beauty in nature that so many people—even, incredibly, scientists—are unable to appreciate. "I marvel that men can spend their whole lives in studying the life of the planet, watching its creatures run the gamut from life to death, bravery to fear, success to failure, life to death, and not . . . be greatly moved."[4]

From Agra a hot and dirty train took them to Lucknow, stopping dead for fifteen suffocating minutes out of respect for King Edward. His death, many Indians felt, had come at a bad time, as there was much unrest and talk of a revolt. Beebe chafed indignantly at the middle-class "Tommies," the English who treated the native people badly. "All take it here very seriously in view of the present rotten government, and many think that the new king will be deposed and a republic set up. Heaven forbid! There is not the slightest doubt but that there will be a terrible revolt in India before long. All the natives we talk with, when they learn we are Americans, make no secret of it. And it will all be due to the pig-

Will and Blair looking toward Everest.

headed Tommies and common English who treat the natives like dogs."[5]

Horsfall joined them in Calcutta, bringing what Will described as a "really good lot of stuff"—specimens for the museum and live birds for the zoo, and many paintings. Blair added to her list of accomplishments winding up accounts, changing money to gold sovereigns, replenishing their "scientific outfit, insect boxes, etc.," leaving mailing orders at Cook's, and writing business letters—all in a torrent of rain, for the monsoon had reached Calcutta. The next day, in stifling heat and humidity, Will won two out of three sets of tennis with a geologist and a vertebrate paleontologist.

Their next destination was Indonesia, for which they left aboard the *Japan* on June 2. On board, Will and Blair found that Horsfall, inflated by his first-class status, had reverted to his annoying ways, unable to resist asserting his importance. Ironically, by trying to assume great dignity, he made himself—and, in Beebe's eyes, all of their party—look ridiculous. Worst of all to Will and Blair was that, like the ignorant Tommies, Horsfall treated the Indians with open contempt.

The *Japan* steamed around the northernmost of the Andaman Islands, along the Mergui Archipelago, and down the peninsula to Penang. In addition to writing to his family and to Colonel Kuser several times and making extensive lists of birds and animals seen in India, Beebe used the time to write a report for Madison Grant and separate letters to young Dryden Kuser, who was showing a precocious interest in birds. He also wrote to his old friends Warrie Mountain and Ferd, and to Hornaday, Osborn, Crandall (who had just been married), the Rices, the Bulls, the artist William Stone, New York Zoological Society treasurer William White Niles, and his publisher and friend Henry Holt. And he worked with Blair on an article about Tibet, which was eventually published in *Harper's*. "I think Blair is going to be a 500% better writer than I have ever been," he wrote Colonel Kuser.[6]

In Sarawak, a British protectorate on the island of Borneo, they were treated to visits from the local dignitaries, including Rajah Brooke, Sarawak's eccentric English potentate. Charles Brooke, then eighty-one, was the second of the "White Rajahs." His uncle, James Brooke, had helped the sultan of Brunei during a rebellion in 1839–40 and had been given the province as a reward. His claim, idiosyncratic as it was, survived bloody native uprisings and many determined legal attacks; in his old age James had adopted his nephew Charles and made him his heir. Charles's own son had just married the young English heiress to the Huntley-Palmer biscuit fortune, and Blair wrote her mother that she wondered "how Miss Millionaire Biscuit will like living in Sarawak."[7]

For Blair and Will, Sarawak was just right. Although American rubber interests were felling the ancient forests, there were still plenty of birds in what remained—though for how long, no one could predict. The country itself was a rich mixture of English, Chinese, and native Indian cultures. The indigenous Dyak Indians, an amalgamation of fierce tribes with a complex animist tradition, were gradually giving way to Chinese Buddhists, Indian Hindus, and ambitious Christian missionaries.

The local architecture, Blair wrote to Mrs. Kuser, "is all Dyak in character, with the defect, unimportant in Dyak eyes, that all the roofs leak and there are no walls. This is unfortunate in a land of heavy rains for people who wear clothes and have possessions. The Dyaks, doing neither, did not consider it in putting up our houses." The unrepentant Anglophile Brooke, however, lent the expedition his "simple" country bungalow, where they enjoyed good food, fine linen and china, and "every convenience." On the outskirts of the town was the menagerie of animals they had collected, which was under Blair's special charge. "It is an interest and a despair at the same time," she wrote, "for I grieve so over the death list."

> We have now alive:
> 1 leopard civet cat
> 2 Exquisite little male wood partridges
> 3 Argus pheasants
> 1 female crested Fireback
> 1 young male or female Bulwers Pheasant.
>
> We had a young Pitta which I longed to keep alive for you, but we could not make it eat and it is now a skinned and labeled specimen. A big moon rat escaped last night.[8]

The house was surrounded by forests rife with pheasants, jungle fowl, and fascinating creatures of every sort. In ten days they collected a hoard of specimens and skins, and prepared to head out into the wilds in search of silver and peacock pheasants. As they left the rajah's wilderness retreat, Beebe noted with frustration the rapidly changing environment: stable for generations, the exquisitely balanced system could be ravaged by man in an instant.

> In the one week time we saw a great change in the surrounding forest. Dozens of splendid trees within view of the verandah were felled, leaving ugly open gaps in the jungle vista, which in time will be burned over and given up to spindling rubber sprouts. If the users of rubber tires and raincoats and overshoes, could see the baby Flying Lemurs and other arboreal jungle creatures fleeing for their lives, or caught and eaten by the Chinamen, they would feel pity for the wild things. In a few score more years all will have become extinct, and civilization have spread to the farthest points of the world. (6.29.1910)

To penetrate farther into the jungle, Will and Blair took passage on a steamer that plied up and down the Rajang River to Sibu. Beebe watched from the bow, fascinated, as a monkey swam competently across—something people had never recorded before. But as he watched, the "brute of a Malay captain" raised his gun to shoot it. Will, who had himself been the nemesis of countless unoffending creatures, unsportingly bumped the man's arm, causing the bullet to sail harmlessly into the air. "It sickens me," he wrote, "to see life taken . . . thus without reason. All my shooting for scientific purposes, when I kill instantly seems very different. This shooting & passing on to let the dying or dead creature drift out to sea is more than bestial." (7.2.1910)

Now a bustling city where modern office blocks shade native markets, Sibu was in 1910 a small town built around an American-run mission where 700 Foochow Chinese immigrants worked a rice mill. The "upriver people" still live in longhouses and stilt houses along the shore of the great Rajang and supply local kitchens with lemurs, snakes, and flying squirrels, as they have done for centuries. Will and Blair hired a seventy-foot dugout and fourteen Dyak and Malay paddlers to take them up the Rajang, Bally, and Mugong rivers.

For the most part they slept in the canoe and worked on shore during the day. Near a particularly fertile pheasant area, they camped on shore for a week and arrived at Fort Kapit armed with Argus, fireback, and Bulwer's pheasants, both dead and alive. Beebe let it be known that he had kerosene, dollars, and tobacco to trade for live pheasants, and natives brought in many wonderful birds. But the Dyaks trapped birds for the pot, not for the amusement of New York zoogoers. Their noose traps snapped the fragile legs, which Beebe strove in vain to repair. As he watched the beautiful birds succumb from their injuries, he despaired of being able to save even one species from extinction.

> What I said about the Himalayan pheasants becoming extinct, applies many fold more to the Bornean birds. The jungles are everywhere being felled for rubber, even as far as we could go in our canoes. Blair has been a treasure and has many times pulled me out of the depths of "blueness" when I would find my best birds dead in the morning. To me it is almost agony to see the poor creatures die.[9]

Blair had more nursing to do than ministering to Will's fits of depression. Working night and day to preserve his numerous specimens, he soon sickened with arsenic poisoning. An important insecticide used liberally in taxidermy, it seeped into his system as he preserved the precious skins and combined with what he was already taking by mouth against mosquitoes.

And always, drawn by fast-spreading rumors that there were white doctors in the area, Indian families materialized out of the forest depths carrying ill, deformed, and injured children, babies weak from malnutrition, elders blind or too feeble to eat. Horrible scrofulous skin diseases were by far the most common ailment. Blair doctored everything with ichthyol, a petroleum-based fish oil ointment common in the United States but unknown in Indonesia. Miraculously, even so simple a remedy made afflicted babies "take heart and eat," recovering flesh visibly in a few days.

By the time they were ready to leave Sarawak, Beebe had secured, in addition to plenty of dead pheasants and a few live ones, eight of the fifteen rare mammals of the region that Osborn was hoping for. He had captured or bought a Bornean wild boar, flying lemur, pangolin, tupaia, porcupine, moon rat, spiny-haired rat, zebra civet, and night monkey. In addition, the Beebes had adopted a tiny little sun-bear cub who charmed everyone with his playful and affectionate nature.

At this point Will finally made the momentous decision to send Horsfall home. His behavior had become dangerously unstable, and he was endangering the expedition. Uncertain of his own status between the topsy-turvy world of the arrogant raj and the natives' Eastern deference to Westerners, Horsfall had befriended servants he should have commanded, and taken liberties with superiors whose politeness encouraged him. Blair, with her plantation-schooled grasp of social rank, suffered almost physical agonies over his gaffes. "He has been the very limit," Will wrote, "and Blair has grown thin and was almost hysterical before he went. He cares neither for his wife nor child, nor work nor opportunity, but only (as I told him) for himself. Always on the look out for fancied slights, but never knowing when he is making an ass of himself, as he always is."[10] When Will dismissed him, Horsfall retorted defiantly that he had been ill treated from the first, and that all his subsequent actions had been with the express purpose of revenging some slights that Will maintained—with perhaps less than complete candor—were imaginary.

Whatever the origins of the feud, a feud it truly had been, and Horsfall's departure breathed new life into the weary travelers. They hired a new taxidermist-interpreter and a reliable, efficient head servant, Haledin, and resolved to make sketches themselves that would provide enough material for artists back home to work from. In mid-August they sailed from Singapore to the port of Batavia, in Java, where

Blair treating a Dyak Indian for an eye infection.

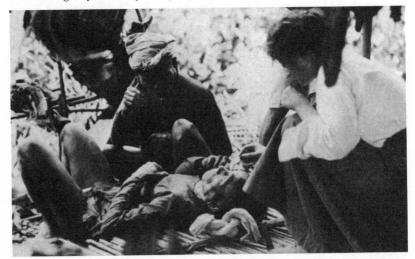

they hoped to add Javan jungle fowl to their list and to investigate claims of undocumented hybrids.

"Even in the biggest hotel in Batavia we may say we lived close to nature," Will wrote.

A mouse ran about now and then; Indian House Sparrows were building in our electric light fixture; Geckos by the half dozen wriggled swiftly over our walls, and a big solemn toad fed on the insects which dropped from our electric light; thus in our one bed-room we had representatives of Mammals, Birds, Reptiles and Amphibians. (8.21.1910)

Some of the geckos in their room were a foot long, and when the lights were put out they would begin making a pleasant *chack-chack* sound. Small bats flitted silently by, brushing their leathery wings against hand or face, drawn by the insects that were attracted to the lights. On the veranda Will managed to catch four huge mole crickets. "Since I first read of them in Thoreau and Gibson, when a little chap in E. Orange I have longed to catch one. Here in far-off Java I have achieved my wish." (8.23.1910)

Will's belief that Java, with its well-established and orderly Dutch government, would be smooth sailing for his mission proved to be optimistic. His guns were impounded in customs, museum administrators refused to help with advice or letters of introduction, and worst of all, in ten days of weary negotiations, no one seemed willing or able to tell them where pheasants of any sort were to be found. Only when he ran afoul of a coterie of voluble German hunters did he discover the seaside haunt of the beautiful green jungle fowl. He secured four cocks and a little hen for Kuser, so tame they ate out of his hand and crowed in the bedroom in the mornings, and skins for the American Museum. He also found the gorgeous Bekisar hybrid and sent Kuser half a dozen in lovely teak cages, handmade for pennies, along with three iridescent Nicobar green pigeons with striking white tails.

With the birds, Will sent specific instructions as to their care and feeding. They needed to be put indoors immediately, as they would arrive in early November and could not withstand cold. They would need careful feeding on whole corn. He tried to impress upon Kuser the birds' value: their unique hybrid characteristics could prove invaluable in the attempt to establish an evolutionary continuum.

Back in Singapore after six weeks in Java, Will found a pile of letters from family and well-wishers. Fame, he wrote Kuser wryly, had at last found him:

> The earwig which I found in the Himalayan lily-pods in the snow, on which the Blood Pheasant feeds, has proved to be a new species and has been named by some great English "bug man" Forficula beebei! Now we can feel that life has been worth living! Blair never ceases jeering at me, especially as I always have spoken so against creatures being named after people; I fear the joke will never die.[11]

Word had not reached him of the grotesque *beebei* slug. By the end of his life Beebe would have eighty-eight animal stepchildren bearing Latinate versions of his family name.

CHAPTER 17

Bearing East

The dawn of a new day had broken in the eastern Himalayas, gilding the first pinnacle of Everest, for so through all the long centuries had the greatest source of light paid homage to the highest point of earth.

William Beebe, *A Monograph of the Pheasants,* Vol. I

I WENT DOWN TO PEARSON'S wharf yesterday to get the launch to visit the boat on which my birds were going. The stairs ahead were crowded with wealthy mostly educated men: Indians, Malays, Chinese, Javanese, Hindus, & tall Sikh policemen. I stood at the top a moment, & the 6 ft. 4, splendid profiled officer swept a half dozen of the men to the wall; shouting, "Out of the way, idlers, do you not see the Sahib!" A Chinaman did not hear & a Malay merchant, perhaps worth many millions of dollars said to me "Kick him into the water, Tuan, it will help his ears." Elderly men, many with brains as good or better than mine for details, yet all without comment or question giving way, or helping to hold the boat while I entered. Such is the white man in the East—everywhere, and he accepts it as unquestioningly as it is offered. We are "it" and no one disputes it.[1]

Will's only preparation for life as a *tuan,* or master, had come from the literature of the raj. Charles and Nettie had reared him in the ideal of an equality of rights under God but a meritocracy of manners and education. He had often visited The Oaks, where the servants were respected and loved, and had been a guest at great houses such as

Osborn's Castle Rock and Kuser's Faircourt, so he had seen the way a well-run staff functioned; but he had not observed the complex methods used to engineer such frictionless running. At the zoo he put a great deal of thought into the way he dealt with the men under him, whose cooperation he needed and whose respect he desired. There, however, he was part of an organization with an entrenched caste system that made all such relations easy to maintain. Here, faced with the sudden necessity of mounting an expedition on which much money and many hopes depended, he had to learn quickly how to make a vast substructure function smoothly.

One of the first lessons he learned was that in order to get anything done, he had to earn the respect of those he commanded. Any slip in presence or control would have been disastrous. Before everyone from shopkeepers to stable boys, he had to preserve the illusion that he was a sahib of importance and means in order to maintain discipline and avoid being cheated or snubbed. It had been this unceasing pressure that made Horsfall such a liability. With Horsfall away and Haledin installed, daily affairs became manageable again, although Will remained wary of his tentative position. His journals at this time reflect his attempts to grapple with the conundrum of authority and power. It was difficult, he wrote Nettie, to know how to treat servants.

> Haledin does something unselfish, of which he has thought and planned, for our comfort—we accept it without comment, or with merely "bayres" (good). He makes an error in counting; he orders a rickshaw half an hour too early & I give it to him loudly, threaten him with all sorts of things, which he accepts meekly, almost tearfully. If I did otherwise, he would class me as a weak fool and leave us for a master whom he could respect, who, verbally and physically, would kick him about and keep him up to the highest level of duty.[2]

Fortunately he had Osborn's and Kuser's example to follow, as well as Blair's Southern belle gentility, to keep him from falling into the ugly snobbery that characterized so many of the middle-class civil servants, catapulted unprepared into positions of power. In addition, he had Rudyard Kipling, whose sympathetic portrayals of native culture and devastating riffs on snobbery reinforced the romantic literature of Will's youth, in which nobility of spirit overcame meanness in savage and civilization alike. The letters and articles he wrote show him to be respectful but undeniably paternalistic to a modern sensibility.

Free of the Horsfall incubus, Will and Blair were much more comfortable performing the social duties of their position. Kuser had briefed them on the necessity of accepting as many of the multitudinous invitations they received as they could, and even returning the favors when possible. "I know you want us to do what is right in such things," he wrote Kuser impatiently, "but it is expensive and I must say I prefer leeches and jungle to some of the empty-headed females I have to take in to dinner."[3]

From Java, Will, Blair, and Haledin took a steamer to the Federated Malay States, where they hoped to find all the Malay pheasant species. In busy Kuala Lumpur, the social duties never stopped. "At 5 P.M.," Will wrote not long after their arrival, "General and Mrs. Watson motored us out to a little club in the botanical gardens where we played clock golf, gossiped with ultra English, and I chafed for my old clothes; jungle sounds drove me wild. Tomorrow ends Governors banquets for a while. If alone I'd beat it to-day." (9.28.1910)

They spent October in the jungles and mountains of Malaysia, following rivers when they could for ease of travel, and trekking up steamy

Will and Blair's houseboat in Malaysia.

mountainsides when the birds refused to be found elsewhere. Their method of going through the jungle, Will wrote Kuser, "was to go in single file and every hundred feet I would give the order Beware— leeches! and we would stop and with sticks and bends scrape or pull off the hideous things. When we stopped for a time, the twigs and leaves all began to shake, although there was no wind, the leeches humping along them toward us." (10.20.1910) Their reward was tangible: they secured every one of the eight Malayan pheasants, including such rarities as the lovely Malay bronze-tailed pheasant and the wonderful crested Argus.

At their jungle *dak* rest house, the cleaning staff were convicts, chaperoned by an armed Sikh guard; the prisoners dragged heavy balls and chains that tangled hopelessly in the bedclothes and around the legs of chairs. At last what Will described as their mournful "anvil chorus" died away as, laden with garbage and weeds, they were herded back into the jungle. Will and Blair watched two members of the chain gang "painfully rolling the tennis-court, carrying their iron balls over their shoulders so as not to damage the court."

Even in remote Kuala Lipis they had to dress for formal dinners, and Will recounts an overwhelmingly grim dinner at the "club," which gave an "impression of coolness that its thermometer refused to second." Like all of its kind in the far-reaching Empire, the clubhouse had old magazines, a little library, a tennis court "a bit weedy and aslant," and a fifty-yard golf course—besides an abundance of whiskey and soda pop. At dinner, sweating turbaned waiters served a full-course English dinner complete with heating alcohol, the atmosphere tense with an undercurrent of wracked nerves as the resident civil servants awaited the deadly cholera epidemic that was heading inexorably upriver toward their "breathless hell." No one could talk without snapping, and snapping was forbidden, so the assembled company begged Will and Blair to talk about anything from the outside world.

The beginning of November found Will, Blair, and Haledin on their way to Burma. After visiting the tourist sites and playing the part of honored guests for the bored ruling class in the British capital of Rangoon (now Yangon), they traveled north to the ancient capital of Mandalay and then on to Myitkyina, where the mountains soared above forests with rushing streams crossed only by fragile rattan bridges. Here the Shan people gave way to the canny Kachin hill tribes, who made strong, agile porters.

The time in Burma made for a fascinating chapter in *Pheasant Jungles,* one of Will's most readable books. The sporadic unrest among the peoples, which was becoming almost endemic in the mountainous area, made travel there quite dangerous, but the pheasants knew no political boundaries and Will was determined to see it through. His government hosts insisted that the small scientific expedition be escorted by a troop of six imposing Gurkhas, an elite corps of native soldiers who were highly disciplined and trained, the only members of the native police force allowed to carry their own weapon—the *kokris,* a short, deadly curved knife. Illustrated by photos of the Gurkhas, whose stern fortitude he admired, along with breathtakingly beautiful land-scapes and the faces of porters, villagers, and enemy tribesmen, Beebe's descriptions of this part of the trek are sometimes humorous, some-times haunting. It is here that he presents an unusually frank and per-sonal account of an attack of depression that threatened to end the expedition.

Will had occasionally confessed to feelings of depression or "the blues" when treasured animals died or hopes were dashed. Less often he acknowledged anxiety, sometimes amounting almost to a panic attack, that would possess him when he was overwhelmed by the responsibilities of his position. These dark moments are part of nearly every explorer's memoirs, terrible narratives begotten of very real dan-ger, unbearable responsibility, and the worst fear—a failure of nerve. Here, at a small government rest house at Pungkatong high in the mountains of northern Burma, Beebe succumbed to a true dark night of the soul. He had arrived feeling exhilarated and fit. But during the first night, he gave way to what he believed to be a nervous breakdown, when "waking fears and sleeping terrors" combined to kill any desire to go on. All he could think about was the shortest route home. To his own amazement, he did not even feel shame at the idea of quitting. For sev-eral days he stayed in the dak, unable to face the prospect of work and utterly incapable of giving orders to others. The burden this put on Blair must have been tremendous.

What saved his sanity, Will wrote later, was blundering into a dank closet in the wayfarers' dak stuffed with cheap paperback mystery novels. For days he read without break, devouring the stories one after another, snapping at his servants—and most likely Blair as well—and refusing to work. At length, he reports, he began to gaze at the distant line of blue mountains without disgust, and when Haledin timidly brought in the first pheasant the local people had trapped, Will reflexively identified it and measured it "before I remembered my recently evolved hatred of them." [4]

Blair remembered the days of Will's "mental convalescence" as a golden time, with "huge appetites, long days together in the jungle, pheasants in abundance, and glorious butterflies." [5] In the jungle, oblivious to the angst of the humans creeping below, trogons—insectivorous birds with brilliant wine- and chestnut-colored feathers—swung through the treetops while monkeys watched the strange new primates curiously, and great hornbills and drongos with impossibly long tails swooped overhead. Beebe added the silver and peacock pheasants he had looked for in Borneo to his collection.

The jungle here was dense with undergrowth; Edenic from a pheasant's point of view, the going was more like hell for humans. Twice at very close range Beebe heard the calls of the pheasants he sought but was unable to see them through the thorny brush. To reach their habitat he had to "caterpillar" along the half-burrowed trails of wild boar, crawling and wriggling his unpadded six feet through thorns and dense scrub. He was adept, however, at locating places where birds were likely to come and drink or scratch for food, and would set up his tiny umbrella observation tent under a tree and become invisible, lying quietly for hours day after day until at last the birds would come.

Blair holding a captured Bornean white-tailed pheasant.

The success I have had is far from being due to my efforts. The Pheasants of one valley have entirely different habits from these near by and when I try to apply the knowledge I have gained I accomplish nothing. But I go out blindly, choose a likely place and efface myself in the leaves and earth of the jungle and wait and pray hard! And within a day or two of 2 or 3 hours of waits, a flock almost always wanders slowly toward me and gives me my chance.[6]

In fact, his skill and experience were what made his luck. Weeks of painstaking research in each area provided him with a good idea of where the birds were to be found, and his own experience as curator at the zoo told him all he needed to know about the birds' predilections and needs. And his habitual use of native trackers, whose knowledge of local habitats he trusted and admired, brought him rich results. Although language was a barrier, the natives could read well the sahib's very real interest and respect, which differed so extravagantly from what they were used to from the British hunters who also relied on their skills.

In addition to collecting live pairs and skins and eggs for the zoo and the American Museum, one of Will's greatest interests was the behavior of each species of bird. While their anatomy was well understood

Pheasant eggs, painted for Beebe's pheasant monograph, by H. Grönvold.

from museum specimens, the lives of the creatures themselves were practically unknown. Live birds in captivity yielded few clues as to the way their wild forebears chiseled out existences in dense jungle undergrowth or frigid mountain forests. Beebe hungered to know everything about these beautiful wild creatures, and went to great lengths to observe them in their native haunts, taking detailed notes and photos when circumstances allowed.

In addition to watching near known water holes, he often looked specifically for the cocks' "dancing arenas"—areas the male birds carefully cleared of brush to display their glossy plumage to advantage for choosy females. Hours of cramped and hungry observation were seldom rewarded with the sight of a displaying bird, though, and Beebe often had to be content with a dropped feather, moldering carcass, or other evidence of the species that frequented the area. After one grueling climb up hillsides upholstered with fearsome stinging nettles, urged on by the distinctive cry of the infinitely rare and desirable Sclater's Monal pheasant, he fell back several feet when he seized a rotting branch. Pulling himself back up inch by painful inch, he grasped a thick smooth trunk that seemed to have an odd rubbery texture. Drawing back, he found he had grasped a large king cobra, fortunately sluggish and headed the other way. His reward for this risky and difficult effort was a sight of the gorgeous bird, pecking at leisure through the leaves on a crumbly bank.

By now Will had compared enough species to venture an opinion on the mechanism by which the sexually dimorphic varieties—those with highly decorative males and plain females—found mates. His conclusion was, for the time, startling; it would be another seventy years before experiments proved him correct.

> It seems to me that the most reasonable explanation of the wonderful performance is of a mental effect upon the hens, not aesthetic, not directly critical or attentional, but a slow indirect influence upon the nerves, the arousing of a soothing, pleasing emotion, which stimulates the wonderful sequence of instincts. This explanation implies no deprecation of the importance of sexual selection. The male who, either by vanquishing his rivals or who by strength and persistence most frequently and effectively displays, will win the hen, regardless of whether the actual process be by aesthetic appreciation or by some subconscious hypnotic-like influence.[7]

Blair accompanied Will on many of his shorter stalks but spent most of her time in camp maintaining order, regulating porters and servants, and acting as chief liaison and physician to the members of the expedition and local visitors. The headman of the tiny cliffside village of Sin-Ma-How and his wife brought gifts and offered to take the group to the valleys where the birds lived. Beebe was impressed by the chief's bearing and delivery; there was nothing in his demeanor, Will felt, that would not have become a minister of state at any diplomatic court.

Other than their native dignity, the villagers of the tiny cluster of houses clinging to the hillside had little to grace their lives. Their huts were filthy and dark, and the village offered no sanitation, so disease was rife. Although Buddhist in culture, rural folk saved their real devotion for the animist spirits they called *nats*. These spirits were malign creatures whose behavior was mischievous at best, and sure to go against Buddhist teachings. They were greedy spirits who required constant appeasement: every meal required a libation, poured on the ground to keep them well fed.

Every dwelling, no matter how mean, bristled with sharp sticks topped with tiny slings, in which the nats would rest preferentially, saving the house and its inhabitants from their evil ways. One of the first signs that the villagers were aware of the strange encampment on the opposite hillside was the overnight appearance of a forest of these nat rests, oriented carefully to provide homes for the nats dislodged by the campsite. "These people are the most stolid and at the same time the most emotional in the world," Will wrote, "and we fall very short of fully understanding them."[8]

Life with the Gurkha guards, too, taught Will much about the religion and culture of the people he was living among, whose intelligence he respected and whose services he depended on. The Hindu Gurkhas, not unlike the Knights Templar of old, maintained rigid discipline and obeyed strict dietary laws. They ate by themselves and used their own dishes. Once after a long trek Will thoughtlessly reached over the pot of rice they were cooking. Haledin whispered to him, too late, to watch the rice.

> Then I remembered, and drew back quickly. All of us laughed together and I went away, but I watched from my tent and saw them throw away the rice, which was polluted by the touch of my shadow. The pot itself

was thoroughly cleansed. They would have to wait hours for the new rice to cook, and they were hungry and tired from a long march. . . . In this incident neither offence nor resentment played a part. It was simply the law and the law is sacred.

The next day he bought a sheep and gave it to the Gurkha corporal to kill and share "in his own way. . . . He saluted, smiled, and we became better friends. If superiority and inferiority entered into this, we divided them equally between us." 9

Back in Singapore, Will and Blair basked in the comforts of mail from home and hotel living. After weeks of hard wear and washing in stony streams, their wardrobes needed replenishing. Will had several suits made for $4 each, including cloth and labor, two "the close-neck collar kind, 3 coats to a suit, which are cool as you wear nothing but an undershirt below." 10 Blair, too, restocked her linen, and made a foray to the port to retrieve a prize—two gowns Mrs. Kuser had commissioned her dressmaker to make for Blair, an evening gown of "lovely rose silk and white lace—a dream" and an afternoon dress in a "heavenly shade of blue silk with an all over pattern of white gold-centered blossoms, lace trimmed, with a sapphire blue velvet girdle and sash." 11

Their next stop was to be China, where they hoped to head for the remote areas of Yunnan and Tibet. Here the heat of summer gave way to cool crisp air, reminding Blair of watching Princeton-Yale football games while wrapped in furs, the scent of chrysanthemums in the air. But their steamer drew into Hankow just as rioting broke out ashore, and they watched in horror as soldiers slashed their way through the crowd. The combination of the uprising and a particularly lethal surge of the bubonic plague made further progress in China impossible. They decided to press on to Japan and return to China when riot and plague had ebbed.

In Japan, Beebe's six-foot height made him feel like a giant. He had to kneel to look into a mirror and kept knocking his head on doorways. Despite the reputation of several Japanese ornithologists he had met in New York and the bright Japanese students he had known at Columbia, Beebe found obstacles to the free exchange of research. The eager students he had known, now respected scientists in their own country, were mired in bureaucracy and making scant progress. They knew nothing of pheasants, nor of the behavior of any birds in the wild. And the wilderness itself was hardly worth the name: Japan was as highly cultivated in terms of land use as it was in manner and tradition.

Chafing against the need for constant socializing and the formidable rituals demanded by politesse, they waited two weeks for permission to study the birds in the Mikado's Imperial Preserves; then, when permission came through at last, they had to wait nearly two more weeks for an actual written and sealed permit to surface. When word came that the plague was receding and the riots calming, the pair fled gratefully back to China, planning to return and make good use of the hard-won permits on their way back to the States.

From Foochow they chartered a houseboat that took them up the Yangtze River, where they found a few pheasants subsisting in the scanty fringes of forest left by a rapidly expanding population and rampant agriculture. Some of the loveliest of the pheasants—the golden and Lady Amherst and the Reeves—lived in deep gorges still wild and uninhabited, and Will secured them only with great difficulty. To find the Manchurian eared pheasant they decided to plunge on into Manchuria, assuring the anxious folks back home that they would use respirator masks as defense against the plague.

In Peking they were impressed by the wide, well-kept streets, the wild-looking Mongolians from the northern deserts, and the mysterious Forbidden City behind its pink walls in the bustling center. The Great Wall mesmerized them, winding "up and downhill like a black serpent. . . . Strings of great shaggy Bactrian camels would come swaying along the trail—a trail which had been in constant use since 200 B.C. Flocks of black-headed sheep, and herds of long-haired giant pigs; then Mongolian ponies, shaggy and tiny." [12]

When they returned to Japan, Will was at last able to study and photograph copper and green pheasants in the Imperial Preserves. The Imperial Household gave him two tall, graceful cranes in exchange for a promised pair of swans, which were unknown there. At the end of his frustratingly brief study, Will found it ironic to be asked by the Japanese scientists to address the Tokyo Zoölogical Society, to tell them about their own birds.

At last, after almost seventeen months, the pheasant expedition drew to a close. Will had secured skins or live specimens of nearly all the pheasants he had sought. He had watched them in their native habitats and had made extensive notes about their behavior. He had gained

experience in leadership and learned hard lessons about his own vulner-
abilities. He had been lifted to heights of ecstasy at the sight of rare
species, and plunged into despair by the effects of human rapacity on
their survival. With Blair, he had fulfilled his mission, and left many
close friends among their British and Asian servants and colleagues.

The voyage back across the Pacific provided sorely needed down-
time after so many months of intense labor and responsibility. As always,
Will found his relaxation in activity. The wonders of the ocean were
endlessly fascinating, and he entered into the roster of deck games
energetically, garnering four firsts in several grueling obstacle races. At
a shipboard fancy dress ball he and Blair won first prize: Blair went as
Martha Washington, and Will borrowed a minstrel mask from the bar-
ber, put it on backward, wore his suit backward with appropriate pad-
ding, and danced the two-step backward with her. There were also a
great many crates of birds with various needs to attend to, and each loss
depressed him. He was particularly distressed when he opened the
cranes' box and found that one of the elegant creatures had suffocated.

*Will and Blair at
the Great Wall, 1911.*

When they reached San Francisco, Charles and Nettie were at the pier to meet them. Will had sent them a check for the cross-country journey, insisting against their frugal inclinations that they should spare no expense. He was determined to show that he was not only solvent but also able to support his parents in the comfort and style to which he had become accustomed. "Don't do any more of the scrimping act. If you want to go to Egypt, I will see to it that the money appears from somewhere. . . . I want you and Papa to enjoy this trip and in comfort and to do that you will need the $300. If there is any left over, well and good, but don't plan for it. Selah! Amen!"[13]

The last hundred or so pages of Will's voluminous journals of this trip are full of the exhaustive lists he loved to keep. They afford an appealing picture of his human side, reflecting as they do his boyhood interests and preoccupations. He lists hundreds of species of mammals and birds he has seen, and 744 skins collected. There are lists of dinner-table tricks he has mastered, people he has met, letters he has written. He lists dates of letters to his parents, to Madison Grant, to Colonel Kuser and young Dryden Kuser, to Osborn and to Crandall. He also wrote to the Journeays in Nova Scotia, to Weary, to Louis Whealton, to his cousin Paul Clark, and to Warrie Mountain, now a physical education teacher at his old alma mater, East Orange High School.

On May 27, 1911, they reached Penn Station, completing their round-the-world trip. They came home to a new house, carefully scouted and rented for them by Nettie and Charles, after as much worried consultation as could occur between parties half a world apart. In November Blair had written Nettie that she would prefer a house, with a porch to sit on in the summer heat. "Of course if you can price a house at $60 . . . –75 a month I should prefer it, and but for the extravagant Will I should never have written the figure $100.—but he is sitting close by and insisting. You know him! And you also know that that is really too much for wisdom to pay for a mere house, when one has the purse of an ornithologist." Blair particularly asked Nettie to try to find a house on Sedgwick Avenue, a pleasant residential boulevard on the water where their friend Robin Niles lived with his father, Robert. The house Nettie had found was 2291 Sedgwick Avenue, next door to Robin, and Will and Blair adored it.

Recalibration

*These things can never be learned in a museum or
zoological park, or by naming a million more species
of organisms.*

William Beebe, *Our Search for a Wilderness*

IT TOOK WILL ALL SUMMER
just to organize the vast amounts of material he had sent back from the
round-the-world pheasant expedition. By August he confessed that he
had done little but classify the 2,400 photographs he had taken and
arrange his new insect collections. Crates of specimens had to be
unpacked with meticulous care and the contents sorted, labeled, and
sent to their ultimate destinations. Most material was for the great
monograph—either skins that had to be stuffed and mounted in lifelike
poses, or samples of plants, mosses, and lichens that would be used by
the artists to make their illustrations accurate.

There were also the live birds to be acclimated to their new existences at the zoo or in Colonel Kuser's spanking-new aviaries at his
northern New Jersey estate. Will's loyalties were torn: his first duty, of
course, was to the zoo, but Kuser had funded the expedition generously
and was daily footing bills related to the infant monograph. And then
there was Osborn at the American Museum, who was directly or indi-
rectly responsible for all the good things that had come to Will in his
career. The American Museum of Natural History was greedy for mate-

rials for its promised dioramas of Himalayan pheasants, and it too needed copies of his notes for accuracy.

The Beebes' small home quickly became a stopover for relatives and friends from all over. Fortunately, they had found a stout housekeeper from Barbados who ruled the house and especially the kitchen like a benevolent matriarch. Blair's fun-loving brothers Roger and Harry stayed for several weeks, "playing golf and tennis, swimming, joking, and enjoying the car." To expedite Will's travel from New York to Faircourt, Colonel Kuser had presented him with the car of his dreams. "Today," Will wrote on August 11, "I got the 4-seater automobile which Col. Kuser has selected from his factory for me. It is a Weber 30-60 horsepower Toy Tonneau and is a beauty." (8.8.1911)

By this time Lee Crandall, now Will's assistant curator, was thoroughly entrenched at the zoo. He had been performing Beebe's duties while Will was off globetrotting, and with a new baby to support was eager to keep his job. Will, meanwhile, had undergone a gradual metamorphosis from curator to distinguished ornithologist/explorer. To his delight, he was made one of the elite honorary members of the prestigious Explorers Club, along with Roosevelt, Shackleton, and Scott— and, later, Charles Lindbergh. He spent weekends with the Osborns at Castle Rock, with the Kusers at Faircourt, and with Teddy and Edith

Sagamore Hill, home of Teddy Roosevelt.

Roosevelt at Sagamore Hill on Long Island. Roosevelt enjoyed the company of the bright, enthusiastic young naturalist whose quick wit and readiness to engage in debate on any subject but politics were a refreshing change from the sycophants he so often had to deal with. Among other things, the two discussed travel and hunting, scientific experiments, the importance of studies of tropical jungles, and ways to penetrate the ocean depths.

Will was constitutionally indifferent to politics. He and Blair delighted in the humorous columns of Finley Peter Dunne, a satirical journalist who, under the sobriquet "Mr. Dooley," had great fun ridiculing the politicians of the day. Roosevelt was a favorite target for Dooley, although Dunne and Roosevelt were good friends. But though Will appreciated the truth behind the humor in Dooley's send-ups of American politics, racism, and imperialism, he had profited greatly from the fruits of British empire in the form of railways and diplomatic facilitation. He had also seen, firsthand, conditions in other parts of the world that he believed needed correction and control. And although he deplored the inhumane treatment of native peoples by both Europeans and other indigenous groups, his greatest concern was for the myriad animal species that were being hunted or marginalized to extinction.

He and Blair had already generated dozens of articles on their various expeditions, mostly for *Harper's,* and they continued to turn them out now they were home. In pieces such as "A Quest in the Himalayas," "Wild Burma," and "With the Dyaks in Borneo," they shared their adventures, discoveries, and insights with a wide audience and gained celebrity as well. Will had to turn down many requests to lecture, accepting only those invitations that most interested him. He gave talks at the New York Academy of Sciences, the American Ornithologists' Union, the exclusive Half Moon Club, the Audubon Society, the American Museum, and the Zoological Society. He also lectured at several colleges, including Columbia, Yale, and City College.

Blair found that she liked being accepted as a writer, and settled into the role with determination. She wrote so much that her eyesight, never strong and long dependent on eyeglasses, began to suffer. She undertook a course of exercises and treatment, and kept on writing.

Will soon realized that he would need to reexamine a few of the great bird collections in museums in Europe—collections he had visited en route but had not been able to canvass in detail—in order to complete his research. He was close to determining a new scheme of evolutionary development for the entire order of pheasants, and needed to make careful observations of the various specimens the museums in London had in their vast storage areas. Existing phylogenies of the pheasants depicted the linear development that was assumed at the time, with new groups branching off from one main line from time to time. Will saw a very different pattern: an explosive burst of diversification long ago, with more typical and gradual changes accumulating over the eons—more like a bush than a tree.

> Further than this it is impossible to go in a linear classification. Whatever may have been the generalized ancestor of the pheasants, the . . . groups in which we are interested have evolved more or less radially, and, considered as terminal living foliage on the tree of evolution, all are equally distant from that common ancestor.[1]

Although now taken for granted, the notion of rapid diversification after some sort of anatomical or physiological breakthrough was a daring conjecture then. To sustain an argument that had never occurred to him before studying the birds in the field, he had to return to the specimen boxes. Luckily, enough funds were left in the expedition account to cover the trip, so he and Blair sailed for London in June 1912 on the Hamburg-American Line's *Victoria Luise*. With them was their friend

Beebe's radial 1913 diagram of pheasant evolution.

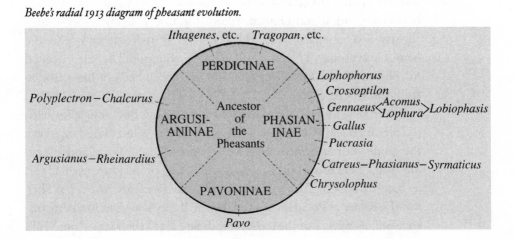

and neighbor Robin Niles, now a fledgling architect, and his father, Robert, a prominent lawyer and zoo board member.

Robin was impulsive, the record clearly shows: irate correspondence from Hornaday to W. W. Niles, for instance, characterizes the nephew as an irresponsible young man who scared the animals by careering noisily around the zoo's newly paved roads in his motorcar. But Will liked his style. After he married Blair, the three had often gone to parties or the theater together. The joint trip to Europe was perfect: Will would be spending most of his days immured in the bowels of the Natural History Museum, but Blair would get to see the sights with Robin and his father.

In London they were welcomed by the museum staff, who set aside a room for Will to work in. Blair was given a room for her writing at the Royal Colonial Club, just across the street from their lodgings at the comfortable Hotel Victoria, where they had stayed so happily at the outset of their trip around the world two years before. After a few days of social engagements and sightseeing, Will settled in to work in earnest, joining the others for dinner and theater in the evenings. The party remained in London for a month, and while Blair and Robin visited castles and museums, Will made great progress in his project to reclassify the Phasianidae, concentrating "every moment of daylight into work." For two solid days he copied the morphological collection labels and specimen descriptions, and made plans to have similar ones made for the zoo.

When the Nileses sailed home in early August, Will and Blair went down to spend a week at Walter Rothschild's quirky museum at Tring, between London and Oxford, where the bird collections were only slightly less extensive than those of the British Museum and the National Museum of Natural History in Washington. Then they headed for Berlin, where the collections held birds Will had not been able to study elsewhere.

Berlin they found to be as modern as New York but far lovelier, with its tree-lined plazas and streets. They visited the Winter Garden, where the vaudeville performers stole Will's impressionable heart, and "everyone drank beer & ate sandwiches & the ceiling of the immense hall was jet black, studded with 100's of electric lights, like stars." (8.27.1912) They saw the palace at Potsdam, visited the small zoo, and went to a performance of *Lohengrin*. But the most incredible part of their trip, from Will's perspective, was a great German military display, complete with cavalry, foot soldiers, artillery, dirigibles, and mono- and biplanes. He was overwhelmed by the discipline and might of the military, and impressed with the martial spirit of the people.

Their last stop was Hamburg, where the Tiergarten held one of the world's finest and most advanced zoological parks, complete with desert and polar environments, a fabulous insect house, and, to Will's great joy, three king penguins. He was able to buy a hornbill that he coveted, as well as several other birds, to take back to the zoo. They also visited the legendary Carl Hagenbeck, animal collector and exporter extraordinaire. This man had managed almost single-handedly to stock most of the great zoos of the world with all manner of rare beasts, and Will paid a visit of respect as a seeker might visit a great spiritual master. The Beebes returned to New York in early September, and Will dug into the monograph in earnest.

The fall passed with intense work on Will's part, studying the notes and drawings he had made and observing the pheasants in Kuser's aviary at Faircourt, now the largest collection of pheasants in the world. He was following up an idea that the order in which different birds molted their feathers could be used to determine their position in the evolutionary line. He frequently spent nights or weekends at Kuser's estate, or at the zoo if he had to work late. The responsibility of producing the massive monograph, which Kuser would pay tens of thousands of dollars to have published, was overwhelming, driving from Will's mind all thoughts of the fieldwork he loved.

CHAPTER 19

Betrayal

In the future . . . I shall be certain of finding the same silence, the same wonderful light, and the waiting trees and the magic. But . . . I shall have to watch alone through my jungle night.

William Beebe, *Jungle Peace*

WILL'S CRYPTIC JOURNAL ENTRY
for January 29, 1913, reads, "Blair left this afternoon, while I was giving a lecture at the Colony Club for Mrs. Osborn." True to his policy of keeping his personal life to himself, Beebe has left nothing but silence surrounding Blair's desertion and their subsequent divorce. Only from screaming news headlines and necessarily lurid divorce records can any hard data be gleaned; only occasional sad, wistful phrases in a few surviving letters to his parents reveal how deeply Will was affected by Blair's abandonment.

Divorce in 1913 still carried a strong social stigma, particularly in states like New York, where the only ground for divorce was adultery. Unhappily married people resorted to various ruses to obtain divorce decrees, from simple bribery to elaborate schemes of entrapment. "Divorce mills" provided conveniently unclothed and compromisingly situated women, doped drinks, and accompanying photographers, private detectives, and legal services. Newspapers had a field day with some of the more egregious cases, and then as now, prominent citizens came in for more than their share of the limelight.

Circumstances like these guaranteed the rise of divorce capitals such as Reno, where divorces could be granted on less compromising grounds and in less time. In fact, Reno had its own sizable "divorce colony" where men and women could, in 1913, obtain a divorce on the grounds of adultery or extreme cruelty after six months' residence. "Extreme cruelty" was infinitely preferable to adultery, as it could be "proved" by the plaintiff's own allegations, properly verified and attested to by compliant friends and legal counsel. To avoid difficulties and scandal, defendants were likely to refuse to fight the suit, allowing whatever allegations the lawyers had worked out to stand with only a token objection.

In the "colonies," plaintiffs could find lodging in any number of hotels, boardinghouses, and private homes. And because of the continual stream of like-minded visitors, a plaintiff could always find company, advice, and sympathy. Reno's economy came increasingly to depend on the colonists, and occasional demands from the conservative element to lengthen the necessary residence time for decency's sake were countered by business interests demanding ever briefer stays to compete with other divorce-mill states. As an anonymous poet in a local paper wrote in 1912,

> If you legislate against the Reno Colony,
> To other fields the fair ones you will drive
>   For ill-advised propriety
>   Brings poverty with piety,
> And some of us would much prefer to thrive.[1]

Blair's sudden departure may have been prompted by one of the periodic drives to extend the state residency requirement. The debate had become so bitter that many alert New Yorkers foretold an imminent change for the worse. Indeed, a massive demonstration in early February resulted in a change in Nevada's law, extending the minimum stay to a year. Blair arrived in Reno just under the rope; six months later, on August 29, 1913, the divorce was granted.

The complaint itself, sad witness to the disintegration of what had been a rare marriage of minds, reads like a multitude of others. To establish "mental cruelty," Blair accused Will of ruining her eyesight by forcing her to sit long hours at the typewriter copying his lengthy manuscripts. She complained that he never liked anything she wore, never praised her, and was hateful to her family and friends; that he made her do all the cooking and housework rather than hiring a servant, and refused to move to a more congenial location where she could entertain friends. She also asserted that

> [t]he nature of the work in which defendant was engaged was such as to make him practically master of his entire time and in consequence many hours each day were consumed by him thoroughly enjoying life and in pursuing whatever line of pleasure he might desire. That, on the contrary, said plaintiff was compelled to be alone the entire time in a lonely flat with neighbors not of her own choice, and not even provided with the convenience of a telephone, and because of her desire to assist defendant, . . . deprived, in large measure, from the enjoyment of reading or sewing or any other occupation which might tend to pass the long hours.[2]

She alleged that he was terribly jealous and repeatedly threatened to kill himself, a threat he knew would cause her particular pain because of

The New York Times
announcement of the
Beebe divorce, August
29, 1913.

NATURALIST WAS CRUEL.

Mrs. Charles W. Beebe Gets Divorce from Well-Known Writer.

Special to The New York Times.

RENO, Aug. 29.—Following her testimony of cruelty and indifference, Mrs. Mary Blair Beebe obtained an absolute divorce to-day from Charles William Beebe, the well-known writer on natural history, connected with the Bronx Zoological Park in New York.

On the stand Mrs. Beebe, whose eyes were shaded by dark glasses, declared that she had assisted her husband in his work until her eyesight failed, and that when she was unable to continue in her help he became cruel and indifferent. Once, she said, he put a revolver in his mouth and threatened to shoot himself in an effort to frighten her. For days at a time, said Mrs. Beebe, her husband refused to speak to her.

The couple were married at Cole's Ferry, Va., Aug. 6, 1902. There was no opposition to the suit to-day.

an uncle's similar fate.[3] In short, he had made her a "nervous and physical wreck." "Naturalist Was Cruel" screamed the *New York Times* headlines. The same accusation was made almost daily against some spurned spouse or other, too thin-skinned to contest the ugly evidence of a marriage gone wrong. The day after the divorce was granted, Blair married Robin Niles.

A week later the papers discovered Blair's remarriage and had a field day with gossipy articles and photos. "Mrs. Beebe Wed Again in Secret; Woman Who Divorced Curator . . . Now the Wife of Robin Niles. Husband Near Neighbor." They loved the news that Blair was thirty-three and Robin a mere twenty-six.

Recently Dr. Beebe praised Mrs. Niles's work so strongly that her action for divorce came as a surprise to the friends of the couple, and her quick remarriage is an equal surprise. The Niles and Beebe families are next door neighbors in Sedgwick Avenue, and have been acquainted for some time. Where Robin Niles and his bride are now is not known outside a very limited circle, and they are not talking.[4]

MRS. BEEBE WED AGAIN IN SECRET

Woman Who Divorced Curator of Birds at New York Zoo Now the Wife of Robin Niles.

HUSBAND NEAR NEIGHBOR

Dr. Beebe Attributed His Success in 50,000-Mile Trip to Study Pheasants to Helpmate's Skill in Dealing with Natives.

The Times *article expressing surprise at Blair's rapid remarriage.*

Despite the lengthy descriptions, for the record, of a marriage that had supposedly been an unmitigated disaster, the break seems to have come suddenly—at least for Will. Despite his keen powers of observation of the animal world, he seems to have been blind to the growing intimacy between Robin and Blair. His candid pleasure in the company of Robin and his father on the trip to Europe, and his joy that Blair would be well looked after while he worked, testifies to a signal myopia.

The divorce weighed heavily on Will, whose idealistic vision of marriage was a rosy hybrid between the Romantic literature that had shaped the rest of his worldview, and his modern vision of a marriage between peers. His concept of divorce, however, had been dictated by the notions of Yankee respectability inculcated by Nettie and the aunts: it simply was not "done" in respectable society. And the circumstances made Blair's betrayal particularly painful: her flight and the marriage that followed so soon after pointed inescapably to cuckolding.

For more than a year after Blair left, Will was heartbroken, ashamed, and despondent to the point of suicide. He dived into his pheasant work with a vengeance; friends had to force him to visit them, and to attend the parties he used to catalyze. Will's friend Weary—Wallace Irwin—moved into the Sedgwick Avenue house as a companion, and Will gave him a small job working on the bird collections, paying him out of his own pocket. Soon Charles and Nettie moved in as well. Will continued to mine his experiences abroad for articles, and gave many lectures up and down the coast. In January 1914, he records that in one week he had written an article for *The Atlantic Monthly* titled "Jelly-fish and Equal Suffrage," half of an article for *National Geographic Magazine,* and two book reviews; given two lectures; "and kept up with my other work."

Loulu Osborn, who had always had a soft spot for Will, made a point of taking him for rides in her open motorcar in Central Park, to make him feel—and to let the world know—that his social status had not been affected by the divorce. She and her husband, the professor, invited him for weekends at Castle Rock, where his acquaintance with Teddy Roosevelt flourished. He was a frequent visitor at Sagamore Hill. "My dear Beebe," Roosevelt wrote in late November 1915,

> Can't you come down on Friday, December 3rd, and spend the night with me? Copley Amory and Jack Coolidge, one of whom had done a

good deal of work in Siberia and the other in Africa, are coming down. They are two young Harvard fellows, whom I really wish you to know.

Mrs. Roosevelt liked your Atlantic piece as much as I did.[5]

Roosevelt encouraged Will to take up aviation, and he leapt at the chance. He learned to fly at Hazelhurst Field, on Long Island's Hempstead Plains, where Roosevelt's son Quentin later earned his wings. Will had always been fascinated with flying and was soon giving lectures on the evolution of flight from archaeopteryx to man.

The pheasant monograph was by this time essentially complete — in manuscript. The publisher had been selected with care: no American publishing house was equal to the mighty task of mounting the elaborate artwork, so George Witherby and Sons, of London, had been chosen, partly on the grounds that they had successfully published Audubon's massive tomes. The completed paintings, photographs, and sketches had been sent to various European houses for reproduction; all Will had to do now was wait for the arrival of proofs to correct.

In April Will and two assistants, G. Inness Hartley and Herbert Atkins, a keeper, left on a six-week trip to Para, Brazil, to collect live animals for the zoo. Will was determined to recover the excitement and high spirits an expedition always inspired. He badgered the stolid captain into letting the men go over the side to collect sargassum weed to examine; with Hartley and Atkins he played deck games and engaged in elaborate contests, such as dressing in increasingly outlandish outfits to gauge the effect on the other passengers. At night they played cards fiendishly but with the port shades down, on account of the German boats that were turning up along the coast, thousands of miles from the Great War that had not yet engulfed America. They passed the *Prinz Eitel* and were preceded into Newport News by the *Prinz Wilhelm*.

The ominous presence of the German boats was exciting, but when their ship sailed past the *Comanche,* the boat that had floated an eager young couple to Veracruz so many years before, Will was devastated. To make matters worse for someone who was supposed to be convalescing, the ship made an unscheduled stop in Norfolk, Virginia, where he and Blair had spent much happy time. As luck would have it, Will ran into her brother Theo almost as soon as he stepped on shore. Although their brief chat was friendly, he wrote Charles and Nettie that the constant reminders of happier times were damaging his hard-won content. "But

I feel all right this morning, & Hartley is good company. I have wonderful work and happiness ahead of me and will try to make you happy."[6] He confessed that nights are "pretty lonely," and that he found himself instinctively looking over the side of the bunk to see Blair in the lower berth, as he had done so often in the past.

"But the future holds still greater happiness than I have known," Will wrote, "and I have so much to do that I shall have no time to think of the past."[7] This sounds suspiciously like a man trying to convince himself as he reassures others; but he was well aware that if he were ever to regain his peace and self-confidence, he would have to stand alone. To ease what must have been a rough parting from Nettie's too-nurturing arms, he writes not so much an apology as a manifesto:

> Please try to trust me. What you wrote about night in the jungle is so silly. You . . . know nothing about the tropics. Drinking water and in malarial districts mosquitoes are the dangers, and night in the jungle save for those is exactly like high noon. I have my work and studying to do, and nothing shall ever interfere with them. . . . I have some big fights ahead of me the next few years and your love will mean a lot to me. I didn't mean to be cross with you but you must realize that I am a man and have faith in me.[8]

The jungle at Kalacoon, Beebe's field station in British Guiana.

The month in Brazil was seminal to Beebe in several ways. At thirty-seven, he was slightly older than Hartley and Atkins. He was also their superior: Hartley was a research assistant and Atkins a mere keeper. More important, it was their first venture into the field, and Will was an experienced hand. He warmed to the role of mentor and assumed a paternal stance toward his juniors that came to characterize his field-work. By early May he wrote his parents that he was getting good work done in the jungle, "though I cannot entice Atkins near it again, Hartley loves it, although woodcraft and he are still far apart."[9] Poor Hartley was to pay for his interest with repeated attacks of malaria. Most impor-tant to Will, he was back in the tropics again, where he felt he belonged. He had no time to make detailed observations or experiments—the monograph called imperiously—but he made plans for a future of worthwhile work. In the short time allotted, the Brazil expedition amassed forty-four mammals, sixty-three birds, and seventeen reptiles —in addition to the myriad insects that Beebe always gathered obses-sively for his own collections.

CHAPTER 20

Wilderness Found

*[The lives of wild creatures reveal] comedy so delicate that
appreciation never reaches laughter, and tragedy so cruel and
needless that it stirs doubts of the very roots of things.*

William Beebe, *Jungle Days*

E VER SINCE HE AND BLAIR
had ventured to South America in 1908 in search of a wilderness, Will
had been looking for just the right place to set up a permanent lab. He
was certain that the only practical way to study the tropics properly was
to have some sort of home base, a station where research could be conducted on a year-round basis, where supplies could be laid in and stored,
and — most important — where the same area could be mined for its
secrets week after week, month after month, year after year, to yield in
the end an in-depth portrait of the whole interdependent ecology of a
region.

The major drawback was economic. What Will wanted was not a
rustic idyll but a functional, modern, well-equipped laboratory and the
staff to go with it. Fortunately, he had what so many of his predecessors
had lacked. Because of his writing, his education, and his social and
professional connections — especially former President Roosevelt —
and because the powerful machine that was the New York Zoological
Society was behind him, Beebe had backers. Treading carefully, so as
not to draw funding from any of the society's other needy projects, the

zoo board managed to get five important and wealthy men—railroad magnates and philanthropists Mortimer Schiff, George Gould, C. Ledyard Blair, and James Hill, and industrialist Cleveland Dodge—to donate $1,000 each for an outpost of the zoo, a tropical research station in British Guiana under Beebe's leadership. Eventually Andrew Carnegie sent in the final $1,000, thus meeting the goal of $6,000 that the board had decided was needed to make a go of the project.

"I take this opportunity for wishing you in the heartiest manner success in what you are doing," Roosevelt wrote to Beebe, in a letter strategically copied to Madison Grant.

> No man is better fit than you are to do the work you have laid out for yourself. There has been no concentrated scientific investigation of the tropics as yet, although there has been a great deal of very valuable collecting work. Your proposal was to establish a proper . . . tropical station where students and observers may study the life histories of animals. No work is more needed at the present moment.[1]

Grant, always mindful of the zoo's purse strings, was notoriously inimical to Beebe, who tended to be a costly asset and whose recent "difficulties" Grant felt sure had ruined him for good work in the future. Ironically, the irascible Hornaday, whose straitlaced Presbyterian values had made him hard on the young phenom to whom so much had

Howes, Beebe, and Hartley at Kalacoon.

seemed to come so easily, was one of Will's staunchest allies during this period, standing up for him against Madison Grant in battles for the research station. With Hornaday and Roosevelt's support, Grant acceded, and Beebe's dream began to take on a habitation and a name: Kalacoon.

Will had decided on Georgetown, British Guiana's capital city, as his base of operations. He was so optimistic that he would find a suitable station that he sailed in January 1916 for Guiana not with a small scouting party but with his entire newly minted staff. Inness Hartley would again be his second in command, as research associate. Paul Howes, a young entomologist, came as research assistant, and Donald Carter, a crack shot, as official collector. Hartley and Howes came as unpaid helpers, as did Inness's sister Rachel Hartley and her friend, Anna Taylor. As staff artists, they would preserve ephemeral flora and fauna, and color or detail too fine for photography, in watercolor or pen and ink.

One of Will's covert purposes in establishing his own field station was to get out from under the crushing weight of having wealthy patrons to please. Whealton had been fun, but Will and Blair had had to waltz to his agenda. Even Osborn, surely the most benign of mentors, had a plan for Will and was never loath to mention it. And Kuser, for all his friendly encouragement, expected immense return on his magnanimous investment, in terms of time and attention and slavish devotion to the treasured pheasant manuscript. With a station funded by a well-meaning but not personally interested consortium, Will felt he would be able to escape from his social and even intellectual bondage.

Locating the ideal site would be easy, he felt sure. British Guiana, like so many of the places he had traveled, had an efficient colonial infrastructure and offered lush jungle within easy reach of civilization. The Wilshires had introduced him to George Withers, who operated a vast rubber plantation on the Mazaruni River just upriver from Georgetown. The plantation had a fine large house overlooking the confluence of the Mazaruni, Essequibo, and Cuyuni rivers, which Withers had offered rent free.

Kalacoon House, as it was known, was ideal for their purposes. Its open situation invited breezes and discouraged mosquitoes; the land around it had been cleared for rubber but some had been allowed to regrow after clearing, so there were samples of various environments

nearby. And not far away was the towering jungle, which Will planned
to study intensely. The researchers moved in at once and began a frenzy
of cleaning, shopping, unpacking, organizing, and hiring staff, getting
all in readiness for their first visitors: Colonel and Mrs. Theodore
Roosevelt.

The visit of the popular ex-president and his gracious wife to British
Guiana was in itself a gigantic social event in the inbred society of the
colony, and the newspapers recorded every dinner and tea they
attended, every visit paid. But as patrons went, Roosevelt was the best
sort: his visit cemented Beebe's status, gained tremendous publicity for
the fledgling research station, and stimulated offers of help, transportation, and supplies—and then he was gone, leaving Beebe without
interference to keep the mechanism he had wound up running
smoothly.

Better still, back home T.R. wrote a characteristically enthusiastic
article for the intellectual *Scribner's Magazine*, touting both the new
research facility and its director's noteworthy new approach to science.
The era of kill and collect was over, he said; it was time to launch a more

The Roosevelts at lunch at Kalacoon.

exacting study of the animals *in their worlds.* It was no longer enough to
make elaborate notes on this or that species, then assert that its protective coloration or lack of it must logically function in some certain way.
It was now time to go forth and observe, to study, to hypothesize, perhaps even to experiment—and Beebe was the man to do it!

> What is now especially needed is restricted intensive observation in
> carefully selected tropical stations, where the teeming animal life can
> be studied fully, and at leisure. The student should be a scientist whose
> training is both broad and specialized. . . . He must be able to see, and
> to understand what he sees; to interpret what he has seen in the light of
> wide knowledge; and finally to record it with comprehensive vividness
> and charm no less than with accurate fidelity to fact. A high ideal! . . .
> [British Guiana] is one of the tropical lands where there is a teeming
> life to be studied; and Mr. William Beebe is one of the scientific men
> who can study it as it ought to be studied.[2]

The perceptive assessment of Will's character is typical of Roosevelt, whose forthright opinions were notorious. But he was also canny in
the ways of men, and his paean stretches to include Hornaday, Osborn,
and Grant, "who are responsible for starting this research station on
the edge of the great tropical wilderness," as well as Withers and the
broad-minded government of British Guiana. And he touts the healthy
situation and comfort of the station—as yet barely established—to
tempt future researchers. "It represents the effort to strike out in a new
line, and the results may be . . . of the utmost value. It always needs
both boldness of conception and very hard work to carry through anything which is entirely original."

Another reason for the Roosevelts' trip to the station had been to
visit Will, who was still depressed and anxious. T.R. and Edith were
more than friends, and their presence was a real gift. The colonel always
had sound, bracing advice, and Edith was kind and perceptive. "It is idle
to attempt even thanks," Will wrote afterward. "You two have pulled
me out of the Valley of the Shadow into sunlight again."[3]

The comfortable, roomy house was ideal for the little working community. Jutting from the top of a modest hillside, it was perched on
brick and stone pillars, providing sheltered storage for all sorts of laboratory paraphernalia underneath. Along the front was a great open
room thirty feet long, which the men used as a dormitory by night and

a lab by day. Will installed his signature shelves, and each staff member was assigned a personal space. The cool nights were miraculously mosquito-free, but mosquito netting was necessary to keep out vampire bats.

The men hacked a trail through the secondary growth to the high jungle but were repeatedly forced back by impenetrable thickets or razor grass whose delicate tracery belied its savage slashing blades. Eventually, with the help of some convicts from the government penal colony across the river, who found the work a welcome change from their usual warden-supervised labors on the prison plantation, they managed to cut two trails into the heart of the pristine forest that Will yearned to know in every detail.

Kalacoon was the answer to his dreams. Although he had loved the excitement and constant novelty of expedition work, his soul yearned for domesticity—on his own terms, in a "domus" of his own choosing. Putting up the shelves was for him a gesture of possession, a sort of aerial root structure that would both tie him to a place and nourish him. Unlike his previous fieldwork, here he would see the birds complete a full cycle of courting, nestbuilding, laying eggs, and fledging young. Here he could watch while an ant colony foraged, fought, and swarmed. Here he could get to know every plant in his study area and what fed on it, lived in it, pollinated it, and killed it. The life history of each inter-

Kalacoon House.

dependent species in the complex jigsaw puzzle of the tropical forest could be worked out; even the lives of the animals that lived in the leaf canopy too far overhead to be observed could be reconstructed through analysis of their stomach contents.

As he would write later, "[T]he joys of exploration are as varied as the numbers and characters of the explorers themselves."

> I can remember the time when my greatest ambition was to be the first to step upon some desert island, or to penetrate to where no white man's foot had ever trod. Then came the period of peripatetic journeys, of covering as much ground as possible in a given time. But I soon found out that the island may be "desert" in very truth, with no return in scientific loot, and the thrill soon passed of encircling a sandy spit and seeing none but one's own footprints. I came to learn that worthwhile observations of birds and animals and insects were great in proportion to the smallness of the territory covered. . . . To be a good naturalist one must be a stroller or a creeper, or better still a squatter in every sense of the word—never a traveller.[4]

Here in his home station he could plant a garden for the first time, and harvest what he planted. Hartley and Howes made perfect companions, good-natured and hardworking, and the "girls," though "uninteresting and with little imagination," were competent and undemanding.

> Our routine is thus: Sam, black butler, housemaid, skinner & general valet comes in at 6 A.M. and we get up; the cots are pushed back, the East India covers put over them & they become fine divans, while the blankets are aired at the windows. We take a shower, dress in our various rooms, & settle the day's plans. Breakfast from 7 to 7:15 A.M. or earlier. We have cereal 3 times a week: bacon & eggs or fish, lime squash, cocoa, or coffee; biscuits, bread, & toast. Then I plan the general line of work, we scatter or go off together, I give flowers or bugs or birds to the girls to draw. We usually return at 12 for lunch. Paul may have gone to one of his wasp holes a few feet from the house. He has been three days photographing one, & has a beautiful print. I had a touch of ptomaine poisoning yesterday so to-day took it easy & only went 10 minutes up the back trail. But it paid as I sat & traced a female orange-headed Manakin to her nest & two eggs. Inness made a wonderful discovery two days ago of a toucan's nest . . . one of the great things I wanted on this trip.[5]

They wrote up the day's notes in the afternoon and had dinner at six-thirty, just after darkness fell. The next day's work was the main topic of conversation. Sleep came readily, with the sound of howler monkeys and sometimes drums in the distance.

> We have many plans for tomorrow. A gardener to engage; a platform to build for vultures & some jars which we have sunk in the ground to examine for rats or small opossums. To-day Inness & I catalogued the birds of the clearing, 57 in all. As soon as the rains start I shall transplant gardenias, roses, etc. to our garden. We have had watermelons, pumpkins, string beans, etc. already. (3.10.1916)

The Mazaruni River, their main road, was a well-traveled thorough-fare used by Indians, gold miners, hunters, and adventurers—bearded, vacant-eyed, indifferently honest men eking out a living panning gold, trapping animals, and trading shabby goods with shadowy provenances. From Georgetown, Kalacoon was accessible by taking a ferry across the Demerara, a train or motorcar to the Essequibo, and a boat to Bartica. Then Withers's beat-up Ford would take them to the foot of Kalacoon hill or to the penal colony across the river. Steamers plied the Mazaruni three days a week, bringing supplies, mail, and ice for the station's "Kalacoon cocktails," usually rum mixed with tropical fruit juices. Will worked hard to keep memory at bay, but inevitably places or incidents would take him back to his time there with Blair.

> In the crescent moonlight I went to the window late at night and listened to a boatful of blacks passing down river and singing as they paddled. Again the harmony came vividly to mind, of our blacks paddling us seven years ago. The steersman setting the words and in high quavering phrase leading the chorus, who took up the words, different each time and repeated them twice, with a sweet pathos which stirred one's deepest feelings. The wild voices coming over the palm tops were filled with a sad wild beauty. (3.12.1916)

In June, Charles and Nettie came down, pleased and shyly proud, to see their son's tropical paradise. "I am going to enjoy showing you all the things which we used to read about in Swiss Family Robinson, Bates & Wallace. Bring khaki short skirt and high laced shoes, and brown shirt waists and felt and straw hats, & for heavens sake, learn to drink a little tea or coffee, for you need something of the kind now & then here," he wrote.[6] It was Nettie's first experience of the tropics, and she was overwhelmed by the heat, the luxuriance of the vegetation, and the incomprehensible variety. She responded by introducing pillows and "proper" towels—with fringes—to the primitive quarters.

When Will proposed the research station to the zoo's board, he had listed several important objectives. First and foremost, to sell the idea to the board, the station had to operate as a collecting and shipping base, where the zoo could obtain choice specimens at low cost. This had become crucial to the life of all zoos nationwide because much of their stock had come from European providers, such as the German animal dealer Carl Hagenbeck. With World War I now decimating European shipments, zoos were powerless to replace dead animals or to acquire new species. Beebe had chosen British Guiana in part because it was a gateway to all the continent's natural riches. He also proposed to collect material for collections at the American Museum and the growing New York Aquarium.

The other stated objective—and Will's obsession—was to provide a facility for stable, intensive scientific studies that would fill gaps in the life histories of species and cast light on their evolution. This he set about with determination. At Kalacoon, hymenopterist Paul Howes was in his element: at least seventeen species of wasps made their

Loading a shipment of animals bound for the Bronx Zoo in New York.

homes in the clearing itself, so he never had to go far to find something worthwhile to study. Inness Hartley was particularly interested in fish, and the river afforded several fascinating and unstudied creatures, including the perai (piranha) and electric eels. His assistants occupied and happy, Will was free to do what he had so long contemplated: dissect a small area of jungle completely, from the canopy top down to the ground, and even below. Intrigued as he watched an ant wren scratching for food in the jungle mold, he dug up a section several inches deep and, with a lens, brush, and dissecting needle, collected every creature he could see.

The groundbreaking scientific paper, "Fauna of Four Square Feet of Jungle Debris,"[7] that he wrote for *Zoologica* proved to be at once thoroughly scholarly and one of his most popular, as he described the minute animals in clear but entertaining ways. His magnifying lens revealed the life histories and intimate connections among tiny forms that generally go unnoticed but whose presence is necessary to the health of the greater jungle life. "Over a steep hill came a horned, ungainly creature with huge proboscis and eight legs, and shining, liver-colored body, all paunch, spotted with a sickly hue of yellow. It was studded with short, stiff bristles, and was apparently as large as a wart hog and much more ugly. It was a mite, one of the biting mites of the tropics, but under the lens a terrible monster." To prove the correspondence, he watched under the lens while it bit him.

The ants were the most numerous fauna; Will found seventeen species. When he sent them to Harvard's famed ant specialist William Morton Wheeler, there proved to be two new genera, undiscovered by other workers or even Wheeler's students, who had spent much time gathering ants in Brazil. "I take it that they did not work in the leaf mould as you did," Wheeler wrote Beebe, "and that probably when other collectors adopt your method an extensive ant fauna will be unearthed."[8]

Will's broad-ranging interests, running counter to the trend toward increasing specialization, made the identification of species difficult. His attempts to identify the organisms he found were frustrated by the tendency of researchers to publish new species in journals so specialized that few libraries stocked them. In his small study area he had found flatworms, roundworms, true worms, myriapods, scorpions, pseudo-scorpions, spiders, termites, roaches, beetles, flies, and many other species, and for the work to be valuable, each had to be scrupulously classified. Nor were there central files or indexes to steer a researcher to the relevant publication. It was partly to counteract this trend that *Zoologica* had been established, in an effort to encourage pan-species investigation.

For his next, more ambitious study, Will decided on an area of jungle about a quarter mile square, to include creatures under, on, and above the ground. The team had cleared several paths to and through this area, and Withers had built roads nearby for his battered Ford, which he allowed the researchers to use when he could spare it. The trek across the cleared land to the jungle's edge was punishing, since the heat in the sun was extreme, but in the heart of the jungle the shade kept the air at a humid, even warmth.

After Will had carefully mapped the ground, he began to look more closely at the animal life. The stability and complexity of the jungle offered many more "niches," more ways of making a living, than did temperate woods. And every niche was occupied. There is one fundamental reality in wild nature, he wrote, "the universal acceptance of opportunity. . . . Evolution has left no chink or crevice unfilled, no probability untried, no possibility unachieved."[9] He quickly discovered that the life of the jungle could be divided into "depths" as accurately as the ocean. For studying birds, there was the floor of the jungle, the lower jungle up to twenty feet; the mid-jungle, which stretched up to seventy feet; and the treetops or canopy zone, which ascended as high as 200 feet. The flora and fauna of these zones, he wrote, were as distinct as the abyssal from the planktonic and from the surface of the ocean. Even the air above the treetops had to be considered, "just as we have sea-birds and flying fish and air-breathing cetaceans in . . . the pelagic simile."[10]

As Will's explorations of the different levels progressed, he became increasingly eager to penetrate the unassailable canopy world, where "yet another continent of life remains to be discovered." His efforts to explore the mysteries of the treetop region were frustrated by every sort of obstacle. The crew tried to pound spikes into one of the huge 200-foot mora trees, the giants of the jungle, to no avail. With ropes and the massive lianas that draped the understory, they could ascend only to its first great branch—a mere fifty feet.

Thus Will could only speculate on the rich harvest that awaited "the naturalist who overcomes the obstacles—gravitation, ants, thorns, rotten trunks,—and mounts to the summits of the jungle trees."[11] He intended himself to be that naturalist, on his next trip south. For the present, "we drag a tiny dredge along the ocean bed, and painfully draw to the surface a few fragmentary organisms, which often burst in our rarefied element. We see a company of fluttering forms high overhead . . . and our guns bring down a swirling, bedraggled fluff which was a bird, whose throat uttered one of the strange songs which we just heard, whose nest and eggs or young are somewhere far aloft. . . . And we realize that until we offset gravitation and establish stations of observation in the tops of some of these giant trees, our ignorance of this roof of the jungle must remain complete."[12]

Although Will fought to focus on the birds, his keen powers of observation and his consuming interest in everything that had life made it impossible not to try to learn all there was to know about every new creature he came across. As in his youth Lattin's catalog had tempted him to collect every curiosity it offered, in the jungle it was hard not to be distracted by every interesting creature. Faced with a crescendo of courtship behavior in the local lizard population, he stopped on his long hikes to try to make some sense of it. The little brown tree lizards would wrestle with each other until the vanquished one offered up the last item he had eaten, on the tip of his long tongue. The victor invariably accepted the offering and released his captive. When Will caught one, thinking of preserving it for later study, the lizard turned its head and held out an ant it had just swallowed. Feeling foolish, he took the ant and let the lizard go.

Using the zoo's hunger for specimens as his imprimatur, he dug pits to catch unwary frogs, toads, beetles, and small mammals, set out dishes of sugar water and gin to attract butterflies, and hung mist nets to collect bats. In addition to 154 birds (as many as 74 species in a single great tree) and untold invertebrates, he quickly accumulated a jaguarundi cub, several pacas, three species of opossums, an agouti, an ocelot, a wild dog, and many species of jungle rats. He came to the astonishing realization that a few acres of tropical forest can contain as many kinds of birds and insects as the entire continental United States.

At the end of May, Will took Hartley and Howes and went in search of hoatzins. These fascinating birds had caught his fancy when he first encountered them on the Abary River with Blair in 1909, when her broken wrist prevented them from capturing any for study. Believed by many to be living fossils, they are awkward in flight but adept in swimming. Hoatzin chicks come armed with a claw on the "thumb" of the wing, which they use to climb, lizardlike, from branch to branch. His fascination with anything that cast light on the evolution of birds or of flight made Will determined to accumulate a body of knowledge about these weird creatures and to establish a colony at the zoo.

Unfortunately, hoatzins are difficult to rear in captivity. Their eccentric lifestyle centers around the mukamuka plant, which grows profusely along tropical riverbanks, and the Bunduri pimpler, whose thorny, locustlike branches provide shelter and defense. Not only is the hoatzin vegetarian, but it eats mukamuka leaves almost exclusively, digesting the resistant cellulose with a fermentation process that produces a musky smell which Will likened to the ripe scent of elephants in an enclosed space.

Up Canje Creek the three men found hoatzins in abundance and collected nests, eggs, and live specimens. It was the height of the breeding period, and Will was able to study their nesting habits, as well as their odd riverine lifestyle, and to add several important facts to the body of

A hoatzin chick using its wing claws to climb.

hoatzin lore. The crew collected the usual background material as well as plenty of dead specimens for a museum exhibit, but were unsuccessful in keeping any alive for the zoo. Will grumbled to Hornaday about how much easier life was for museum collectors, who were not expected to come back with living creatures or new research data. "When I am old & infirm & on the Park pension list," Beebe wrote, "I shall ask for active employment at the Museum!" [13]

In August, the rainy season now well under way, Will reluctantly closed Kalacoon for the year and returned to New York to fulfill his obligations to the zoo and to drum up support for what he now believed to be a viable research station. Kalacoon had proven the value of concentrated study of a particular area. Now all he had to do was to convince the zoo board, its wealthy donors, the academic community, and the reading public of the practicability, desirability, and intellectual necessity of such research.

To the zoo board he had written long, chatty letters detailing the perfection of the situation and discussing the 300 living specimens he had been able to bring back to the zoo. For the zoo's members and the other scientists he hoped to lure down, he wrote scholarly papers for *Zoologica*. Roosevelt wrote congratulating Hornaday on Will's paper on the hoatzins. "Taking into account the scientific interest of its subject, its scientific insight and grasp with which it has been treated, and its literary power, clearness and interest of its treatment, I regard it as the best bit of bird biography I have ever seen." [14]

To recruit researchers, Will inserted in almost everything he wrote evangelical paeans to the healthful situation of British Guiana and the immensely practical arrangements for in-depth research on every creature the tropics could boast. For the public he wrote articles for the intellectual triumvirate of *Harper's, Scribner's Magazine,* and *The Atlantic Monthly,* and for his growing number of camp followers in *Ladies' Home Journal.*

The book that came from the Kalacoon experience was *Tropical Wildlife in British Guiana.* Published in 1917, it opened with an astute introduction by Roosevelt lauding the novel way of studying wildlife that Beebe employed, the beginning of a "wholly new type of biological work." The book, the sort of natural history–cum–travelogue at which he was so adept, included his scientific analysis of the bird life of the area, along with descriptions of the local wasps by Howes and the fish by Hartley. *Tropical Wildlife* spurred researchers to consider the tropics as a practicable place to work. It inspired many scientists to plan trips to Kalacoon or to begin field research stations of their own; the Smithsonian, for instance, would open a station on Barro Colorado Island in the Panama Canal in 1923, where Will's colleague Frank Chapman would study birds and which continues to provide an invaluable window into tropical ecology for students and researchers.

CHAPTER 21

Jungle Peace

Slip quietly and receptively into the life of the jungle, . . . accept all things as worthy and reasonable; . . . . sense the beauty, the joy, the majestic serenity of this age-old fraternity of nature, into whose sanctuary man's entrance is unnoticed, his absence unregretted. The peace of the jungle is beyond all telling.

William Beebe, *Jungle Peace*

Back home the war was on everyone's mind, and feeling was high. The deadly battles at Verdun and the Somme in 1916 had demoralized the Allies, and by 1917 German submarines were attacking American merchant ships ruthlessly. Relief efforts monopolized New York society, and even the ivory towers of academe resounded with bellicose threats. At the zoo, ranks were being thinned as staff members joined up; the remaining keepers were formed into a uniformed brigade.

Will was desperately eager to go to France, but at forty he was not considered for regular service. Roosevelt, chafing to fight himself, had secured a position for Beebe training American pilots for an esquadrille on Long Island, and Will hoped it would lead to a commission. For several months he juggled aviation with writing and work on the pheasant monograph, whose publication was being delayed by the war. One set of proofs from the publisher in London had been sunk in transit by a German U-boat, and many of the plates, consigned to a Viennese press, were virtually inaccessible. Then disaster of another sort struck, dashing his combat hopes: while performing aerial maneuvers for some visiting Italian officers, Will crashed on landing as he veered to avoid a photographer who ran in front of his plane to get a daring shot. Three bones in his right wrist were broken, and the tendons were damaged severely. His spirits foundered with his hopes, and again he plummeted into serious depression.

The bones healed rapidly, but the soft tissue damage needed long, careful tending. He had to sit out the United States' April 1917 entry into the Great War, benched by what he saw as an inglorious accident. Its staff depleted and even its animals on short rations, the zoo was unwilling to underwrite a season of research in the tropics. In August Will managed to arrange a quick trip to Kalacoon to check on the state of the house and the equipment he had left stored there, and found to his horror that the jungle he had studied so intently was gone. The great demand for rubber had spurred Withers to clear every inch of his holding for rubber trees, right up to the edge of the house. Realizing that Kalacoon's usefulness had been irreversibly compromised by the wholesale destruction of the area, he began the melancholy job of packing up.

When he returned to New York in October Will was determined to go to France, if only to see for himself what the pilots he had trained were accomplishing. Several friends were there working with relief groups, and *The Atlantic Monthly* commissioned him to write articles on what he found. The Zoological Society board granted him a leave of absence "for war duty and in connection with the Kuser Pheasant Monograph." Roosevelt busied himself writing letters of introduction to friends in the service, including an adjutant of General Pershing, that would get Will to the front lines:

> The bearer Mr. William Beebe is an old and close friend of mine. He has done a great deal of flying. He would have had a commission in my division if I had been allowed to raise troops. I vouch for him in every way—for his judgement, courage and loyalty—and bespeak every courtesy for him.
>
> Faithfully yours,
> Theodore Roosevelt[1]

Will wound up sailing for Europe with his faithful shadow, the perennially unemployed Weary Irwin, his friend from childhood summers in the Pocono mountains. They were met in Paris by Will's friend

Carita Spencer, who was vice chairman of the Women's Surgical Dress-ing Committee, a group of American women who made and distributed bandages to hospitals on the Continent. Organized and run by influen-tial New Yorkers, the society was one of the few really effective relief organizations. Carita was one of the reasons, with a resilient "can-do" attitude that took her from one hospital to another to determine needs and distribution.

Between his own friends and Carita's, Will had plenty of opportuni-ties to assess the situation. Eleanor Roosevelt, T.R.'s niece, whom Will had met at Sagamore Hill, was in Paris at the time; they dined together and talked of T.R., who was ill, and his sons Quentin and Archie. He frequented the various canteens to talk to the men and the worn-out doctors and clergymen who were trying to help the dazed, battle-weary soldiers. "The eyes of the soldiers haunt me. They laugh with their mouths but their eyes never change."[2]

While he was in Paris Will never ceased hoping to get to the front, to see it for himself and describe it for readers back home. At last T.R.'s

Beebe in France, 1917.

letters took effect and he had his chance: he spent several weeks in the thick of battle, flying with the great aerial photography planes that lumbered out over German artillery to snap a few quick shots of gun emplacements and trundle back, escorted by a bevy of agile little fighters. The unwieldy planes, protected by their small escorts, reminded him of large parasitic cuckoo or cowbird nestlings being fed and cared for by their tiny warbler hosts. He spent time in trenches and even accompanied a Canadian Indian platoon on a night raid.

The articles he sent to *Scribner's Magazine* and *The Atlantic Monthly* were uncharacteristic of Will's usually enthusiastic style: dispassionate, factual accounts, they told of aerial photography, raids, and the war's effects on the natural world—hedgerows leveled, with all their varied animal life; fields trenched and turned to muddy killing fields; wildlife displaced and slaughtered.[3] Referring to an article about the Indian platoon's nighttime raid, Edith Roosevelt wrote Nettie that "Mr. Roosevelt asks me to say that he has seen nothing finer of its kind than your son's last article in 'The Atlantic.'"[4]

Will returned safely in February, unsatisfied but at least with an educated imagination. The experience had driven depression away temporarily, but returning to the zoo renewed his misery over the loss of Kalacoon and of Blair, which continued to haunt him. "In confidential conversation with Beebe," Osborn wrote Madison Grant, "I find that he is worried and far from well. I want to talk with you about this. We cannot work a star of the first magnitude as we would a cart-horse; we must realize that we have in Beebe a star and the making of one of the greatest naturalists of our time. Without telling him so, we must take care of him."[5]

The appearance of the long-awaited first volume of the pheasant monograph in the fall of 1918, heralded by reverential reviews, was elating, especially when it was awarded the prestigious Elliott Medal of the National Academy of Sciences. British convoys had made the publication possible, but the future of the remaining three volumes, complete as to text but with their plates immured in Vienna, remained uncertain. Of the first volume, however, there was unanimous approval. "This work is believed to be the most perfect zoological publication as to illustrations, printing, and text ever issued," wrote the Zoological Society *Bulletin*. "It differs from an ordinary monograph in being an extraordinarily readable and interesting book."[6]

Roosevelt, who had felt that Will's talents were wasted on a study of a single family when he had such important ideas about the whole range of animal life, still admired the monograph's accuracy and scope and its surprisingly engaging style. But in the future, he advised his friend to "put more of the 'I' in," to personalize his accounts, in the tradition of adventure writing, as Roosevelt himself did. (9.6.1918)

Jungle Peace, which came out just after the monograph's first volume, was the kind of book T.R. wanted Beebe to write. Using the theme of the jungle as a refuge from war's alarms, *Jungle Peace* reprinted several essays that had appeared in *The Atlantic Monthly* and added a few more, including Will's painstaking analysis of a square yard of leaf litter hastily scooped up from a Brazilian jungle. With a magnifying lens and the boredom of a slow cruise back to New York, Beebe discovered armies of minute creatures, all leading lives of adventure and peril.

The book opens with a declaration of Beebe's subversive agenda as a scientist and writer: after studying his subject objectively, transcribing facts and observations with technical accuracy, the writer shifts to an oblique view of the same subjects, "observing them as actors and companions rather than as species and varieties; softening facts with quiet meditation, leavening science with thoughts of the sheer joy of existence. It should be possible occasionally to achieve this and yet to return to science enriched and with enthusiasm."[7]

The conclusion of an essay called "Islands" shows nature defeating the ship's elaborate blackout precautions:

> When we steamed away from shore that night, no lights . . . were allowed. Yet the path of the vessel made a mockery of this concealment. . . . The outward curve of the water from the bow was a long slender scimitar of phosphorescence, and from its cutting edge and tip flashed bits of flame and brilliant steely sparks. . . . Alongside was a steady ribbon of dull green luminescence, while, rolling and drifting along through this path of light came now and then great balls of clear, pure fire touched with emerald flames, some huge jelly or fish, or sargasso weed incrusted with noctiluca. . . . Perhaps to some unimaginably distant and unknown god, our world system may appear as fleeting.[8]

T.R.'s review in the *New York Times* lauds Beebe's service in France and describes him as "sick of the carnage which has turned the soil of

Northern France into a red desert of horror. To him the jungle seemed peaceful, and the underlying war among its furtive dwellers but a small thing compared to the awful contest raging among the most highly civilized of the nations of mankind."[9] Readers loved the book, whose message acknowledged the chaos of war but used it to highlight the overarching peace of wild places. Six editions were printed in the first three months.

In early January 1919, long blind in one eye from a boxing injury and plagued by painful resurgences of chronic parasitic infections picked up from various exotic locales, Roosevelt died unexpectedly from a blood clot at the age of sixty. Will had been so close to his mentor and friend that his grief was tremendous. "Word has just come of Colonel Roosevelt's death at 4:15 this morning," he wrote on January 6, 1919, still numb from the shock. "It is in my mind, the second thing in my life, Blair being the first. I fear I shall only drift now, or live in their memory." (1.6.1919) From his hospital bed, T.R. had written to Will what would be the last letter he ever wrote, a congratulation and comment on the pheasant monograph.

Ever solicitous of Will's state of mind and shaken himself by the death of his old friend, Osborn worked assiduously to negotiate a new lease on life for the tropical research station. With offers of several inviting sites in his pocket, in March he swept Will away on an expedition to find a new research station in British Guiana. The Osborns invited several of their friends along, so the SS *Guiana* carried a rowdy group of would-be explorers as it set sail. "There are many funny people on board," Will wrote Nettie, "& our own party as usual runs the upper deck."

Prof. Osborn & I had many talks & I think I shall do some splendid things at the Station. Mrs. Osborn stood on her head in my cabin on two of my cushions, as a Doctor told her it was good for her colon. So you will probably want to do it. She made us all do it, although fortunately for some of the party it was not rough. . . . I caught weed & found crabs & things so the days have gone, with rest & sleep. . . . The first several days are always hard, & each morning before I am quite awake I reach down my hand to the lower berth for Blair to take. But the hearts gradually heal and the ocean is always blue, & my friends are fine & I love Papa & you lots & lots."[10]

The outcome of the "exploring" party was never really in doubt: Beebe had been offered Kartabo Point, an outpost of a New York–based mining corporation upstream from Kalacoon, and was enthusiastic about its possibilities. Osborn's support and encouragement was invaluable, and the prospect of a new station rising from Kalacoon's ashes gave Will new life and hope. With resurgent optimism he decided that Kartabo, at the junction of the Cuyuni and the Mazaruni rivers two miles up from the ghost of Kalacoon, was in every way superior.

With Will at Kartabo that first season was his new assistant, John Tee-Van. John had come up from the ranks at the zoo, the son of a widowed Irish keeper with a large motherless family to keep afloat. John had begun work in 1911 cleaning cages but spent his evenings studying architectural drafting in night school. Impressed by the young man's careful work habits and studious demeanor, Will asked him to draw a bird bone and was so pleased with the result that he made him his assistant. When John "graduated" to his employ, it was serendipity for both men.

Isabel Cooper's watercolor of Kartabo House, British Guiana,
Beebe's field station from 1917 to 1927.

Beebe loved to teach everything from woodcraft to deck tennis, and Tee-Van was the perfect student. During the early years of their association, Will took on John's education as if he were an adopted son. He presented the young man with the *Encyclopedia Britannica,* whereupon John worked his way through the entire set, his sharp mind remembering, questioning, categorizing. With Will as his role model, the self-effacing Tee-Van refined his diction and manners and gained self-confidence and ambition. Beebe's tutelage would eventually propel John Tee-Van to the directorship of the Zoological Society.

On these first expeditions John acted as general factotum while he learned the ropes. With Osborn and his lively guests out of their hair, they were free to get the new station under way. Will's first report to Hornaday was full of excitement—as well as careful justification of every expense he undertook, for Madison Grant's parsimonious eyes.

The new place will exceed all expectation both as to situation, convenience to the jungle and enormous supply of organisms. As I sit in the glass-covered laboratory I have just seen and heard three species of kingfishers on the stakes which moor the boats. A colony of caciques has just commenced to build in the bamboos behind. I get up and go down to the river au naturel and dive into the water which at this hour is very cool. There are mangroves farther down, reeds nearer, and pure sand in front, so we have all varieties of beach. For all of this week we must keep at the roof which we are covering with tar paper. Last week we fitted in glass windows in a solid line to the gallery.[11]

In addition to helping Will with the construction, John helped Isabel Cooper with the sketching. A young Bryn Mawr and Columbia graduate who had been teaching science and art in the city schools, Isabel signed on eagerly as artist for the expeditions. She was joined by Helen Damrosch, also an artist and the daughter of New York orchestral conductor Frank Damrosch. Ruth Rose, a Broadway actress who had driven official cars at the front during the war, acted as household manager, learning zoology as she went. Mabel Satterlee, a studious young granddaughter of J. Pierpont Morgan, joined the staff as a general assistant.

Inness Hartley was again willing to come at his own expense, as was Albert Reese, a zoology professor who had written a book on alligators.

Alfred Emerson, a graduate student from Cornell, and Clifford Pope, a graduate student in herpetology at the University of Virginia, came along for the experience as unpaid research assistants. In those days—as indeed today—students whose families could afford the expense would affiliate themselves with scientific expeditions, paying their own way and often subsidizing some valuable piece of equipment in compensation. Pope, for instance, came bearing a wooden canoe as dowry. He eventually achieved success as a curator at the Field Museum and author of *The Reptile World*. Alfred Emerson continued to work at Kartabo for several seasons; he would become a professor at the University of Chicago and a respected expert on termites—a project suggested by Beebe that became Emerson's life work.

These students usually received good value for their money, as Will was a natural teacher and loved steering a quick mind onto a new, promising path. His multifarious interests practically guaranteed that there would be some fascinating question that needed a solution, or a problem that required careful, controlled data collection. He put them to work at once on a project of special interest to them, or on someone else's if they had no chosen field of work. Most of them went on to distinguish themselves in their various areas, but some fell by the wayside when they discovered how hard Will expected every team member to work.

The practical routine established at Kalacoon worked just as well at Kartabo. At dusk, the entire party would gather on the patio to watch

Local skinners at work in the jungle lab, Kartabo.

the sun set, sip the rum and fruit juice swizzles that were so integral a part of British colonial life in the tropics, and discuss the day's work. As vampire bats flitted softly by and birdsong gave way to the nighthawk's plaintive cry and the buzz of the cicadas, Will would bring out his ukelele or banjo and they would sing, talk, and drink until the servants announced dinner—usually prepared from some bush meat the Indian hunter had procured: agouti, tinamou, or peccary.

Will rose before sunrise to take a swim, and breakfast was early; there was always oatmeal and the dry cereal Nettie believed to be the cornerstone of good nutrition, along with toast, jam, and fruit. Lunch was quick and followed by a brief time-out while the sun was at its fiercest. Then it was back to the jungle for a few more hours, until it was time to swim and dress for dinner—not the stifling black tie affected by the British officials, but clean, loose garments, often pajamas, which were at that time accepted "at home" wear. Often Will dived back into the jungle after dinner for a last hour or two of observation before bed.

In the field, Will had settled into a costume that worked for him. In the steamy jungles the topee helmet of the sahib had given way to a worn brown tennis hat, the much-loved leather "leggins" to the cheapest tennis sneakers he could procure. He wore thin woolen shirts and khaki trousers, which he changed once a week. Everything was subject to sudden, constant drenching, but a few moments in the sun would dry all but the shoes. The women remained skirted or demurely knickered in khaki or serge, but were beginning to loosen some buttons.

The house quickly sprouted a little community of tents to accommodate staff and visitors. There was of course no running water, so facilities were rudimentary, but the tents had wooden floors, shelves, and beds. One particularly dry tent was reserved for "important" visitors—those who were not scientists but who were likely to provide support of a more solid and financial nature. There was a capable staff to keep the household running—a butler, a housekeeper and cook, a caretaker, a "man of all work," and a boy, little Richard, who "does what his size permits." With Will's extravagant love of parties, birthdays—especially his—were causes for complete disruption of every schedule as the entire staff spent untold hours and creativity dreaming up entertainments and costumes and zany practical jokes.

Kartabo's first season was a roaring success, and the 1920 season proved still better. Thanks to Will's evangelical tracts, word was spreading, and the distinguished scientists he had hoped to attract began to arrive—tentatively at first, as the tropics' reputation for discomfort and disease was a powerful stigma to overcome. One of the pioneers was William Morton Wheeler, the eminent ant man from Harvard who had identified the ants Beebe's four square feet of jungle debris had yielded. Will liked and respected Wheeler, who returned for several seasons.

With jungle looming all around and the river running sluggishly just down the hill, Kartabo's location was better adapted than Kalacoon's to the needs of researchers. The diversity and abundance of life forms were dazzling. When Wheeler went to work on a huge tree he had felled, he discovered sixty-four species of ants living in or on it. Will cut an imaginary column a yard square from the treetops down through the leaf litter and into the ground, and tried to conduct a census of every plant and animal that lived within it. He cataloged many hundreds in the soil alone.

Convicts from the penal colony cleared a trail from the old boat dock through to the jungle, as they had at Kalacoon. Like other local workers, they brought interesting finds to "the Doctor" for identification and sometimes reward. By Kartabo's second season an established network of trails cut through the jungle for miles, winding through clearings where jaguars stalked and herds of peccaries rooted for food, leaving their strong, musky scent hanging in the air like a cloud. Trails wound through dense underbrush where shy birds such as tinamou and nighthawks hid, to the feet of gigantic jungle trees draped with lianas, whose branches formed a canopy so rich with life that Will believed a scientist could spend his whole life and never scratch the surface of what there was to be learned. The insects alone were overwhelming in their individual complexity and in the intricacy of their interactions.

When Beebe and Tee-Van scoured the zoological literature for information about the great number of species Kartabo offered, they were surprised to find how little any of these creatures had been studied. Many of them were actually new to science. References to known species were few and, where they existed, usually dealt with preserved specimens, often of only one sex. There was no information at all on environment, courtship and breeding habits, diet, methods of offense and defense, voice, color, seasonal changes, enemies, or instincts. All the things that were crucial to an understanding of how the whole forest community worked were lacking. Will's mission at Kartabo was increasing in scope and importance.

Many facts about a species' behavior were gained by accident or serendipity. Bats flitted through the bedrooms and tents, so John and Will slept with toes exposed, hoping to lure the vampires to bite, to discover if stories of their painless bloodletting were true. One day a giant anteater came swimming past the dock, trying to escape a group of miners who were chasing it in a boat. The whole crew sprang into action, determined to capture it alive for the zoo. It put up a terrific fight, striking out fiercely with its strong, curved claws adapted to digging termites and ants out of hard mud. It was at length subdued, however, and eventually became quite tame. When approached, it would curl into a tight ball and wrap its great fluffy tail around its body, peeking out occasionally to see if the coast was clear.

Will made a great effort to downplay the perils of the region in his reports and articles, and was wont to assert that a walk through Central Park was just as dangerous as a trek through the jungle—and this was in the early 1900s. But there were at least three more venomous snakes in the area than Central Park could boast. Although not abundant, there were huge bushmasters as well as the small, deadly fer-de-lance and an occasional coral snake. Boa constrictors and anacondas were common. Snakes—the deadlier the better—were a perennial favorite with zoo visitors, so every effort was made to catch them alive. Once when he was hunkered down to study a patch of forest, Will felt something strike the edge of his soft brown tennis hat. It was a fer-de-lance, and he had nothing in his hand but a frail butterfly net. Fortunately, the snake had recently swallowed a large frog and so couldn't strike as hard as usual; a lucky blow with the net killed it.

The vast rivers, a mile wide at Kartabo Point, yielded plenty of food for thought as well as dinner. In the 1922 season Will began an intensive study of the river creatures and their interdependence, setting out seines of various meshes to catch both fish and the plankton they consumed. He compiled an exhaustive survey of the river like the one he was amassing of the jungle around Kartabo. Great primitive catfish lurked in the shallows, armored like their distant Devonian ancestors, and the electric eels were too tempting to leave behind: Beebe and Tee-Van endured shock after shock trying to capture the great fish, some as long as four feet. The New York Aquarium was hungry for specimens, and Ulric Dahlgren at Princeton was eager to study their electric force. The river was host to two species of freshwater rays, both covered with toxic slime. But despite the dangers, a swim was a daily ritual for all the permanent staff, the warm water and absence of leeches more than making up for an occasional water snake or piranha. Beebe believed that swimming in the clear brown water was what kept them all fit despite the rigors and malaises of tropical research.

Although visiting researchers always had their areas of interest clearly mapped out, Will remained omnivorous. Every living thing truly interested him, and he collected data on many fronts at once. Birds, of course, were always on his mind, and he was engaged in an ambitious course of study of the vocal apparatus of tropical birds. He removed the syrinx of every bird brought in dead and shot specimens of those he needed for comparison. One day he discovered, almost accidentally, that the throat muscles of a decapitated bird could sometimes be manipulated to produce sounds very similar to the call of the bird when alive.

He was still fascinated by mimicry and camouflage, which were subjects of lively scientific debate. Why so many creatures, particularly

View up the Essequibo River near Kartabo, with typical jungle vegetation.

insects, should be colored either protectively, as camouflage, or vividly, as if to advertise their presence, spurred discussion of evolution and adaptation, and prompted lively if not always genial dinner conversation. Beebe's broad interests and love of fun led him to be tolerant, if not always reverent, of most points of view. He thought of himself as the host of an ongoing party, and the mandate of noblesse oblige carried him through many sticky moments. But when people used science to reinforce political agendas, such as using social ants to prove the superior efficiency of a Socialist system, he bristled; if he couldn't divert the conversation, he forced himself to walk away.

Among the creatures that captivated him was the three-toed sloth, an animal which moves so slowly that its shaggy coat is often green with algae and mold. To study sloths and other animals intensively, he dug an elaborate habitat island surrounded by a dry moat. The tree-bound sloths were unable to escape, so Beebe could observe aspects of sloth behavior that had gone unsuspected, as they were normally so hard to observe. Their eyes enthralled him by their complete vacancy, and their inability to deal with problems left them at a real disadvantage. "Psychologically," he wrote, "they are either a mystery or are beyond belief simple and dull; which is perhaps another way of saying that I am not able to put myself in their position and get their point of view on life."

A three-toed sloth, a favorite research subject.

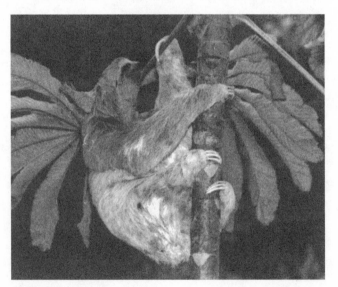

He calculated that out of 169 hours, one sloth spent 11 hours feeding on cecropia leaves, 18 hours climbing aimlessly, 10 hours resting and scratching, and 129 hours sleeping.[12]

Will's dream was thus coming to pass: he had created a place where scientists and researchers of many stamps could live and work comfortably in the midst of the almost unimaginable richness of the jungle, sharing ideas and theories, proposing new avenues of research, and challenging old dogmas. And he had the stability he craved—a place he could feel at home, without the stress of family but with all the accoutrements of the scientific life. The great joy of returning, he wrote, was "to find things in their same places—the nail for my shaving mirror or to hang my forceps upon . . . where a single spadeful of muck will produce one or more fresh water Nereis. . . ." (3.4.1922) In his spartan room he had a camp bed and shelves of specimens and chemicals, including the medicines for the whole camp, and souvenirs from earlier trips. It was at once office, library, lab, and home. Thanks to his responsibilities as director, the constant companionship, and his almost frenetic devotion to work, the cloud of depression was lifting, and Beebe was schooling himself in the ways of forgetting.

> Little petty jealousies and irritations and selfishnesses arise, but I am learning how silly it all is to let small things fret. . . . I am infinitely happier here with only a few of the necessities than in New York with all my 100's of odds and ends which . . . are only of temporary value.[13]

PART III

Marine
Biologist

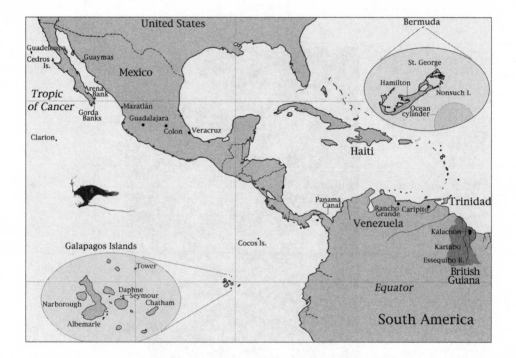

CHAPTER 22

The Encantadas

*[There is] one fundamental reality in wild nature—the universal
acceptance of opportunity. . . . Evolution has left no chink or crevice
unfilled, no probability untried, no possibility unachieved.*

William Beebe, *Edge of the Jungle*

BACK IN NEW YORK, WILL
settled into the bachelor life he had made for himself. He had taken
possession of an apartment in a new building on 67th Street, just west
of Central Park, designed by the impressionist painter Childe Hassam
to accommodate the literary and artistic set. It was all studio apart-
ments, and the residents formed a good-humored, party-prone group.
Will's neighbors were writers, theater people, and artists. Few had "day
jobs." Will was unattached and had a fund of good stories and a ready
laugh, so he was quickly absorbed into the clubby atmosphere.

As honorary curator of birds and director of the newly created
Department of Tropical Research, he now had no animal maintenance
duties to hamper his writing or his research. Lee Crandall had taken the
reins as curator, although he still relied on Will's keen eye for illness
among the birds and help in nurturing the exotic species Beebe brought
back from the tropics. Writing and planning for the coming expedition
—whatever it was—were Will's labors of love, but the worldly necessity
of fundraising was constantly demanding. Not only did he have to find
"angels" to fund every project he undertook, he also had to convince the

zoo board that his aims were complementary to theirs. Toward this end he wrote popular articles, gave interviews, and promoted the zoo and its research and conservation goals incessantly; cultivated wealthy donors he had met at Sagamore Hill, Faircourt, and Castle Rock; and talked up new ones as he traveled on the lecture circuit.

Harrison Williams, a particularly generous contributor, had made a reputed $680 million in the utilities boom, investing on a much grander scale than Kuser had. Williams was captivated by Beebe's account of the outstanding scientific possibilities of Melville's *Encantadas,* the wild and unexplored Galápagos Islands off the coast of Ecuador. Will had always dreamed of going to these uncharted isles, where Darwin had stumbled across the finches and giant tortoises that sowed the seeds of evolutionary theory in his mind. Will speculated that, given the current state of scientific knowledge and technology and a competent staff, he would be able to amass the hard data that Darwin, on his brief, solitary visit, had been unable to collect. Only real, visible evidence would put the world's nagging doubts about natural selection to rest, and Will imagined that he could get it.

The islands' inhospitable barrenness had largely protected them from damage by settlers and sailors, and had also kept them unknown to science. The absence of water or food, along with the rough and jagged coastline, made any sort of long-term exploration impractical. Astoundingly wealthy and always adventurous, Williams was perfectly willing to finance the outrageously expensive expedition, with the proviso that in addition to Beebe's staff, Williams could invite his own non-scientific party of game-fishing friends. After a great deal of discussion, wrangling, and crystal ball gazing, Williams, Beebe, Osborn, and the all-important zoo board agreed on a trip to the Galápagos to collect animals, materials for several exhibits at the American Museum, any incidental research Will was able to fit in—and a great deal of sportfishing.

Will's staff was composed of his Kartabo contingent as well as several others whose skills and personalities he deemed complementary. Bill Merriam, a good-natured and ingenious adventurer, signed on as "Chief Hunter" and would later prove his worth by devising solutions to intractable problems. Harry Hoffman and Isabel Cooper were the team artists, with John Tee-Van as official photographer. Ruth Rose, who had cut her teeth with Will at Kartabo in 1922, again proved as flexible as she was indispensable. She was a fine writer and researcher with a witty, sophisticated style, utterly fearless and endlessly enterprising. She managed the unexciting household details as well as the considerable menagerie, and was always game no matter what danger or fun was afoot.

Williams's party comprised a cheerful, adventurous set of wealthy professional men, each provided with an honorary expeditionary title. There were a surgeon, a dentist, a lawyer, and a congressman and former assistant secretary of the treasury (designated curator of dredging and diving); but sadly, Williams himself, the official "Patron," had to leave the ship on urgent business before reaching the islands.

Williams chartered a 250-foot steam yacht called the *Noma* for the expedition, and the small force left New York on March 1, 1923. The ship had its quirks, which everyone—especially Beebe—made an effort to take lightly. James Curtis, the congressman, kept a humorous log of the journey. He describes the "principal occupations and favorite indoor sports" of the first days as "the issuance of complaints by all hands."

> First: that too much water was leaking from the pipes. Second: that no water was coming from the pipes. Third: that the salt water was too salty to shave with, and Poland water was too expensive, except for the

The Noma *staff, 1923.*

Patron. The indoor sports were reading and sleeping, until the evening session at which the Professor translated the works of Darwin, Beck, Rothschild, and others, into language understandable to the various curators, all concerning the Galápagos Islands. . . . At breakfast, a learned but not entirely clear recitation from the Professor as to "What is a Boobie." [1]

Finally the engines passed muster and the vessel headed south. The yacht proceeded slowly, the stokers becoming conveniently seasick when asked to work faster. Several crew members deserted at each port, as they began to understand the desolate destination of the vessel. The passengers, however, passed the time agreeably. Their passage through the Sargasso Sea had not produced the dense mats of sargassum weed that Will had hoped for, but even the occasional clumps were rich with life. He spent his days scooping weed from the surface, and the whole party convened to watch the staff comb carefully through for interesting creatures, which soon occupied every spare bathtub, basin, and cocktail glass. Will, alias "the Professor," devised an infinite variety of deck games, and there was a nonstop poker game in the parlor. Rotting vegetables floating away on the tide gave their lives as targets for rifle practice.

He read or lectured to a receptive—and well-oiled—audience every night, and the exotic creatures salvaged from the weed mats sent even the most hardened nonscientist scurrying to the tiny but select library for information. Minute filefish and crabs so elegantly camouflaged as to be invisible until rudely shaken out, larval stages and eggs of myriad sea creatures hiding their vulnerability in the weed's embrace, minuscule jellyfish feeding on microscopic plankton—all were admired and looked up and sketched and placed in makeshift aquaria by men accustomed to judging fish by the pound and delegating note-taking to the secretarial pool. By the time the motley group reached the Pacific islands, all were caught up in the thrill of the chase.

Will quickly realized the difficulty of work in the Galápagos. The "heaps of cinders in a vacant lot" Melville had described were immeasurably rich in animal life, but the islands guarded their riches jealously. Spawned by a ring of ancient volcanoes, their surface was rough, razor-sharp lava that crumbled and slid underfoot, so that slopes that looked gentle were savagely unclimbable. The scrubby brush that grew on the barren slopes was thorny and shed burrs that pierced clothing and worked their way through shoes and into feet. And then there were the spiders.

> One of the worst things were the cob-webs strung everywhere between the whitened, half-dead shrubs. The spiders were the same zigzagged backed species found on all islands, but here very large with enormous and very strong webbing. . . . When you are panting and dripping from every pore, with a heavy gun & tripod in one hand, a box of plates in the other, a game bag hanging over sun-burned shoulders, gripping sharp boulders, which slide back 1 in 4 feet, with spines from cactus and some kind of ground juniper threatening every careless step or slip— then to have 100s of sticky webs covering eyes and ears and face, with spiders crawling everywhere from hat to knees is . . . the last straw. (4.6.1923)

With the usual mandate to bring back animals live for the zoo and aquarium and preserved for the American Museum, Will set everyone to collecting. Anchored well off the islands' rocky shores, the *Noma* sent out boats of explorers every morning and returned for them in the afternoon. They divided into search parties and took on different sections of each island in turn, focusing on the smaller islands that offered more chance of grasping the interactions of their denizens, so tame that catching them seldom required guns. Huge iguanas yawned and shifted lazily inches away from bulky camera lenses; a hawk perched on the barrel of a gun that had been aimed at him. The penguins waddled

The Noma, *Beebe's floating research station in the Galápagos.*

around the deck getting stepped on and were nearly impossible to photograph because they refused to stay a decent distance from the camera. The penguins and the sea lion pup were inordinately fond of the
phonograph, and clambered up and over the fine leather furniture of
the smoking room to get near it.[2]

To Will's dismay, almost as soon as they arrived in this paradise, the
ship's captain began to fret over the lack of fresh water for the boiler.
Although the navigation charts promised water at several of the islands,
each promise proved false, and Will had to watch his long-awaited,
meticulously planned expedition turn into a million-dollar scavenger
hunt. At each island, possible sources of water became their primary
goal, and studying or collecting animals something to do hurriedly en
route. Ironically, the ship's supply of coal became perilously low as they
steamed from place to place in search of water. "We had now attained
the perfection of the vicious circle; we must have fresh water in order to
go on burning coal, but if we went on burning coal we should not have
enough coal to go on looking for fresh water. The brain reeled."[3]

Albemarle, the largest island in the chain, promised the best compromise between the possibility of water and interesting fauna.[4] Large
schools of porpoises and flocks of shearwaters kept them company as
they steamed toward the one fitting landing place, Tagus Cove. They
found none of the water that appeared on the charts, and the captain
gave them only a few early morning hours before they would have to
leave for the next island. Beebe was desperate to collect specimens of
the rare Galápagos flightless cormorant, which he expected would be
extinct soon, and of the small Galápagos penguins. His group headed
for a seabird colony on top of an imposing cliff. Scaling the loose, slippery clinkers with a forty-pound movie camera in one hand and a three-
barreled shotgun and collecting bag in the other, Will noted that his
feelings were "far from science" when his footing gave way. On the way
back he watched Bill Merriam descending with the unwieldy camera
tripod, whose legs seemed alive, and the game bag full of live boobies,
"miraculous in their ability to get a wing free at a critical moment."[5]
They found the cormorants nesting as well, and took movies and photographs from all angles. Will made hurried notes on their behavior, in case
they should not survive the trip to New York. Then the ship's whistle
blew, and they had to scramble rapidly back to the beach.

After a tantalizing ten days among the islands, the captain ordered an immediate return to Panama, where, if Will could extort more funding, they could refill the boilers and restock the coal supply. Not knowing whether he would be able to return, Will dispatched three parties for a final frantic day of collecting on Eden, the island richest in specimens. One group caught fish and crabs and assembled cacti, shells, and anything that could be used to make realistic groupings back home. Another team went in search of sea lions, and Tee-Van and Beebe headed for the best place to catch marine iguanas and crabs. Will hated "with my whole soul" to leave; if water supplies had been adequate, he would have had at least two more weeks.

In Panama, Will radioed Osborn and explained the problem. Osborn, who had used his considerable influence to set up the expedition in the first place, was loath to go begging for further funds from an already munificent donor—particularly when Williams had been unable to take part in the expedition he was financing. However, he cabled Williams for another $64,000—and got it.

The sportsmen were nearly as elated as the staff, for they had enjoyed playing scientist, and the extracurricular game fishing had been memorable. They caught and released huge mackerel, jack, grouper, and bonita, and still had plenty for the table. And they had stalked and wrestled and become acquainted with creatures from iguanas to octopi to giant tortoises that would provide them with cocktail party conversation for several lifetimes.

They also joined in the lesser sport of seining for smaller fish to feed the growing shipboard menagerie, which by the time they left included, in addition to an assemblage of fish and other marine creatures, three penguins and four sea lion pups, several monkeys, a hawk, two flightless cormorants, twenty-five large land and marine iguanas, a Galápagos tortoise, a sloth, a grison weasel, a pair of small red doves, a pair of jays—and a puppy and kitten that Will had adopted in Panama.

Watered and coaled, from Panama the *Noma* steamed back to the Galápagos. The $64,000 had bought Beebe another ten days, and he determined to do the collecting he had to do for the zoo and the American Museum, and to do as much research as he could. On each island he collected specimens of the amazing variety of finchlike birds. Darwin had discovered that nearly all the birds on the Galápagos, whatever they looked like and whatever they ate, were actually finches. It was this discovery that jump-started the idea of natural selection. On volcanic islands only a few million years old, a single species of finch had radiated into thirteen distinct niches. There were seed-eating finches, a woodpecker finch, and even a finch that drank blood. But the problem that intrigued Beebe was that three different finches lived on seeds that fell to the ground, and each type had a different beak. How could natural selection account for three separate coexisting solutions of the same problem?

Since the 1859 publication of *The Origin of Species by Natural Selection,* several expeditions had collected specimens and had anatomized them thoroughly. But none had recorded adequately the stomach contents of each species, and thus none had been able to determine the extent to which the finches' differing sizes, bills, and behavior might be affecting what each species could eat. Beebe knew he could do this investigation methodically and well. He was expert at deciphering the contents of bird stomachs, and hoped to find evidence of diets sufficiently diverse to account for the differences in the beaks of the various species.

They spent one or two days on each of several islands. Off Tower Island a huge flock of ungainly frigate birds swept over the boat, all trying to land on the wireless crosspiece or on the small rounded top of the mast. Big red-footed boobies dashed about, just out of arm's reach, and graceful shearwaters skimmed the surface. "Giant sharks swam past, and huge devil fish rose & fell with the shock of a barn door being dropped on the surface." (4.18.1923) On Chatham the terrible burr-thorns took their toll, working their way in and under socks, making walking on the lava in sneakers hell. At the end of one blisteringly hot day, the whole crew dragged in, too exhausted even for poker.

Between the islands they trawled for sea life, using nets that reached sometimes a half mile down. Strange masses of jelly that once had been creatures whose shape could only be imagined came up, flattened and exploded by the deadly difference in pressure. Closer to the surface the creatures were equally ephemeral, but they often lived long enough for description, sometimes even for a rapid watercolor to be made, which caught the delicate shadings more accurately than the black and white of photography. Most were the minute creatures that make up the free-floating oceanic plankton which nourish larger marine animals.

Squatting in the companion platform at 6:15 I scooped up one of the beautiful blue glowing organisms which drift singly past the ship all the morning. He went out at the instant of catching, but glowed in the net with a purplish sheen. In a vial I found it was a giant Copepod, perhaps Sapphirina, perfectly transparent and colorless.

One large arrow worm, absolutely clear, had two parasitic worms in its gut: one a roundworm, the other a flatworm. "Think of being so invisible that nothing but one's parasites showed!" (4.25.1923)

Daphne, which was actually two islands, one very tiny, held a closely guarded secret. Beebe dubbed the pair Daphne Major and Daphne Minor, and described them as looking like the mouth of a submerged bottle and its stopper. After an arduous scramble up the sides of the large crater that was the mouth of the bottle, the crew discovered that the crater's floor was carpeted with boobies, nesting thickly over the sandy bottom by the hundreds. On Seymour Island Beebe bagged a moray eel, which turned out to be quite rare. He had noticed it peering myopically out of a crab hole several feet above the current tide line, and it put up a mean fight as he wrestled it determinedly into the snake bag. He studied the insect life on Seymour as well, noting hordes of brightly colored grasshoppers and migrating butterflies, and finding

A crater booby nursery on Daphne.

many burying beetles on a dead pelican in limbo "between decay and dessication."

The journey back to New York was lightened by frequent hauls from the sea. Will had devised a "pulpit" for himself—an iron cage affixed to the bow of the ship that enabled its occupant to be as near the ocean surface as practicable. He had also built a thirty-foot "boom walk" that jutted out from the side of the ship, convenient for netting drifting weed, floating animals, or intriguing detritus. The trip gave him time to study some of the specimens they had collected on the Galápagos. As he worked up the results of his study of the finches, Will was disappointed to find that most of the birds, despite their various beaks, ate similar diets. He had been sure that he would find a correlation between diet and beak size.

Thanks to years of study by Peter and Rosemary Grant of Princeton University, we now know that the finches' beaks do radically affect the birds' survival, but that the system is, like so many natural phenomena, at once more complex and more elegantly straightforward than the hypotheses. Natural selection is indeed at work, but it is weather dependent. Climatic shifts brought on by the powerful El Niño and La Niña currents, unknown in Beebe's time, affect the island's vegetation radically, for years at a time. All the finches eat small, easily opened

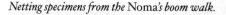

Netting specimens from the Noma's *boom walk.*

seeds, but the group with small beaks handles them more rapidly. In times of extended drought, plants produce fewer of these accessible seeds, so the birds that can open the thick hulls of the larger, tougher seeds left by the others have more options. Unaware of this powerful long-term cycle, Beebe was downcast at his failure to vindicate natural selection and determined to look for evidence of it elsewhere.

CHAPTER 23

New York Aerie

New York City, to several million of us, is the focus of the universe. . . .
Like cave men of old we clamber up to our little cubicles, tunneled out of
the street canyons, there to be happy or miserable, successes or failures, to
live and to die.

William Beebe, *Unseen Life of New York*

NEW YORK MEANT A RETURN
to writing and to the unending drain of zoo politics. Like any other educational community, it pitted unworldly specialists and researchers
against hardheaded businessmen who knew exactly how much it took
to keep those academics fed, clothed, and housed. Every penny had to
be cadged from well-worn membership rolls or wheedled from the limited herd of wealthy benefactors who were being begged for the same
funds by every other nonprofit organization.

To the benevolent but sharp-eyed men of the zoo board, Beebe was
a cash cow. With the manners of one born to wealth and the reputation
of a Columbia graduate—a reputation he did nothing to quell—he
remained playful and unassuming, the perfect addition to almost any
gathering. His popular books were on so many shelves, his photo in so
many rotogravure sections, his caricature in so many glossy magazines,
that his bald head and thin, eager features were known wherever he
went. Of course, this favored-son status excited the envy and ire of his
fellow curators, whose education and devotion to their work had
never paid off in such coin. Hornaday and the irascible Grant—more

edgy than ever as his health was failing—had to deal with the inevitable friction.[1]

Galápagos: World's End, the 1924 book that came out of the *Noma* expedition, was an instant best-seller and remained on the *New York Times* top ten list for months. With his customary light touch Will told the story of the expedition, diving fearlessly into scientific detail along the way, educating and flattering his readers without losing their interest. He described a wild, fruitless hunt along the Chagras River in Panama for the yapock, an aquatic opossum the zoo had wanted, poking fun at his missteps and surprises and managing to convey the natural history of the Panama Canal at the same time. He wrote of each island they visited and its characteristic creatures, bringing each discovery— whether giant lizards, tame sea lions, a crater of boobies, or transparent creatures of the tide pools—alive for his audience. He included a chapter by Ruth Rose on the history of the Galápagos, a long chapter on the dwindling population of giant Galápagos tortoises and the urgent need to protect them from sailors who prized the meat and sold the shells to collectors, and a surprisingly entertaining chapter by a member of the crew on game fishing among the islands. Will and Ruth wrote an account of the vain search for water, rendered humorous and breathlessly sus-

Charles and Nettie Beebe's house in Pelham, New York.

penseful by their witty rapid-fire prose. Reviews were overwhelmingly favorable, and George Palmer Putnam, by now one of Will's greatest supporters and best friends as well as his publisher, had to print a new batch of books after three weeks.

When he had moved from the Sedgwick Avenue house into his own apartment on Sixty-Seventh Street, Will had built a house for his parents in Pelham, a leafy and settled residential suburb north of Manhattan. Nettie's health was failing, but she was devoted to what she saw as the rational system of thought of the Christian Scientists, and she steadfastly refused to call on medical aid. Will implored her to seek the help of reliable men such as his old friend Joe Fobes, who had become an excellent physician, but she remained adamant. There is no record of what she suffered from, but on February 5, 1924, Will wrote, "Mama died at 8:20 this evening." Nettie was seventy-one.

Desolated, Charles and Will accompanied her body to Glens Falls, where she was buried in the family plot next to her parents and the redoubtable aunts. Her death had not been unexpected, but Nettie's gentle and steadfast manner had been the rock that had anchored both men, and her absence was agonizing. Will hated to leave his grieving father, but he had little choice: less than a week later, with Charles in the care of close friends and relatives, Will sailed again with his colleagues for Kartabo.

With Beebe on board this time were his regular Kartabo staff, including John Tee-Van as his invaluable assistant and the now married Helen Damrosch Tee-Van as one of the artists. Ruth Rose came as "historian" and assistant, and Bill Merriam brought engineering skills that were constantly being tapped. Isabel Cooper and Harry Hoffman joined Helen as the team's artists, as before. Alfred Emerson came to continue his work on termites, and there was an exuberant new naturalist and collector, the Russian immigrant Serge Chetyrkin. Chiriqui, a Cebus monkey Beebe had bought in Panama, was endlessly entertaining and helped keep him from the depths of grief over his mother's death.

Returning to Kartabo was a panacea for Will's spirit. There was the sense of returning to the place he truly belonged, as well as the stimulus of constant labor and company. With this crew there was always entertainment. Aside from the ritual cocktail hour, there were the almost religious observations of birthdays. Harry Hoffman was given an elaborate fancy-dress party in March to which Will came as a girl—having sacrificed his beloved mustache for effect. At a charade party in April, Ruth and Isabel acted out Mayhem and Dynasty, a visiting couple were bedbugs, and Will was a chigger. The servants put on their own party and entertained the staff with a variety of strange instruments, a joke speech that invoked the name of every staff member, a love song, a flute solo, and a Chinese mock-sermon.

Will's notes on his "outfit for general jungle collecting" succinctly indicate his life and methods in this period:

Three barrel gun w/ insertion barrel
#10 and #6 shells and 32 cal shot cartridges.
Short-handled net stuck in belt w/net part behind.
Marble ax in belt (sometimes).
Haversack w/smaller one attached.
In larger one paper, cotton, shells, notebook, compass (sometimes forceps and small brush).
Upper left-hand pocket of shirt, #3 stereos.
Left trousers pocket, #6 shells, pitch pipe.
Right ditto, knife, pencil, #10 shells, 32 cal cartridges.
Left hip pocket, small notebook, toilet paper.
Right ditto, handkerchief.
Watch pocket, watch, cotton wad.
Tennis cap,
BVD's and socks,
Khaki trousers,
Woolen shirt,
Sneakers.

His continuing object, as at Kalacoon, was to make as complete a record as he could of every order of animal in the jungle around Kartabo, to study each species as far as he was able in order to get an idea of its life, its habits, and its niche, and to begin to understand where each creature fit into the vast fabric of the Kartabo ecosystem. He was convinced that every animal and plant, even in the richness of the jungle, was so interdependent on every other that no "disturbance in the force," such as the extinction of even one insignificant species or the clearing of one patch of jungle, would be without repercussions.

As had been the case with the finch study, Will was well suited for this monumental endeavor. Omnivorous in his interests, he had studied nearly every order of creature. He had worked at the zoo long enough to know the importance of small variations in diet and environment to an animal's health and reproductive fitness. And perhaps most important, he had the patience, and finally the opportunity, to see an intensive, long-term study through.

The field station was a continuing joy. He loved the easy camaraderie, the intellectual stimulation from staff and visiting scientists alike, and the incomparable wealth of life in the river and forest around him—and often in the bedrooms and residential tents as well. In one typical day he recorded, collected, or banded several swallows, a lavender jay, a goldbird, iridescent caddisflies, mollusks, sponges, damselfly nymphs, howler monkeys, tinamous, woodpeckers, parasites in a termite nest, tapirs, peccaries, a snakebird, and several species of beetles. Bill Merriam had even rigged a way of shooting a line up and over branches sixty feet high so Will and his observation gear could be pulled up into the canopy in a sling—not as high as he would have liked, but it was a start. In the 1924 season he wrote several scholarly papers as well as articles for his popular audience back home, but spent most of his time working on his massive analysis of Kartabo's entire ecosystem. This paper would be the first major documentation of the nascent science of tropical ecology, the crucially important pursuit that steps back and looks at the entire flora and fauna of the world's richest and most important biome.

The article, "Studies of a Tropical Jungle: One Quarter of a Square Mile of Jungle at Kartabo, British Guiana," which came out in *Zoologica* in 1925,[2] remains a monument to one man's almost supernatural grasp of the animate world. The breadth of Beebe's vision was breathtaking. In nearly 200 pages, he gives a history of the area around Kartabo Point, a rapid-fire view of the geology and soil, a month-by-month analysis of the climate (more than 100 inches of rain, average humidity 84 percent, mean temperature 79 degrees Fahrenheit), an abbreviated but informative digest of the plant life, and an exhaustive rundown of every species of animal that had, in thirty-six months of study over five years, been positively identified. The descriptions include the number of individuals, their mode of living, their predators, and their prey. The article reflects countless hours of careful collection, dissection, preservation, and study; of stomach contents analyzed both chemically and microscopically; of parasites identified and cataloged and frequency of occurrence noted. The most crucial thing it reveals is the infinite time spent simply observing, watching the creatures go about their business in a habitat where "the race of man exists only by hearsay."

Longer and more detailed scientific papers put the meat on the bones of these brief descriptions. For the public—*Ladies' Home Journal,* for instance—Beebe wrote of the effect of the death of a single tree on the life of the jungle; for *The Atlantic Monthly* he described a chain of life: a protozoan inside a frog inside a fish which was eaten by a snake which was killed by a bird and found by a naturalist in the jungle. And as always, the Zoological Society's monthly *Bulletin* was provided with articles to entice the zoogoing audience.

When the crew arrived home in July, they were greeted by an article in the *New York Times* about the capture by Kartabo's intrepid female staff members of a lovely rainbow boa, and a wonderful cartoon by Ralph Barton of Will masquerading as a palm to avoid being eaten by an "unusually good specimen" of giant lizard.[3] In March he had been featured in *Vanity Fair*'s Hall of Fame along with violinist Jascha Heifetz, the writers Rebecca West and Van Wyck Brooks, and humorist Will Rogers. Ruth, who had been on the stage, introduced him to her acting friends, and his old love of the theater reasserted itself. The actor

Ralph Barton cartoon from the New York Times, *1924.*

Roland Young became a good friend; one night Beebe had dinner with him and then went with him to see him act in *Beggar on Horseback* on Broadway. Movies were still silent but full of possibility, and he was intrigued by the mechanics of film. He went with Ruth to see *The Thief of Bagdad*—"a marvelous movie"—and the next evening saw Cecil B. DeMille's epic *The Ten Commandments*.

In the little time Will spent alone he was prey to depression, which he abhorred in himself. Nettie's death was still fresh, and one evening he wrote that he became "rather low, looking at the old hair earrings of Grandma Younglove." (8.4.1924) (Jewelry made from the hair of departed loved ones, often quite intricate and lovely, was a Victorian custom, serving as both a sweet remembrance and a memento mori for the living.) When he visited his father at the Pelham house for the first time since his mother had died, he tried to dwell on how happy she had been there, in the house he had built for them.

He also still brooded over Blair, whose loss remained painful. "There is a big part of me that is very lonely and vacant," he had written once to Nettie. "I did love Blair so & it seems as if I should have to wait for another incarnation before I could forget & start again." [4] Friends such as the Osborns and the Roosevelts, who had adopted him into their families and made him feel less lonely, had assuaged a certain part of the pain, and his cuckolded pride had long since ceased to rankle. But Will was naturally affectionate, and family life was dear to him; despite friends and despite the quasi-family of his devoted staff, he missed the constant companionship of a partner. And he still considered Blair the love of his life.

Blair, however, had blossomed into her own life of adventure. As Blair Niles, she was traveling widely and writing articles about those travels. Her life with Will had sharpened her taste for travel in exotic places, and her interest in the people gave her a different slant on the places she visited. The amenable Robin was apparently better suited to be her traveling companion and aide-de-camp than Will, with his driven personality and take-charge approach. Robin's job as an architect gave him plenty of freedom for travel, and his family fortune freed him from having to pay suit to patrons. He developed an interest in photography and willingly toted the cumbersome photographic gear through steaming jungles and arid deserts for his adventurous wife, taking the pictures that would illustrate her articles and books.

In 1923 Blair published her first book, *Casual Wanderings in Ecuador,* and her second, *Colombia, Land of Miracles,* came out in 1924. *Black Haiti* followed in 1926, and 1927's *Condemned to Devil's Island* was made into a motion picture. Her magazine contributions were written with verve and dash, and she was becoming a sought-after lecturer herself. In bohemian New York, her divorce was actually an asset. Her association with the avant-garde literary establishment flourished, and in 1931 she would publish *Strange Brother,* a forward-looking novel about a gay musician who finds tolerance and friendship in the freewheeling atmosphere of Harlem's cultural Renaissance.

Will's bohemian friends moved in a different world. He was living in an elegant apartment at 33 West 67th Street, just a few steps from Central Park, where he used to eat his solitary lunch when visiting the American Museum as a child. But when the sculptor Emil Fuchs showed him his studio in the newest "artsy" building in the city, the Hotel des Artistes just down the street, Will was captivated and almost immediately claimed the penthouse there as a sun-drenched writing studio.

The Hotel des Artistes boasted sumptuous two-story flats with tall windows and balconies that extended all the way around to form a second floor. The restaurant downstairs catered to its domestically challenged residents, whose minute kitchens were barely functional, by

Beebe in his studio in the Hotel des Artistes.

cooking their meals for them and sending them up in an elaborate system of dumbwaiters. Will's gregarious neighbors were the same well-to-do artistic set that inhabited the other building.

He kept the larger flat as a residence, where he gave popular theme parties with his friend Ezra Winter, painter of such monumental and fantastic scenes as *Quest for the Fountain of Youth,* which dominates the foyer of Radio City Music Hall. Winter's mammoth studio in Grand Central Station was the scene of numerous Beebe-Winter galas where music, cocktails, and fancy dress were mandatory. Cartoonist Rube Goldberg was a frequent guest, harmonizing with Will on banjo or mandolin. Upstairs neighbor Fannie Hurst's parties catered to a more literary and theatrical set, and there Will met writers for the budding movie industry, film heartthrobs Douglas Fairbanks and Mary Pickford, actress Katharine Hepburn, and the famed dance team of Alfred Lunt and Lynn Fontanne.

Will in turn often invited friends to his studio to watch movies of his expeditions. One young screenwriter at these parties was writing a story about an adventurer in the remote Himalayas. Elswyth Thane's first novel, *Riders of the Wind,* was to be a romantic tale of a young woman who abandons her stodgy husband to travel through the myste-

Monkeying around with Rube Goldberg.

rious East with a dashing older man. Much of the tale had been influenced by Will's published accounts of the pheasant trip, and he was more than willing to give encouragement and advice. In the end, the book—the first of a long series of popular novels Thane would write—was dedicated to William Beebe.

Almost as soon as he returned from the Galápagos in the summer of 1923, Will had begun planning his next major expedition. The completion of the article on the ecology of Kartabo added impetus to his drive to search new places for evidence of natural selection at work. Henry Whiton, a wealthy and philanthropic businessman whose avocation was yachting, volunteered to provide Will with a steamship to make a thorough exploration of the Sargasso Sea, whose weed-choked depths had conjured a reputation for mystery and peril. The relative failure of this part of the Williams expedition had made Whiton that much more eager to back his own expedition.

Will, however, had experienced the vagaries of the so-called Sargasso. He was more interested in the Galápagos, which had truly enchanted him. The idea of returning, with a properly fitted-out scientific research vessel to trawl and dredge the ocean's depths, had become all-consuming. Whiton was at first reluctant, fearing that the costly expedition that should make his name famous would become simply an appendage to the Williams expedition. He was also sympathetic to Dr. C. H. Townsend, the director of the New York Aquarium, then in Battery Park, which had been the zoo's undervalued younger sibling since its acquisition by the New York Zoological Society in 1902. Townsend knew that a Sargasso Sea exhibit would be a big draw for the aquarium and wanted that to be the expedition's first priority.

The negotiations took more than a year, but at last, with additional capital from Vincent Astor and Marshall Field, and another hefty donation from Harrison Williams, the expedition began to take shape. The zoo agreed to pay the salaries of Will and his staff, and they began to fit out a ship for a voyage to the Sargasso, which they hoped to find in place this time, and then the Galápagos Islands again, by way of the intriguing Humboldt Current, whose strong temperature gradients as it sweeps north from the Antarctic make it home to a great diversity of animals.

The Arcturus Adventure

Over six miles below the warm, sunlit surface . . . weird worms,
fish, crabs—uncouth and unearthly, live out their lives in the midst
of eternal silence, blackness and quiet—feeding on the refrigerated
remains of animals, which fall from unimagined regions overhead.
William Beebe, *The Arcturus Adventure*

B EEBE'S INTEREST IN THE SEA,
born in the rocky pools of the Bay of Fundy's intemperate tides, had
never left him. Every expedition had begun and ended with long sea
voyages, and the observing, trawling, and netting that passed the time
on these transits were building to a strong urge to understand what lay
beneath the ocean's vast surface.

People had been trying in more or less clumsy fashion to penetrate
the depths since time immemorial. Aristotle and Pliny record attempts
to decipher the ocean's mysteries; Alexander the Great is said to have
been a formidable diver. In more recent times, technology had allowed
expeditions to map the ocean floor for the great commercial venture
that was to result in the transatlantic cable. Sophisticated dredges
aboard scientific research vessels had plumbed the ocean floor in search
of new life forms, hoping to find, in what was then imagined to be the
uniform environment of the deep sea, remnants of the "living fossils"
Darwin had postulated—fossils that would supply links between
extinct and extant species. The *Challenger* expedition of the 1870s, still
in the slow-moving age of sail, had proven that there was indeed life in

the frigid depths beneath the level at which light could penetrate, although the explorers found only one living fossil, a strange squid with a partially internalized shell that seemed to bridge the gap between the nautilus and the modern squid.

A Norwegian vessel, the *Michael Sars,* had done extensive trawling around the Atlantic in 1910, and its leaders, Sir John Murray and Johan Hjort, had written an analysis of their methods and results, *The Depths of the Ocean,* which Will had practically memorized. Its detailed descriptions of their equipment, with its successes and drawbacks, and of the creatures they pulled from the depths were invaluable as both information and muse. There were tables of temperatures and depths, along with photographs and drawings of deep-sea fish, coelenterates, crustaceans, and even microscopic animals, all carefully classified. The *Sars* crew had used nets at many different levels, so they had gathered data on the distribution of creatures as well. But the fine silk nets had to be drawn with painstaking slowness, and the larger meshes let much escape. Weather had limited their trawling severely, and the very vastness of their subject—the Atlantic—made even their most determined efforts little more than random sampling. Thus they had not produced a satisfying picture of ocean life.

Will hoped that with his powerful new ship and his energetic, informed young staff, he would be able to succeed where other expeditions had failed, and bring in species new to science that would answer some of the many questions his fertile mind was forever devising. Pairing emergent technology with his gift for imagining new ways to get around old problems, he dreamed of setting Darwin's theories, ignominiously on trial in Tennessee, beyond doubt. He hoped that with increased speed and better winches they would be able to ensnare species that had been able to outswim the *Challenger* and the *Sars,* and that with modern preservation techniques they might manage to catch at least glimpses of the ephemeral ones. At the same time, he had a mandate to collect specimens and scientific data for the American Museum, the zoo, and the aquarium, and to make enough spectacular finds to content his patrons.

With Whiton's knowledgeable aid, they generated a design for an oceanographic research vessel that would be able to trawl and dredge at the greatest depths then possible, and would provide the stability for microscopic and dissection work that had been lacking in the *Noma.*

Using the *Challenger* as a model, they came up with a strange-looking but practical steam yacht of 2,400 tons, suitable for extended periods of voyaging. In 1924 Whiton had an existing steamship (with a greater water and coal storage capability than that of the *Noma*) completely refitted to these specifications, and Loulu Osborn christened her the *Arcturus*.

The backbone of the *Arcturus* staff was the core of workers Will had come to rely upon in Kartabo and on the *Noma:* John Tee-Van had become indispensable as both a general aide-de-camp and a friend, and had also studied ichthyology intently; Helen would be joining the team later. Isabel Cooper and Harry Hoffman came as scientific artists. Isabel deserved special credit, as she endured repeated bouts of malaria and never ceased to grow hopelessly seasick whenever there was a swell. Ruth Rose, the actress turned writer, researcher, collector, and manager of creatures and crew, was fearless and endlessly creative when problems arose. Bill Merriam returned as the ingenious deviser of solutions to whatever logistical, electrical, or plumbing problems arose, and he was also a crack shot. Serge Chetyrkin, whose language skills were still rudimentary, made up for his inability to communicate with his zeal in learning taxidermy and collecting whatever specimens the director desired.

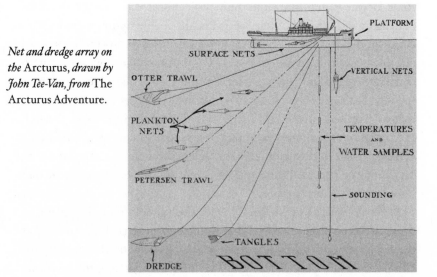

Net and dredge array on the Arcturus, *drawn by John Tee-Van, from* The Arcturus Adventure.

Joining this seasoned crew were Charles Fish and his wife, Marie, "on loan" from their positions as government ichthyologists. William Gregory, an associate of Osborn's at Columbia and the American Museum, planned to pursue his own research into the evolution of fishes. Ernest Schoedsack, a cinematographer of broad experience whose masterful documentary, *Grass,* had followed a nomadic tribe through Persia, was along to record the expedition on film. Schoedsack, who at a genial six-foot-six was known affectionately as Shorty, was intrigued by the challenge of filming in the Galápagos and under water, and needed capital for the next film he and his friend Merian Cooper had in mind.

In biting cold the *Arcturus* sailed out of Brooklyn harbor on February 10, 1925, waved off by a host of friends, well-wishers, and curious crowds drawn by the spectacle of the strange vessel. Outfitted with towering winches and arms to hoist the immense trawls, it could have been a cargo ship, but for its several stories of deck cabins for passengers. The *Arcturus* carried a dozen small boats, two with glass bottoms. In lieu of a figurehead, its bow featured Will's pulpit: the iron cage he had lovingly designed for the *Noma,* which could be lowered to the water's edge when the ship was under way, allowing its occupants to scoop surface nets or scratch the backs of accompanying dolphins. Over one side swung the "boom walk," also designed for the *Noma,* which allowed people to walk out from the deck over thirty feet of ocean. These additions helped overcome the frustrating distance from the deck to the water. The captain's original skepticism turned to enthusiasm as he dis-

The Arcturus, *under way.*

covered how easy it was to take soundings from the boom walk, well away from the ship's roiling wake.

Several of the forward cabins had been turned into a large laboratory with cages, tanks, and wells for live animals; microscopes, chemicals and vials for preserving; and a darkroom for developing film and studying the luminescent creatures they hoped to draw from the depths. With his own money Will amassed a library of reference books that lined the walls of the lab on neat shelves with retaining boards for rough weather. He stocked the ship's library with Darwin, history, and oceanography as well as the fantasy and mysteries he needed for entertainment and escape. He was particularly devoted to the works of Lord Dunsany, an eccentric Irish peer whose popular plays and tales were both fanciful and grotesque. Beebe and Dunsany had met and become friends during the war, and they continued to exchange visits and correspond for decades. Will loved to escape into the worlds of the gods and goddesses Dunsany imagined, and appreciated his dry wit and sophisticated philosophy. Everyone aboard packed sparingly for the cramped quarters of a ship cabin but found space for everything necessary for the inevitable formal parties onshore, as well as the equally inevitable charade and costume parties on board.

They headed out into a gale, and many were ill. On top of that, Ruth came down with an attack of malaria, and Will soon followed. These attacks, which sent their temperatures soaring to 103 degrees, were accompanied by intense headache and nausea. They passed quickly but left their victims weak for a few days afterward. Will had previously denied having malaria: news of it would have dealt a blow to his carefully limned portrait of the healthy tropical climate, a vision that was successfully recruiting scientists to the research station. In fact, he had truly believed that his malarial attacks, which were in his case accompanied by frightful nightmares of his experiences during the war, were simply "flying nerves," the product of his own mental weakness. Presented with Ruth's illness, though, and his own similar symptoms, he at last had to admit that he had succumbed to this scourge of the tropics.

In Bermuda they picked up the necessary ship's cat for pest control and as company for Chiriqui, Will's mischievous monkey, and headed out into a sea finally calm enough to trawl. Before the crew grew accustomed to the work they lost several of the precious silk nets, but they eventually caught first a half-inch squid and then, after dark, a myriad of wonderful creatures. There were "1000s of organisms of all kinds," Will wrote, high on success. "Tunicates, larval fish, including 20 of the Leptocephalus eel larvae, wonderful feather-tailed copepods, pteropods & heteropods, sea-worms, in fact a solid month of investigation would not exhaust this one haul." (2.23.1925)

The weight and stability of the *Arcturus* made deep dredging feasible, but the Atlantic was uncooperative. Distant storms produced deep rolling swells that made life miserable enough, but as soon as the captain stopped to make the detailed soundings necessary for dredging, the *Arcturus* turned across the swells and wallowed miserably, sending Isabel to her cabin with plenty of company. In the lab the rolling made work impossible. Specimens slid across the floor, microscopes scooted across desks, and a bottle of red ink splashed against a wall, making gory patches against the white. Anyone caught off guard walking with a delicate specimen wound up slipping and sliding in a desperate effort to keep upright.

Having managed to make a compensating tripod for his movie camera, Shorty was able to do some filming despite the restless ocean; meanwhile, Will resigned himself to cataloging specimens while lying flat on the deck. One all-day dredge came up bulging with marvelous possibilities, but a lurch caused the net to split before their eyes, dumping the entire load back into the sea. Confined to the companionway by the high waves, the collectors couldn't gather much, but Serge caught a squid that promptly squirted him with sepia and then bit him on the palm of the hand with its sharp beak, while Will dispassionately recorded its color changes.

When the sea calmed a bit they put a little rowboat over the side, but it was immediately dwarfed by vast swells that towered over it, sending it into bottomless valleys, then up almost as high as the ship deck. Will was so eager to study the sargassum weed that he dived off the boat and went down underneath as far as he could. "I dived and saw an ultramarine world with sprigs of amber drifting near the ceiling of that world— and then I tried to imagine what was down, down, down beneath me, two miles and more." (3.3.1925) The weed was scattered, and they were once again unable to find the thick mats that would signal the true Sargasso. Deciding that the sea must have seasons of its own, they determined to cross to the Pacific and look for the Sargasso on the way back, months later.

In Panama they were joined by Dorothy Putnam, wife of Will's friend and publisher George Putnam, and their twelve-year-old son, David. David was to keep a diary of the voyage to be made into an adventure book for boys. On his birthday, the crew threw an elaborate pirate party. They traversed the Panama Canal to the Pacific on March 28 and headed for the Humboldt Current, where they planned to take meticulous temperature and location measurements and study the organisms that thrived on the nutrient-rich cold waters of the Antarctic stream.

The Pacific lived up to its name: on its calm, blessedly smooth surface they were able to perform what Will considered the first real work of the trip. The Humboldt Current, however, proved as elusive as the Sargasso Sea. There was no trace of it. What they did find was an astounding current rip, a junction of two currents of very different temperatures that formed a distinct line in mid-ocean. It made a slick along its length many feet wide, unbelievably full of life. Water samples yielded a brown soup of plankton, and hordes of fish were feeding on it. Tropic birds, boobies, and frigate birds circled over, dropping precipitously to seize a fish or to steal one from another bird. Vast schools of dolphins raced through at lightning speed, feeding and leaping, and

David Putnam's pirate birthday party aboard the Arcturus.

brightly colored sea snakes weaved back and forth, making vivid yellow and green squiggles where they passed. The fauna on one side of the line was typical of the warm waters of the western Pacific, while the other side teemed with the cold-water organisms of the eastern ocean.

Substantial logs of coconut, bamboo, and even a forty-foot cecropia tree floated along the slick, caught on the boundary between the two currents, each supporting its own thriving community of creatures. Herculean efforts by the whole staff, some hanging by their knees from the boom walk, succeeded in landing one of these log cities, and even the hardened sailors gathered to watch the animals that came creeping out—triggerfish, crabs, long worms with bright red feathery tentacles, grotesque sargassum fish, odd little blennies that skipped across the deck on their fins, staring curiously up through goggle eyes. These fish were able to withstand exposure to the air, and were popular pets in the laboratory aquaria.

Will at first imagined that one of the currents was the elusive Humboldt, but the temperatures made that impossible. Both sides were too warm for the Antarctic flow. He decided to follow the current rip for a couple of days to chart it, measuring the temperatures as they steamed slowly along. The more he thought about it, the more convinced he became that this rip could be the cause of the unusual climate that South America had been experiencing, with unprecedented temperatures, rain, and disastrous flooding. In fact, this is probably the first study of the phenomenon now called El Niño, which arises from unusual movement of a warm current through the Pacific.

As they neared the Galápagos, Will's excitement increased. When he boarded the *Arcturus* he had brought with him a brand-new salvager's diving helmet and pump. The heavy, awkward helmets were used to recover goods from ships that had foundered in shallow waters. His eccentric friend Zarh Pritchard had been using a full diving suit to paint exotic underwater scenes—under water—and Will realized that this apparatus might give him the very window he wanted onto the world of the creatures he could only imagine in their native element. He could not bear the thought of the cumbersome diving rig, though, and opted instead for the bare helmet, which would leave him more exposed to pressure and current but much freer to move and feel. He remained skeptical: the tall copper cylinder weighed in at sixty pounds, and its air had to come through a hose tethered to a hand pump aboard a boat overhead.

As soon as they anchored near Darwin Bay off Tower Island, Will prepped the equipment and planned his first dive. With John and Ruth he set off in a small boat and went searching for a calm, shallow spot near shore where the boat could anchor safely. Just under the looming cliffs they found an ideal place. The pump handle was attached, a leather washer screwed in; the hose was looped carefully and its line attached to the top of the helmet that sat in readiness, its double panes of glass coated with glycerin to keep them free of condensation. Lead weights were attached to the shoulder flanges to keep the diver on the bottom. They lowered the collapsible metal ladder, and Will stepped over the side. He descended a few rungs, until his shoulders were just level with the surface, and paused while John fitted the heavy helmet onto his shoulders. As he lowered himself slowly, rung by rung, the

Beebe descending in the diving helmet, 1925.

almost unbearable weight of the helmet lifted, and the water began to rise inside. Sound deadened gradually until only the rush of bubbles surging out beneath the helmet to rise to the surface was audible.

The air that Ruth and John were pumping in turn through the hose kept the water inside the helmet safely below chin level, and provided plenty of oxygen for normal breathing. He began to feel pressure against his ears, though—first a dull ache, mounting rapidly to a sharp stabbing pain as he descended. From his flying days, Will knew the trick of swallowing to equalize the pressure against the eardrums, and the pain passed. Feeling in control once again, he stepped off the ladder onto the firm lava surface of the ocean bottom, twenty feet beneath the surface.

An almost disappointing feeling of being too comfortable—too safe, too calm, too dry—assailed him. But looking at his hands, he saw that the skin of the fingers was wrinkled, and knew that he was indeed at home but in an alien element. The angled panes of glass that met at the front of the knightlike helmet provided a strange, skewed vision of the world. His first impression was of color—endless, limitless shades of ultramarine.

Then he became aware of fish. As he focused on the coral reef in front of him he saw movement, which resolved into a myriad of sea creatures going about their daily business as if nothing miraculous were taking place at all. Bright parrot fish chased each other through weird arches and turns, snaky wrasses weaved in and out of recesses, and bril-

A coral reef, photographed for study.

liantly colored crabs scuttled along the bottom. The cylindrical construction of the headpiece made it impossible to turn his head to see
what was behind, and Beebe became uncomfortably aware of fish nibbling the backs of his legs, without being able to discern whether they
were three-inch angel fish or six-foot sharks. Slowly he lumbered across
the bottom, his air hose floating eerily behind. Then, stepping off a
coral ledge that to his skewed vision had seemed shallow, he found himself floating lightly through space, free of the limiting pull of gravity.

Although Will Beebe was not the first to go down in a diving helmet,
there were no books that described the experience, no photographs to
bring the wonders of the reef to life. Snorkeling and scuba gear were
still the subject of fantasy. And though salvagers had plumbed the relative deeps for booty and divers for pearls and other treasures, scientists
had penetrated the surface only with water glasses and stationary bells.
One other scientist had tried a helmet, but he never published his work
in any but the driest of scientific terms. As far as Beebe knew, he was
the first to study the rich life of the coral reefs up close with knowledge,
curiosity, and a well-stocked lab and staff behind him.

So, awash in the wonder of a silent world that spoke with infinitely
subtle colors, bizarre creatures, and grace beyond all imagination, Will
rose reluctantly from that first dive, in love with the ocean and determined to learn as many of its ageless secrets as he was able—for as far
and as long as he could stretch his own strength and knowledge, the
patience and persistence of his staff—and the pockets of his good-
natured patrons at the Zoological Society.

CHAPTER 25

Fire and Water

A stoic indeed must he be, who is not deeply moved by such a sight;
the ancient peak, so cold, so dead, and yet at the center so vibrant
with the everlasting fires of earth. It is the most awe-inspiring—
the most beautiful sight in the world.

William Beebe, *Two Bird Lovers in Mexico*

SHORTLY AFTER MIDNIGHT ON
April 11, the captain woke Will to look at a rosy glow on the horizon
that was too broad to be a ship on fire. Both men hypothesized that the
light arose from volcanic action in the vicinity of Albemarle Island.
Will's first thought was to drop everything and head for the strange
glow—"the exact effect of a fire alarm on a school-boy."

They cleared Darwin Bay by ten that morning and steamed toward
Albemarle, the largest island of the Galápagos chain. Will eased his very
slight guilt by stopping frequently to take soundings for the contour
map they were developing, along with occasional deep hauls of sea creatures. During the day the glow disappeared, but as darkness fell and the
sunset dimmed on the other islands in the chain, the earthly glow
replaced the sun's. Through his binoculars Will could just make out distinct bright spots of flame.

On the morning of Easter Sunday, Will and John found a spot calm
enough to land and went ashore, climbing to the closest small crater to
make observations and take temperature measurements. No lava was
flowing on this part of the island; the center of the eruption seemed to

be an area between two peaks on its extreme northern end. To distinguish them, they named the peaks Mount Whiton and Mount Williams.

At first the going was relatively easy, as they walked along an eons-old river of lava that had cooled smooth and hard. But soon it became almost impossibly rough as the lava turned to jagged petrified scoria, razor-edged froth that could look solid but crumble at a touch. Slow cooling of the ancient flow had allowed trapped gases to escape, blowing the congealing lava into thin crusts and "skeleton rocks" as it forced its way out.

> The metal soil of this great ploughing was piled in pinnacles and mounds, brittle, sharp as knife-points and varying in size from a needle to a house. At every step we crashed down through the mass as one might tread upon hill-sides of delicate glass, or we leaped unsuspectingly on a harder, steely stratum only to slip sideways or in turn bring down a lava slide upon legs or body. [1]

The heat from the lava when they stood still was unbearable, and even a moment's rest was impossible, as the scorching rocks were too hot to sit or stand on. They stumbled on, determined to reach at least one of the fumaroles—deep holes where gases and steam broke through—to get a glimpse of the boiling lava beneath. Will managed to scale the crumbling walls of a small hill and found himself perilously

Volcanic activity in the Galápagos, 1925.

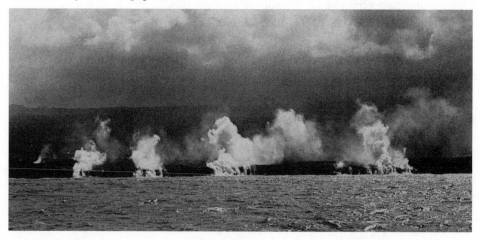

close to a crater that was venting sulfurous steam. Handkerchief over
mouth, he fought dizziness and nausea as he battled his way back,
humbled in his ambition but triumphantly clutching a few small bits of
newly hardened lava.

Leaving the excitement of Albemarle's eruption behind, the *Arcturus*
steamed back to Tower Island and peaceful Darwin Bay. The expedition
was half over, and a recoaling trip to Panama was unavoidable. In Balboa,
they discovered to their surprise that they had briefly been given up for
lost. Unbeknownst to the crew, there had been a period of radio silence
when the ship had been en route to the Galápagos from Panama in
March, and the papers and newsreels had played the story to the hilt.
"Sargasso Explorers Lost?" asked the *New York Evening Graphic,* above a
photo of Beebe, Ruth Rose, and Ernest Schoedsack aboard the supposedly ill-fated *Arcturus.* "What strange fate has overtaken the Arcturus,
the largest vessel ever fitted out for deep-sea exploration, which
departed from New York February 10 with William Beebe and a group
of scientists aboard for a six months' trip to the mysterious Sargasso
Sea?" The *New York Times* of April 10 had reported "Fear Felt for Sea
Explorers and the Steamship Arcturus.—No Word From Them in 11
Days, Although They Reported Daily by Wireless Before Heading Into
the Pacific." Anxious cables were exchanged between the Zoological

The Arcturus
off the Galápagos.

Society board and the sponsors of the expedition, some worrying that the ship's wooden structure might have allowed it to burn, and speculating on the advisability of sending the military out to check. Meanwhile, of course, the party aboard ship had continued their work, nowhere near the Sargasso Sea, in blissful ignorance of anything but a temporary disruption of communication on account of too much static. After several days of breathless headlines, the papers were forced to admit that "Message Reports Beebe Safe," and the world relaxed, some of it no doubt disappointed at the anticlimax.

As the *Arcturus* steamed back toward the Galápagos from Panama, Will worked tirelessly on his writing and on several research interests. One night he worked until midnight on the luminescence of the eerie lantern fish they had dredged up from the depths, whose photophores gave off a cool blue light even as they expired on the deck, and he also continued his studies of fish and parasite interactions. The expedition spent ten days off the hospitable shores of Cocos Island, where dives in the cumbersome helmet allowed unprecedented access to the flora and fauna of the coral reef communities. The most vigorous pumping allowed only one diver at a time; Will and John and Ruth took the pumping in turn, using the exercise to warm up after the chilling dives. John Tee-Van was growing increasingly adept at sketching under water and was also becoming an expert on reef fish. Dr. Gregory went down occasionally and was making a thorough study of the anatomy and metabolism of reef fish that would eventually win him the position of curator of fish at the American Museum. Will encouraged any of the staff who wanted to dive to go down at least once, so that ideas were continually forthcoming and excitement was kept at fever pitch.

In the spirit of his favored research approach, which brought the focus of study down to a narrow isolated area but included all actions and interactions of flora and fauna within that small space, Will demarcated an aquatic "island" in mid-ocean, sixty miles south of Cocos. For ten days they steamed in a tight circle about an area with depths ranging from 500 to 800 fathoms,[2] catching fish from the surface and making hauls every 100 feet down, until the great dredges were scraping the bottom for strange and bizarre creatures. The last haul yielded 242 fish of 20 different species, including deep-sea miniature sharks, blind fish, squat quadruped-like batfish, and fish with green and purple "lanterns" hung over them. They also found the egg case of a huge stingray that

included a young ray about to hatch. In all, Station 74, Will's oceanic
island of water, yielded a collection of 3,776 fish of 136 species, many
of them new to science. Beyond the discovery of new forms, the cumu-
lative net hauls of the next decade would elucidate the complex development of many mysterious fish, and even allow Will to "undiscover"
several species whose larval and adult forms differed so greatly that
they had been classified as belonging to different groups.

A day of diving along the shore of Narborough Island was exciting, as
divers were accompanied by fearless sea lions, penguins, turtles, cormorants, rays, groupers, sharks, and the usual coral reef denizens. Will
was awed by the aquatic grace of creatures he had previously seen only
lumbering about on land. An 800-pound turtle floated as lightly as
thistledown over his head; flightless cormorants that on land climbed
awkwardly by hooking their necks over rocks flew through the water;
and great hulks of sea lions pirouetted like ballerinas.

> It was now my privilege to see these same creatures in their chosen ele-
> ment, graceful, glorified reincarnations of their terrestrial activities.
> . . . Whereas here at the sea bottom I sprawl awkwardly, clutching at
> waving weeds to keep from being washed away by the gentle swell,
> peering out of a metal case infinitely more ugly than the turtle's head.[3]

Sorting a net haul.

Another month of sounding, dredging, and seining around the Galápagos filled in much of the data Will felt had been missing from previous studies. In June, he reluctantly left the Galápagos for the Sargasso,
hoping but not expecting to find it weed-choked and fertile with animal
life. As they steamed past Albemarle nine weeks after their last visit,
the dense lava rivers were reaching the sea, crawling down the cliffs
"like the tentacles of some huge scarlet octopus." When the molten
lava reached the water it hardened rapidly into gas-filled bombs that
exploded and shot a rain of projectiles trailing fire, gas, and water. Great
boulders of lava toppled in, making the ocean bubble and seethe.

With the glow of Albemarle still visible on the horizon, the steering
mechanism broke without warning, and the condenser soon after.
Feeling lucky to have escaped the stiff onshore wind that would have
tossed a powerless ship back into the boiling waters, they limped
back into Panama, where Will's father, Charles, had arrived with Helen
Tee-Van.

The stresses of leadership had been gnawing at Beebe; sleepless
nights had engendered more depression and fatigue. Disgusted with
himself, he set to work to overcome the blue devils of melancholy he
inwardly considered the wages of vanity and self-absorption. Although
Will was not religious, his moral philosophy preached that character
arose from self-control. To allow depression to seize the upper hand was
to be weak. Worse yet, it decreased his productivity. He played endless
games of tennis, and forced himself to relax and enjoy his father's gentle
companionship and the hot baths at the hotel. After a few days, Charles
Beebe went back to the States while the *Arcturus* steamed on through
the Panama Canal into the Atlantic, en route to the Sargasso Sea.

The Sargasso, when they reached it, was little more concentrated
than it had been in February. Clumps of weed dotted the surface, but
hardly in the dense mats that had legendarily entangled ships and
dragged hapless sailors down to Atlantis. Will had undertaken the voyage with the understanding that Whiton wanted the Sargasso research
to be his finest hour, but the weed was refusing to cooperate. They
studied what they could, made the situation sound as encouraging as
possible in their messages, and started home.

Pounding his brain for a way to please his backers, Will thought of
the Hudson Gorge, where the outflow of the great river had cut deep
into the ocean floor not far out from New York City. What could be

more compelling to the folks back home than an introduction to the deep-sea world, not as it is in unimaginable tropical locales, but as it looked at their own front door? Exploration of this unknown frontier might even draw enough media attention to soothe Whiton's disappointment at the delinquent Sargasso.

Will and the captain carefully plotted a dredging station—the expedition's 113th—that would place them just at the edge of the continental shelf, where the depth dropped off rapidly. The Hudson had once been a mighty cataract high above sea level that poured over cliffs half a mile high into the sea. As the land subsided, the cliffs sank to the palisades that frame the river now, and the Hudson became more of an inland tidal fjord than a powerful rush of waters. The gorge that the rushing waters had dug through the resistant rock was still there, though submerged in fathoms of sea water, and was now presumably home to at least a sampling of those fascinating creatures that the *Arcturus* had been collecting in faraway waters. Will trusted his powers of language to imbue the gorge with interest and intrigue.

When they reached the area of the gorge, the surface yielded wonders of its own. Schools of tunnies passed, looking like violet torpedoes, and dolphins rocketed by, leaping and falling back with loud reports. Small blackfish whales swam by in what Will described as a "dignified, elephantine manner," sending up sighing spouts of mist. Great Cyanea jellies throbbed slowly along, concealing beneath their pulsating amber umbrellas schools of small butterfish protected by the "medusa tangle" of poisonous, stinging tentacles.[4]

Drifting clumps of sargassum weed are common in the area, and yielded the usual suspects—pipefish, sea horses, filefish, and sargassum fish, the comical fish that mimicked the weed in every way. A small male triggerfish was caught navigating alone through the vast waters. He was the despair of the artists, as he could change color with every emotion that surged through his body, from ultramarine through purple and black to an astounding yellow. "This little Joseph of the sea was one of my greatest delights," wrote Will, "and in his scant two inches I saw and respected what . . . typified fearlessness, dignity, poise, adaptation, besides incredibly kaleidoscopic beauty."[5]

At night the researchers shone a strong light from the gangway, drawing a crowd of footlong squid that darted about, first in curiosity and then after prey attracted by the light.

They shot back and forth across the circle of light, now scarlet, now pale rose, now white, and when we scooped them up in nets and transferred them to our big tanks neither their activity nor their shift of kaleidoscopic colors ever ceased. Once, and only once, there came to the light a great silver-armored, fang-jawed snake mackerel, headed straight for the squids. Instantly, the keen eyes of those molluscs perceived him, their bodies became colorless and they melted into the blackness of the nocturnal sea.[6]

From the pulpit they harpooned specimens of the various sharks that lazed by, on the lookout for garbage. And garbage there was. The first day netted the following haul: one rubber nipple from a baby bottle, four cardboard milk-bottle tops, one milk of magnesia bottle, one leg from a rubber doll, one piece of bathtub, one large wooden spigot, some tin cans, and an empty Gordon's gin bottle (which Will confessed might have come from the *Arcturus*). The pristine Galápagos the Hudson Gorge was not.

The first dredge actually caught on the steep slope of the undersea shelf and sustained some damage. Succeeding dredges and trawls netted long chains of salpae—common, transparent jellyfish-like tunicates—but among them were a respectable assortment of satisfyingly grotesque deep-sea fauna. The most abundant were the little cyclothones,

The lab aboard the Arcturus.

"delicate as tissue paper, with series of lights along the body" and enormous mouths with which they engulfed still tinier prey. When brought up they looked like bits of string stuck to the net, but when they were eased out into water their intricate structure became visible, including their minute light-generating organs.[7]

Beneath 400 fathoms they took from the black depths lantern fish with gloriously brilliant gill covers and scarlet or green eyes, and spiny scarlet and wine scorpion fish. Farther down they found stomias, with huge mouths and sinister teeth, and spectral snipe eels: "pale, slender eel wraiths, with inconceivably evanescent fins, large staring eyes, and the most absurd and useless jaws imaginable."[8]

There were so many creatures of so many types, many of which had been thought native only to tropical waters, that the press and the public were thrilled, and the donors were appeased. Popular articles about the amazing creatures to be found in the Hudson Gorge captured New Yorkers' imaginations, gave rise to exhibits at the zoo and the aquarium, and inspired innumerable school trips and science projects. The huge popularity of Will's book about the expedition, *The Arcturus Adventure,* was reassuring to all of his backers, and its influence caused a swell of concern about preserving the Galápagos environment. Scientific papers originating from the expedition continued to appear in a fairly steady stream until the mid-forties.

CHAPTER 26

Elswyth

*Too thin, and somewhat weather-beaten. . . . A quiet, low-voiced man
with little crinkles round his eyes when he laughed. And then you caught
the lean line of his jaw, or a glance like a scalpel on the soul.*

Elswyth Thane, *Riders of the Wind*

At the end of July 1925,
Will returned to his studio on Central Park West to great acclaim, a
bevy of well-wishers, and Elswyth Thane.

Elswyth, whose first novel, *Riders of the Wind,* was about to be published, had been introduced to Will before he left on the *Arcturus.* Born
Helen Ricker in Burlington, Iowa, in 1899, she had grown up in Des
Moines, where her father was a high school principal. Trained in the
natural sciences, Maurice Ricker was one of the first educators to realize the potential of visual aids in teaching. He became an avid photographer and filmmaker, specializing in zoological subjects. When a last-
minute opening occurred for a photographer on a University of Iowa
expedition to the West Indies in 1918, he secured an "extended leave of
absence" from the Des Moines school board and left hurriedly for the
start of a new career.

Helen was eager to shake the Iowa dust from her heels and join the
ranks of hopeful young writers in New York City. She and her mother
accompanied Maurice as far as New York and settled in an apartment in
Manhattan. There may be an interesting story behind their sudden

departure from Iowa: a frantic scramble ensued to find a new principal, and Helen is not listed among the graduates of her high school.

On the strength of some articles she had written for the *Des Moines Register* and other papers, Helen managed to land a job in the fledgling world of screenwriting in New York, but she longed to write novels. She decided that her Germanic patronymic, Ricker, would not endear her to readers in those belligerent times, and that her Iowa background put her at a disadvantage with snobbish New Yorkers. With characteristic determination she invented a new identity for herself: "Elswyth Thane" was romantic and unplaceable.

Upon his return from the Antilles, Maurice Ricker adopted a new identity of his own. Rather than return to Iowa, he took a government position, working on ways to implement visual presentations in education. He and Edith moved into a Brooklyn apartment, but soon separated.

From childhood Elswyth had been enthralled by the literature of adventure and exploration. Her father, an avid student and teacher of natural history, would certainly have introduced her to Beebe's books, with their mixture of science and adventure. When she wrote *Riders of*

The young Helen Ricker.

the Wind, with a hero strongly resembling the adventurous Will Beebe of the pheasant expedition, he was the logical person to consult.

The heroine of *Riders of the Wind* is a restive young lady with "Viking blood," unhappily married to a stuffy housebound husband with a possessive mother. Its hero (for some unaccountable reason called Dodo), a dashing explorer in the Himalayas much older than she, is captivated by the "slim grace of her body, the sweet line of her averted cheek, her meek, shining hair in its great coil on the nape of her neck." But the demure appearance cloaks an ardent, coltish spirit.

> "If I were a boy," she said slowly, . . . "I'd run away and go with you. I'd ask nothing better than one day of burning sand—one night of tormenting cold—one hour of a great uncertainty between life and death! Oh! Just one little glimpse of the things that make just existing from one hour to the next a great adventure!"[1]

Boy or not, she does run away with him. After many perilous and romantic adventures, however, he sorrowfully prepares for a life alone, certain that he is too old for her. Of course he never has a chance. She

Elswyth Thane.

manages to convey her willingness; she obtains a divorce and the two are united.

There is little evidence to reconstruct the progress of Will and Elswyth's relationship. His journals are mum, and no correspondence survives. But Will made it his business to get Elswyth off to an ambitious start. She soon found herself invited to parties given by Noël Coward and Fannie Hurst, taken to the theater to meet actors and writers, and introduced to publishers who might never have glanced at a manuscript by an unknown Iowa girl. During the *Arcturus* expedition she had the loan of Will's elegant studio apartment on Central Park West, and mingled with the elite coterie of writers and artists nearby. Her next book, *Echo Answers,* came out in 1927 and was well received.

Beebe, now forty-nine, responded cautiously to the silent adulation of this quiet, serious girl with the large eyes and Pre-Raphaelite mane of uncut hair coiled on her neck. He idealized marriage, and Blair's defection had made him shy purposefully away. He had grown close to Ruth, but she had fallen in love with Shorty Schoedsack aboard the *Arcturus* and would soon leave with him for Hollywood, where they would marry and become part of the emergent movie empire. She would accompany Schoedsack and his coproducer, Merian Cooper, to Laos to shoot their masterpiece, *Chang.* Having honed her writing abilities with Beebe, Ruth would also write the screenplay for their movies *King Kong* and *Mighty Joe Young,* two other Schoedsack and Cooper ventures.

Will was, as he had always been, appealing to women. He was now well heeled, well connected, and divorced. His career had an aura of romance, and his energy and drive gave him a magnetic attraction. Charles Shaw, the noted celebrity profiler, wrote a personal sketch of Beebe which said that "though half a century old he can still run a mile in five minutes, clear the bar at five feet, go to sleep in five seconds, beat any world's record escaping from an infuriated water buffalo, and make the shortest after-dinner speech in the world."[2] In addition, Will declared that "marriage . . . is the most wonderful thing in the world." Not surprisingly, he never lacked for casual female companionship.

Despite his popularity with cocktail society, Will and Elswyth were together a great deal after his return from the Pacific. His free time was limited, however; he and his staff were overwhelmed with all the specimens they had brought back. The variety was so great that each creature had to be identified, experts consulted, the scientific literature searched for previous work. Again Will's broad interests were stymied by the lack of museum facilities or comprehensive keys to identification. Literature searches at that time were tedious; libraries did not routinely stock obscure journals of ichthyology or carcinology. Some specimens were new to science: they matched no description given anywhere, and no fish or crab or sponge or snail expert they consulted knew anything about them. The lab shelves, floors, and aquaria overflowed, and Beebe and Tee-Van in particular were never idle.

Will turned out scientific and popular articles literally by the score on such diverse topics as the volcano in the Galápagos, bioluminescence, sloths, the life cycle of the eel, the Sargasso Sea, and the swallow-winged puffbird. *The Arcturus Adventure* included several chapters authored by Ruth Rose and illustrations by all of the staff artists. Will felt strongly that all the members of the Department of Tropical Research (aka the DTR) should share in what glory and profit there was in publication.

In addition, Colonel Kuser had requested that he publish a condensed, popular version of the pheasant monograph. Although loath to take any time away from his analysis of specimens, he could hardly refuse. As he worked on *Pheasants: Their Lives and Homes,* he was reminded of the great stories from the pheasant expedition that had not found a home in the monograph, and began writing a new book. That he enjoyed telling his adventures to a wider audience is evident in the book that resulted: *Pheasant Jungles* is one of his richest, most exciting works.

This unexpected writing task and the extensive lab work the *Arcturus* expedition had generated kept the DTR in New York during 1926. Kartabo had been "sublet" to the University of Pittsburgh, where Will's colleague Charles Fish was a professor. Under the university's leadership, students from several American colleges spent the summer there, learning conservation and field study techniques. Several visiting researchers were also working there, so Will felt that the station was serving its purpose.

In 1927 he was offered the chance to lead yet another expedition. This one would be to Haiti, where the colonial governor, Sir John Russell, was eager to add to the knowledge of the colony's resources. The coral reefs just off the Haitian shore were reputed to be extensive, and there were several intriguing and largely unexplored inland lakes and caves. The Zoological Society came up with $10,000 plus staff salaries, and Will chartered a four-masted schooner, the *Lieutenant,* picking up the tab for an extra $5,000 himself. Charles Beebe, Elswyth, and her mother, Edith, joined the expedition for several weeks.

For four months the *Lieutenant* lay at anchor in Port-au-Prince Bay, where Will and John Tee-Van studied the reef fish. Living in nine tents set up on the deck, the crew spent every seaworthy minute mapping and studying Haiti's unspoiled reefs. With the diving helmet, Will had great hopes of coming to understand the intricate interrelationships of the creatures that made their homes in and among the baroque growths of coral. In addition, he had worked with Elswyth's father, Maurice Ricker, and an engineer to develop a watertight brass box and a weighted tripod for the motion picture camera. Various time exposures allowed them to get breathtaking underwater footage that Will would use in lectures and newsreels to heighten awareness of the treasures of the reef communities and, incidentally, to raise interest among potential financial donors to the DTR, whose funds he now regularly shored up from his own salary.

Bell Telephone Labs designed and donated a telephone outfit to be used with the diving helmet. Although it took several trials and much readjustment, it eventually proved its worth in freeing the divers from the cumbersome zinc tablets they had been using to take notes, allowing them to dictate instead to an assistant on board. He had also been given a specially designed air rifle with which to obtain specimens under water, and in return lent his endorsement to the Daisy Manufacturing Company. It never worked very well, but made a great ad.

Will and Elswyth were married on September 22, 1927, aboard Harrison Williams's yacht the *Warrior,* anchored off the Long Island coast near Sagamore Hill. There is characteristically little information in Will's journals or correspondence about the four months between his return from Haiti and the wedding. It was not a spur-of-the-moment decision, however: Edith Roosevelt offered her minister for the occasion and helped Will with the arrangements, and the *Warrior* had been scheduled well in advance.

The select guests included Harrison and Mona Williams; General and Mrs. John Russell, High Commissioner of Haiti; Henry Fairfield Osborn and his wife, Loulu; the Kuser family; and the Herbert Satterlees, whose daughter Mabel had worked with Will in Kartabo. Headlines read "William Beebe Weds Novelist: Explorer and Miss Elswyth Thane Married on Yacht Warrior off Oyster Bay. Notables at Ceremony. Bridegroom 50 and Bride 24." Elswyth was actually twenty-eight; she always fudged her age by at least a year. Photos show a tall, grave young woman in a demure light-colored pleated skirt and middy blouse, a long scarf tied around her neck. Red tape at City Hall had left them too little time to change into their wedding finery. After the ceremony they sailed for Bermuda, where the impeccable British atmosphere pleased the anglophilic Elswyth, and the beautiful beaches and clear water promised good research possibilities for Will.

They stayed in Bermuda for a couple of months. On previous brief stays en route to British Guiana, Will had met many prominent British residents and had lectured to crowds of expatriates at Government House, so the pleasures of society were assured for his bride. His father and Elswyth's mother joined them barely a week after their arrival, and John and Helen and the rest of the staff settled in soon after. Charles eagerly made his first dive wearing the helmet. Still wiry and spry, and game as ever, he found the experience thrilling and went down frequently.

Will and Elswyth in Pelham after the wedding, 1927.

Elswyth moved in with Will, into the large apartment at 33 West 67th Street; their first winter together as a married couple was packed with parties, dinners, lectures, theater, and intense writing. Will was working on *Beneath Tropic Seas,* a slight work on the Haiti trip during which his attention had been diverted by Elswyth's presence. It was the first work of his to receive less than enthusiastic reviews, mainly on account of its fragmented, episodic structure. Elswyth complained that Putnam had pressured him into submitting it before it was ready, thus sabotaging its chances. Will and John also published jointly a lengthy paper on Haiti's fishes, which had never been cataloged. Thanks to their dives, they had managed to document 284 species, 6 of them new to science and another 43 never before seen in the area. John had applied himself to a meticulous presentation of their findings, and the result was scholarly and complete. As Will had hoped, Tee-Van's professionalism and expertise was winning him a reputation as a solid ichthyologist, even without a degree. Elswyth's mother, now calling herself Edith Thane, was an artist, and made several of the illustrations for *Beneath Tropic Seas.*

Haiti and the extended honeymoon on Bermuda taught Elswyth that she had no taste for the unglamorous drudgery of collection and classification, and still less for the discomforts of fieldwork that were routine and even pleasurable for Will. Cold sandy labs with no indoor plumbing and people who spent their days hunched over microscopes, growing ecstatic over foul-smelling bits of jelly, were hardly the stuff of romance. What she longed for was the distant glamour and imaginary order of history, and particularly of English history. Having been to Britain as a child, she cherished memories of its castles and keeps. She was further intrigued to discover that the Reading Room of the British Museum held unpublished treasures of history—treasures that could be mined and transmuted into fiction.

Will was fascinated by Britain as well and had many friends there, so in May 1928 he took Elswyth on an expedition of her own. They lunched with his friends the A. A. Milnes, and visited Lord and Lady Dunsany at their Irish castle. Will had met Rudyard Kipling, whose work he had long admired, years earlier, and the Kiplings had the Beebes to tea. Elswyth adapted readily to the society of literary aristocrats, and soon found herself at home at the British Museum while Will talked diving with his colleagues at the Natural History Museum.

When he returned to the States at the end of May to receive an honorary doctor of science degree from Tufts University and a doctor of
letters from Colgate, Elswyth stayed in London to do research on the
life of Elizabeth I.

Will, now officially the Dr. Beebe that people had been calling him
for years—though still technically lacking a bachelor's degree—was
making plans of his own. His general staff was undergoing some change
at this time: Ruth had left with Shorty for Hollywood and the glories of
King Kong, and financial strictures at the zoo had cut many from the
support staff. Tee-Van remained in an official position, and Will paid
Bill Merriam out of his own pocket. The artists had always worked on
a volunteer basis, and William Gregory and other researchers were
happy to analyze the specimens in their particular fields and to publish
the results. Will was pleased to share his finds with specialists in the
best position to analyze them—and pleased, too, to share the glory
with them, as long as the interests of science were served.

Just as John Tee-Van was beginning to find his own career in ichthyology and needed more time to devote to his own pursuits, a stroke of
serendipity brought Gloria Hollister to Will's office at the zoo. Gloria
was an intense, athletic young woman who was also a gifted scientist.
She had majored in zoology at Connecticut College, earned a master's
degree at Columbia under William Gregory and other colleagues of
Beebe's, and then worked for three years with Dr. Alex Carrell, a pio-

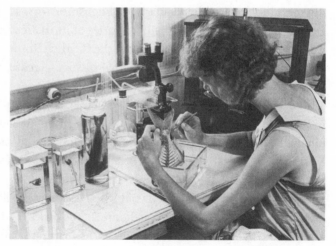

*Gloria Hollister using
her staining technique.*

neer in the use of tissue culture in cancer research at the Rockefeller Institute. Because her first love was the natural world, though, laboratory work was seeming ever more claustrophobic to her. When she heard· through her Columbia connections of the legendary William Beebe's need for a skilled naturalist for an oceanographic expedition, she applied at once for the job.

Will found Gloria to be an ideal assistant. Her background and training had prepared her to shoulder much of the technical work, and her thoroughness and high energy level matched his. From the outset he encouraged her to develop her own interests, as he had John. She had been staining cells in her previous job, and with Will devised a system of staining fish with potassium hydroxide so that their skin and internal organs were made transparent and their skeletons revealed under ultraviolet light much more clearly than previous methods, which had used bleach. This discovery made her reputation in science.

Gloria accompanied the first real expedition that Will mounted to Bermuda, in 1928. His previous stopover visits and his extended honeymoon there had introduced him to the advantages of Bermuda as a study site, but he had not done any sustained work there. Because Bermuda's coral reefs are the most northerly in the world, the water temperature was not as warm as he liked to dive in, but Elswyth did not like the heat and disorder of more southerly areas with better reefs, and Bermuda seemed a good compromise. She emphatically refused to accompany him to Kartabo; he never returned there.

Bermuda in the late twenties was a sociable place. A British colony, it had a governor and all the attendant infrastructure. So far automobiles had been kept out, and transportation from one end of the long island to the other was by horse-drawn carriage, bicycle, or boat. Elswyth, Will, and his staff and crew of volunteers took part in Bermuda's lively social life, attending dances, teas, and cocktail parties. To foster good will, Beebe gave talks illustrated with slides and films to every group that asked, making friends for the project and inspiring the donations of goods, services, and money that made every expedition run. He was introduced by the governor, Sir Louis Bols, to Prince George, who was cruising on the royal yacht, reluctantly fulfilling his duty by progressing from one British island to another, from one elegant dinner to another equally elegant tea. The prince was delighted to meet Will, whose books he had read with fascination. They talked long into the night, the prince first speaking little, then becoming fluent as he warmed to his subject and his audience. He was intrigued by the idea of diving and, after several trips out with Will and the crew, was determined to go down himself.

Will was pleased at the prince's eagerness but appalled by the responsibility of taking him diving. In fact, when George was down at one of the crew's favorite reefs, the boat began to drift a bit and the royal air hose became entangled in the ladder. Panic ensued on the boat, but George, enraptured by the world below, remained blissfully unaware of his danger. Will managed to get him up—clinging to the wrong side of the ladder—and the prince joked later that perhaps Beebe would have done well to let him drown for the sake of the publicity.

George ventured out with Will again, this time accompanied by Governor Bols. They visited Nonsuch, a twenty-five-acre island off the east coast of Bermuda that had once housed a hospital. Its two large buildings had not been used for nineteen years, but a caretaker lived there with his family and kept the facilities in perfect repair. Prince George and Governor Bols offered the island to Beebe for a research station. Along with the use of the island he proffered a seagoing tug, the *Gladisfen,* with a crew, and a small fleet of motorboats to make access easy.

The offer came at a perfect time. Bermuda was not the tropical jungle Will loved, nor were its reefs as lush as those of warmer waters. But he had been longing to do a thorough vertical study of a defined

Nonsuch Island, Bermuda.

area of ocean, and Bermuda sat on top of an extinct and sunken volcano with steeply sloping sides. Its progression of shallow to deep water would be ideal for such a study, he thought. And perhaps more important, it was a place where Elswyth could be content. Will accepted the offer, and the prince, with the zeal of a new convert, asked Will to order him a helmet of his own, so he could join in whenever he was in Bermuda.

Two of Will's oldest patrons and friends, Harrison Williams and Mortimer Schiff, put up $25,000 to cover the expenses of an intensive study of ocean life in the way Beebe had pioneered: using diving helmets, trawls, and dredges, they would study a carefully delineated eight-mile-square area with unprecedented thoroughness. From a depth of two miles to the surface, both in the open ocean and along the steep slopes of the submerged mountain whose peak formed the island of Bermuda, they would dive, dredge, and scavenge. They would classify every living creature they came across, and collect specimens wherever possible for the zoo, the American Museum, the aquarium, and subject-hungry researchers. They would map the reefs and the bottom, and work out the complex interdependence of temperature, depth, light, substrate, and plant and animal life—all from a research station on their own private island.

Beebe's eight-mile cylinder, trawled repeatedly by nets. Beebe hoped to canvass the life of a small sector of ocean as thoroughly as he had the jungle.

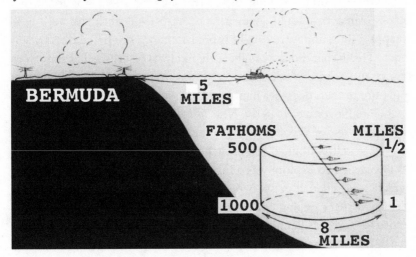

CHAPTER 27

Bermuda Diary

*I realized where the net was—a full mile beneath the ship and sunlight,
a region which for power over the human imagination and for utter
inaccessibility compares only with interstellar space.*

William Beebe, *The Arcturus Adventure*

NO MATTER HOW THOROUGHLY
they combed and sifted the eight-mile cylinder of water Will planned
to study, though, they could never really know what was down there.
Trawling nets operated under tremendous disadvantages. Trailing the
delicate webs on vastly long lines, a trawler's top speed was about two
knots. Most fish could swim in and out at will until the net's jaws
snapped shut, and were caught more by chance or choice than by science. Those that did get caught were likely to be crushed as they were
drawn up, or to explode from the difference between the astounding
pressure of their deep-sea haunts and the surface atmosphere.

In late November 1928 the *New York Times* carried a story, complete
with pictures, of William Beebe's plans for an underwater exploration
device. A self-contained vessel that could be lowered into the depths
Will yearned to explore was a logical extension of the diving helmet,
which had given him such tempting glimpses into the mysterious
depths. In fact, diving bells and submarines had been attempted for
centuries, even millennia, without great success, and Will's own ideas,

the offspring of a mind versed in almost every sort of knowledge except mechanics, were totally impractical.

The article detailing Beebe's ambitious project caught the eye of a young engineer named Otis Barton, who had long shared the same dream. Barton was studying natural history at Columbia with the hope of a career as an explorer-naturalist like his idol, William Beebe. He set himself to designing a truly practical diving structure, and at length presented his detailed plans to a skeptical Will.

Barton's first attempts to get through to Beebe were stymied. As he wrote in his 1953 book, *The World Beneath the Sea,*

> To my bombardment of letters he was serenely indifferent. Later, after I met him, I realized why; he was harassed constantly by crackpots with fantastic, half-baked ideas for machines that were supposedly designed to fit his particular needs. Every other person who was unemployed, he said, spent his time on a park bench designing a diving tank.[1]

A mutual friend obtained an audience for Otis, and Will was immediately struck with the simple elegance of Barton's design. The spheri-

Beebe with Otis Barton and the bathysphere, 1932.

cal shape solved the problem of pressure that had plagued previous schemes, including Will's own, which was based on a cylinder. In a simple helmet, pressure made breathing difficult by even the 60-foot depths Will had reached. By 150 feet, 2,160 square inches of body surface would be subjected to a pressure of 144,072 pounds. Even in a pressurized suit, human divers could not endure depths much greater than 300 feet—500 if immobilized in an armored suit—and Will wanted to go much, much deeper than that.

Years earlier, Will had spent happy hours at Sagamore Hill talking to T.R. about submersibles, but not until 1927 did he seriously begin to consider the possibility of making one. With John he worked diligently on the math, but no cylinder he designed could withstand the pressure, which increases by one atmosphere (14.7 pounds to the square inch) each 33 feet. By half a mile down, the pressure is more than half a ton on each square inch of surface. No flat surface could resist the crushing weight of water.

A ball is the ideal shape for distributing pressure: with water pressing equally on every bit of surface, each inch of a sphere pushes against its neighbor, which pushes back with equal force, strengthening instead of weakening the structure. With Will's blessing, Barton took his ideas—and a considerable part of the inheritance his grandfather had left him—to an engineering design firm to work out the details. One of the company's engineers, John Butler, was able to take Barton's sketchy plans and turn them into solid reality. "Never for a moment," Will wrote euphorically, "did any of us admit the possibility of failure—Barton and his associates were sustained by their thorough knowledge of the mechanical margins of safety, while my hopes of seeing a new world of life left no opportunity for worry about possible defects." [2]

While Barton wrestled with the ungainly steel ball and its endless engineering problems, the Twelfth Expedition of the Department of Tropical Research of the New York Zoological Society sailed for Bermuda. On March 15 Will, Elswyth, Gloria, John Tee-Van (Helen would arrive later), and Bill Merriam arrived at the new research station on Nonsuch Island with 196 large cases of freight and a mountain of cabin baggage. Even landing on the island was a challenge—the rocks were rugged, the surf was difficult, and the cluster of low white buildings perched high above the ocean.

The island itself was dense with dark green cedar trees, and low scrub provided shelter for birds. The view from almost everywhere was glorious, either out toward the open sea, where Will's cylinder of deep ocean lay, or toward Bermuda itself, lovely and low on the horizon. They set to work reclaiming the long-deserted buildings, stowing gear, and arranging everything on shelves; they closed in the open veranda and turned it into a fine lab with marvelous light. Chiriqui the monkey, who had been wintering with Will's friend Helen Ives on Bermuda— possibly exiled from his usual West Side winter quarters by Elswyth— arrived in a downpour, making the expedition official. Will soon added two black kid goats as pets. To him the situation was ideal, and he actively enjoyed the ordered chaos of setting up a lab. But Elswyth, unable to work amid such distraction, decamped to the mainland until things were more civilized on the island.

When Otis Barton arrived at the new station on Nonsuch Island later in the spring, the diving tank was still being cast at a foundry in New Jersey. His first introduction to the island and its crew paints a vivid picture both of Barton and of the DTR. He viewed the encampment as if it were a team of curious Martians.

Every morning when the weather was good, [Beebe's] crew went to sea in his trawler, let down nets more than a mile, and by evening were back in the harbor. Specimens caught in the net were brought to the

The dock at Nonsuch.

laboratory on Nonesuch *[sic]* Island which, together with dining quarters and a small colony of living huts, harbored Dr. Beebe and his expedition. Here John Tee Van, his artist wife Helen, and Gloria Hollister, a
golden-haired scientist of Amazonian stature, worked all day and most
of the evening first trawling and then classifying their deep-sea catch.

Life on the island revolved around Dr. Beebe who spent his free
moments on a ledge studying the heavens through a telescope. Before
dinner he gave each one of us one whiskey and soda, and later, in his
bunk, he read us *Barrack Room Ballads*.[3]

Never really a scientist and averse to poetry in all forms, Otis soon
decided to move to the more congenial St. George Hotel in the town of
St. George's on the mainland, with its clean sheets and cozy pub, where
he could confer more easily with Captain Sylvester, who would be in
charge of things maritime when the sphere went down. Will had salvaged the heavy-duty winches and reels from the *Arcturus* and planned
to mount them on a derelict lighter, the *Ready,* which could be towed
out to sea by the *Gladisfen,* the tug that was on loan from the government. "All of this I was financing single-handed," Barton wrote, "but we
were still in the plush twenties, and I was spending my inheritance
doing what I wanted to do more than anything else in the world."[4]

Unfortunately, when Sylvester and the dockyard engineers calculated the stresses, they realized that the five-ton tank was more than
the winches and reels could handle. Poor Barton had to order the steel
ball melted down and recast. He spent the summer redrawing the plans,
while doing some diving and underwater photography of his own. With
characteristic impetuosity he ordered a waterproof camera and a bulky
pressurized suit of the type Will eschewed, deciding that his next venture would be to make underwater motion pictures, in the vein of
Verne's popular *20,000 Leagues under the Sea,* which had been made into
a spectacularly successful film in 1916.

The little enclave on Nonsuch never did become civilized enough
for Elswyth. She had a little octagonal building of her own that served
as writing studio and bedroom, and the crew would often assemble
there in the evenings to play and sing and read aloud. But Elswyth was a
night owl: she chafed under the rules that required the generator to be
turned off early, to save power for the multiple refrigeration systems
and aquaria that kept specimens alive, or at least preserved. To con-

tinue reading, she had to resort to a kerosene light that was smelly and not really strong enough, and then she missed breakfast and the excitement of the early morning catch. Always an early riser, Will was up by dawn to get the most out of every hour. It is hardly surprising that Elswyth, like Barton, was never really happy on Nonsuch and often stayed with the "nobs" at Government House in Hamilton, the capital, or at the St. George.

Elswyth and Will never ceased to be fond of one another, but Elswyth preferred intellectual companionship and the seductions of wit to close physical relationships. She was jealous of her privacy and intolerant of intimacy. She was, in addition, almost chronically ill. She had suffered all her life from stomach distress and menstrual disorders, and was subject to recurrent respiratory problems. These illnesses may have accounted at least in part for what she referred to as her "never very lighthearted childhood."[5] She was not at all athletic or energetic, except in her strict self-discipline.

Will, too, was a private person, but he was never happy without sympathetic female company. Losing Blair had been a blow not only to his pride but also to his need for companionship. After propinquity coupled with similarity of interests brought him together with Gloria Hollister, the two were soon on intimate terms, although they adhered to a strict code of discretion. The press was always interested in Beebe's exploits, and there was little privacy on any expedition. In his journals, the few comments that concerned their relationship were entered in Will's singular runic code, which he employed whenever he wanted to record something personal. It was based on a simple system of letter

Coded journal entry; reads "I kissed Gloria and she loves me."

substitution, and replaced the awkward German of his adolescent effusions. He and Gloria explored the island, sheltering in its many caves during the frequent showers that swept the coast and watching the extravagant double rainbows that followed in the rain's wake. They cataloged every species of plant they found and scavenged creatures to study from the lush tide pools. Occasionally they exchanged visits at night, but only with a caution that heightened the experience. They would maintain their intimacy for a decade.

Will and Elswyth may well have had some sort of agreement, even before their marriage. In an interview in *Today's Woman* in the forties, Elswyth said she and Will had a very "modern" marriage and that she enjoyed the knowledge that he was attractive to women.[6] They complemented each other in many ways and delighted in each other's company and fame. His letters to her and hers to him from abroad are full of wry anecdotes and humorous asides that make it clear that they had much in common, whether or not they shared a bed. They always began with a "Dearest" and closed with an "I love you" that rings true decades later. How much Elswyth knew of Will's relationship with Gloria, if anything, she never wrote.

On special days such as Easter weekend, Will took the whole crew across to St. George's for the luxury of warm baths, and Elswyth joined them for the Sunday church service. The quaint White Horse pub on the quayside was an inviting place to drink rum swizzles, and the locals flocked to hear about the team's discoveries. Will's childhood friend Warrie Mountain, now a teacher in East Orange, came for a visit, and Will took him diving. As competitive as when they were writing essays for Nettie, Will went down so far that his ear began to spurt blood, but "it was wonderful down there."

The Bermuda government was pleased to have the researchers there, especially with the prospect of worldwide publicity when the diving sphere arrived, and expressed its appreciation in many useful ways. Though old, the *Gladisfen* was a sturdy little boat, up to the task of hauling up the heavily laden nets that rose from a mile or two below the surface. But rough weather, in addition to flooring the entire crew with seasickness (with the exception of Will), made it difficult for them to land their loads at the Nonsuch dock. A sunken barge, the *Sea Fern,* was hauled to the island and reincarnated as a breakwater—and eventually developed her own thriving coral reef community. The Tucker family, who had been maintaining the island for the government, continued on Nonsuch, and another large family, the Foxes, tended to any odd jobs that arose, from doing laundry to collecting sea slugs, with energy and zest.

Winter in Bermuda brings cool temperatures and water too rough to trawl, so the DTR went home to New York as that season approached. Will spent the winter of 1929–30 in his office at the zoo reading every book and talking to every authority he could find on oceanography and marine science. Barton was confident that the new sphere—dubbed the bathysphere, from the Greek *bathys,* or deep—would be perfect. After lengthy consultation, he and Will had decided on fused quartz for the three-inch-thick windows because of its inherent strength and clarity, and because it would pass all wavelengths of light. One of the questions Will hoped to answer was the rate at which light levels fall off with depth, and which wavelengths penetrate farthest; ordinary glass filters out some violet and red, and much of the ultraviolet. Reams of correspondence with experts on chemistry, light, respiration, electronics, and telecommunications—every field related to the bathysphere's mission—accumulated on the big rolltop desk that had been a present from Hornaday.

In early April 1930 Will and Elswyth set off for Bermuda again, this time with John, Gloria and her fox terrier Trumps, Gloria's mother, and three college volunteers, Howard Barnes, Jack Connery, and Phil Crouch. Three more "boys," John Cannon and Jack Guernsey and Perkins Bass, joined them in June. The boys were a great help in getting the lab set up and running, and their high spirits were infectious. While they labored, the "adults" went to tea with the Kiplings, who were visiting, or stayed up all night at parties and dances in Hamilton with Prince George, who was again eager to dive. They often returned on their power launch, the *Skink,* in time for breakfast at seven, dress shoes in hand. At fifty-two, Will seemed to have more energy than ever. His lithe, wiry frame never carried an ounce of fat—in fact, his doctors were constantly trying to make him put on weight—and his weathered features and perennially bald head seemed ageless.

Elswyth left in May for a summer of research in England. Barton arrived with the sphere, which was painted a bright white outside, its interior black to make any outside illumination more perceptible. The second casting had produced a sphere of a mere 5,400 pounds, its walls an inch and a half thick throughout. With Captain Sylvester, they mounted it carefully on the barge and fastened its heavy cable to the winch. The cable, a specially constructed seven-eighths-inch steel-core nontwisting wire made by the Roebling Company, was 3,500 feet long and would support twenty-nine tons; under water the cable alone weighed in at two tons. The crew of the *Gladisfen* spent several days making final measurements and adjustments, and finally ran a test dive. The sphere passed its test: it came up dry.

On May 27, Will, John, and Jack Connery took the *Skink* to St. George's harbor, where the sphere lay in readiness. Otis and Will clambered in the tiny hatch and curled themselves uncomfortably on the cramped concave floor, leaning against the curved, clammy sides. The inside diameter was fifty-four inches. There was no flooring, only the cold steel bowl where any condensation or leakage pooled and sloshed. The noise as Tee-Van hammered the 400-pound hatch into place was almost unbearable, and the close darkness pressed in on all sides. Only the small quartz windows provided a view of sky and air, and the anxious faces of friends.

The bathysphere being lowered from the Ready.

Finally the last bolt was hammered into place and the winch came to life, swinging the ball wildly on its tether. Will later recalled that "Otis was very nervous at first but got over it."

> The light and the telephone were working. . . . We had an oxygen tank with a valve which allowed a litre an hour to escape (2 litres as there were two of us), also two racks, one covered with sodium to absorb our CO_2, and the other to absorb moisture. When all was ready we signalled and suddenly the whole affair gently tilted and lifted. I had the same feeling as if I had started in a cylinder of the future to Mars. . . .

They sank slowly down; the water lapped at the glass of the window and then covered it.

> I noticed that very soon after all color left our faces—our skin looked almost as it does in ultra violet light. We could find only a small bit of printing in red, & this showed as blue black in the light from the window, but perfectly red in my flash-light.[7]

They went down only forty-five feet on this first dive, and the water so near the harbor was too murky to see through; after fifteen minutes Will signaled to be taken up. The sphere had functioned perfectly: the only water inside was from condensation.

CHAPTER 28

Out of the Depths

Before we have the complete solution of the whys and wherefores of herding and flocking and schooling, there must be a great deal of uncomfortable climbing and diving, hiding in unpleasant places, getting wet and hot and cramped and weary.

William Beebe, *Nonsuch: Land of Water*

WITH THAT FIRST SHALLOW and murky dive, Will's life was changed. The *Ready,* with its load of steel, was towed into the lee of Nonsuch, and that became the base of operations. The next calm day they steamed out to the eight-mile-diameter circle that they had been crisscrossing with methodical multi-net trawls at every depth, gradually accumulating as complete a picture of what was down below as they could. Now, armed with that data—and Murray and Hjort's exhaustive volume, *The Depths of the Ocean,* from the 1910 *Sars* expedition, as bible and field guide—Will would go in search of the animals in their element. On June 3 they sent the sphere down empty for its first really deep plunge. When they tried to haul it up, they discovered that the rubber hose that carried the electrical and telephone connections had twisted wildly around the cable. When it was finally unsnarled, they found that it had wrapped itself around forty-five times.

Beebe and Barton figured that the cable must have twisted imperceptibly as it had been wound from its spool onto the great winch, then took the hose with it as it unwound. Miraculously, the electric and communications wires still functioned perfectly. The lighter's crew worked around the clock unwrapping the Medusa-like tangles and then rewinding the stretched and straightened cable back on the winch.

June 6 was the next calm day, and after sending the empty sphere down 1,500 feet as a test, they reeled it in twist-free and with only a quart of water inside. The deepest descent at that time had been made by a Navy diver in a prototype armored suit, who had been lowered to 563 feet. Primitive windowless submarines had descended to 383 feet. No definite plans had been made for this dive, but Beebe and Barton hoped eventually to go down 2,000 feet.

> Finally, we were all ready and I looked around at the sea and sky, the boats and my friends, and not being able to think of any pithy saying which might echo down the ages, I said nothing, crawled painfully over the steel bolts, fell inside and curled up on the cold, hard bottom of the sphere.[1]

He had to have been scared. They were about to be sealed in a spherical steel coffin and thrown into the ocean. The bottom there was 8,000 feet—not that it mattered. If there were a leak, they'd be dead— shot through by arrows of pressurized water—before they drowned. If

Beebe looking out of the bathysphere.

the winch malfunctioned or the cable broke, there would be no rescue. Perhaps he was simply too single-minded to see the risk beyond the goal; more likely, he embraced it. He needed excitement to stay fully alive, to keep working at fever pitch. Will used exhilaration as a drug, to keep his "blue devils" at bay—diving into Nova Scotia's icy waters with a rock to go deeper each time, confronting the wilderness of the Himalayas, flying in France during the war. If he couldn't risk his own safety he read mysteries, getting thrills vicariously, as he had in that dark time in the Indian mountains. The anticipation of the grab bags from Lattin's seductive catalog long ago; the tantalizing net hauls from the depths, any of which might hold some unknown creature; the thrill of tracking a bird or a jaguar—these quests kept him whole. The bathysphere, like one of Lattin's geodes, was ripe with the prospect of treasure, of things no one had ever seen before.

The sphere might be a tomb, then, but that was an acceptable alternative to life without risk, without an attempt to experience something marvelous. Beebe and Barton settled themselves as best they could on the cold steel, Barton taking charge of the earphones and sitting where he could keep a wary eye on the various instruments that measured oxygen levels and humidity, and Will up against the only viewing window. Although five had been cast, two had cracked or broken as they were ground down, and one was being kept on reserve; the bathysphere's

The bathysphere on the deck of the Ready.

third "eye" had been plugged with steel. A strong searchlight pierced the gloom from one of the two remaining windows, partially obscuring the view.

After the heavy hatch was hoisted and clanged into place, the voyagers shook hands all round through the four-inch round opening in the door's center.

> Then this mighty bolt was screwed in place, and there began the most infernal racket I have ever heard. It was necessary, not only to screw the nuts down hard, but to pound the wrenches with hammers to take up all possible slack. I was sure the windows would be cracked, but having forcibly expressed our feelings we gradually got used to the ear-shattering reverberations. Then utter silence settled down.[2]

They saw themselves being swung first up from the *Ready*'s deck, then away from the side, where they hung swaying for a long moment. Twice in tests the crew had misjudged the swell and crashed the sphere into the half-rotten side of the *Ready*, and to Will it appeared through the tiny window to be happening again. "Gloria wants to know why the Director is swearing so," came down the telephone line. Then down slowly, "until a froth of foam and bubbles surged up over the glass and our chamber was dimmed to a pleasant green."

The landscape was familiar from many dives in the helmet—"a transitory, swaying reef with waving banners of seaweed, long tubular sponges, jet black blobs of ascidians and tissue-thin plates of rough-spined pearl shells. Then the keel passed slowly upward, becoming one with the green water overhead."[3]

The hull of the *Ready* passed slowly out of sight, and word came down the hose that they were at fifty feet. "I looked out at the brilliant bluish-green haze and could not realize that this was almost my limit in the diving helmet. Then '100 feet' was called out, and still the only change was a slight twilighting and chilling of the green." Then a slight pause, while the first of the 200-foot clamps was attached on board the *Ready,* both marking their depth and linking the communications wires in their hose to the heavy cable, to prevent the wires from breaking under their own weight.

Otis gave a sudden cry and pointed to a slow trickle of water from the door. Will was sure that the increased pressure at greater depth would stop the leak, and instead of giving up called for a more rapid descent. Two more minutes took them to 400 feet, then 500, 600, and 700. The leak remained a trickle, and the men relaxed. "Only dead men have sunk below this," Will wrote.

The fish he saw at this depth were familiar from hauls with the nets, but what occupied Will's mind was the light. Both men knew that it was decreasing, yet it seemed to be increasing in brilliance. Barton was so enraptured by its gleam that he was certain he could read by it, yet when Beebe brought out a page, it appeared completely blank.

> Only by shutting my eyes and opening them again could I realize the terrible slowness of the change from dark blue to blacker blue. On the earth at night in moonlight I can always imagine the yellow of sunshine, the scarlet of invisible blossoms, but here, when the searchlight was off, yellow and orange and red were unthinkable. The blue which filled all space admitted no thought of other colors.[4]

At 800 feet they halted for a few minutes of quiet observation. Otis regulated the oxygen canisters to a careful two liters a minute, and both men waved their palm-leaf fans. Will watched the slow dance of myriad tiny lights outside the window, deducing from the patterns of light what luminescent creatures he was seeing. When he switched the searchlight on he could see the animals themselves as they swam through its beam, appearing and disappearing as suddenly as though they existed only for that instant of time.

Although everything was functioning, Beebe decided to call a halt. He had long ago learned to listen to his inner monitors, and something warned him to stop. He signaled to the crew above to begin hoisting the sphere. "Coming up to the surface and through it was like hitting a hard ceiling—I unconsciously ducked, ready for the impact, but there followed only a slather of foam and bubbles, and the rest was sky." More "boiler-factory pounding" freed the men, who began to emerge, only to realize that an hour's immobility had taken its toll.

> I started to follow and suddenly realized how the human body could be completely subordinated to the mind. For a full hour I had sat in almost the same position with no thought either of comfort or discomfort, and now I had severally to untwist my feet and legs and bring them to

life. The sweater which was to have served as cushion, I found reposing on one of the chemical racks, while I had sat on the hard cold steel in a good-sized puddle of greasy water. I also bore the distinct imprint of a monkey wrench for several days. I followed Barton out on deck into the glaring sunshine, whose yellowness can never hereafter be as wonderful as blue can be.[5]

The strange brilliance of the darkness exerted an eerie fascination. On subsequent dives Beebe used a spectroscope to map the gradual disappearance of the visible spectrum from the depths. Red was the first wavelength to leave, followed by orange, yellow, green. At 800 feet nothing was visible but a faint tinge of violet.

On later dives they discovered that by 2,000 feet, every trace of light disappeared. Beebe was entranced by the phenomenon and amazed at the impossibility of describing it adequately in prose. "Brilliance that was not brilliance" and "an indefinable translucent blue quite unlike anything I have ever seen in the upper world" left him reaching for new words, for metaphors that simply did not exist. He was certain that earthbound human optic nerves had never developed a way of coping

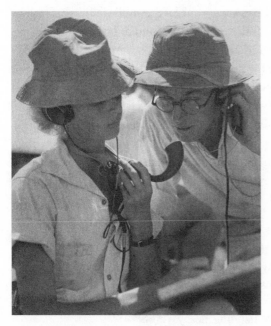

Hollister and Tee-Van talking to Beebe aboard the bathysphere.

with such light, and that consequently human language had no apt vocabulary to describe it.

On one early dive the telephone line, which carried not only Will's ecstatic descriptions but also the crucial check to reassure John and Gloria, broke. The nervous crew hauled the divers up after only 250 feet. Disappointed at losing the dive, Will nevertheless admitted that when the line went dead he had felt completely adrift, as if they would float on forever as a sort of rusty seaborne plankton.

On June 10 they sent the sphere down empty to 2,000 feet, and it came up blessedly intact. Several feet of hose had pushed in through its gasket from the pressure, a problem that would continue to plague them unpredictably, but it caused only inconvenience as the sphere's occupants fought off the rubbery tentacles. The first time the coils writhed themselves around Otis, Will reminded him of the death of Laocoön in the coils of the snake. An unappreciative Barton later wrote that "there isn't much choice between a rubberized serpent and a real one when you're being squeezed to death." [6]

After the first awe-inspiring dive, Will focused his attention on the creatures he had come to see. For succeeding dives he wrapped a "very ancient squid" or lobster in cheesecloth and hung it just below the window to serve as bait. For the next dive, Will was determined to go deeper. The first animals he saw were the large Aurelia jellyfish, familiar from his earliest voyages, and at 100 feet a cloud of brown thimble jellyfish quivered past the window. The tiny jellies were the first demonstration of the incomparable clarity of the fused quartz window: Beebe saw them coming from twenty feet away, and could judge distance and size with accuracy to forty-five feet.

At 200 feet, a strange fish came to the bait. From its size and shape it resembled a pilot fish, but the color was all wrong: instead of a rich blue, it was a "ghost of a pilot-fish"—pure white, with eight wide black bands. Whether its lack of color was the result of the light or of the pilot fish's ability to change color, Beebe did not know.

> Long strings of siphonophores drifted past, lovely as the finest lace, and schools of jellyfish throbbed on their directionless but energetic road through life. Small vibrating motes passed in clouds, wholly mysterious until I could focus exactly and knew them for pteropods, or flying snails, each of which lived within a delicate, tissue shell, and flew through life with a pair of flapping, fleshy wings.[7]

At 400 feet he saw the first true deep-sea fish—cyclothones and lantern fish. The cyclothones, or roundmouths, were the first he had ever seen alive, and though he had netted thousands, he did not at first recognize the living fish. The lantern fish he saw for the first time with light organs intact, in their full armor of iridescence. He saw big bronze eels nosing around the bait, and also their larva, the leptocephalus,

> a pale ribbon of transparent gelatine with only the two iridescent eyes to indicate its arrival. As it moved I could see the outline faintly—ten inches long at least, and as it passed close, even the parted jaws were visible. . . . Pale shrimps drifted by, their transparency almost removing them from vision. Now and then came a flash as from an opal, probably the strange, flat crustacean, well-named *Sapphirina*.[8]

At 700 feet, small shrimps and flying snails drifted past "like flakes of unheard-of storms." A large jellyfish bumped up against the window, so transparent that its stomach showed, full of luminous food. The powerful beam that shone out of the second window cast a narrow corridor of light, and the intriguing little constellations that marked so many of the deepwater fish would blink out at its border, to be resolved as suddenly

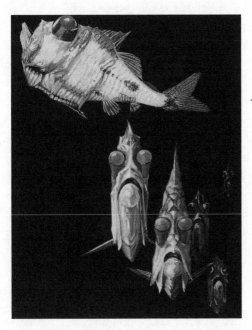

Bostelmann's silver hatchet fish (Argyropelecus).

into startled and curious fish, as if a photographic negative had been exchanged for a positive image in the blink of an eye. At 1,050 feet he saw a shower of sparks, a series of luminous, colored dots moving independently. The light revealed them to be a school of silvery hatchet fish, *Argyropelecus,* gleaming like tinsel. At the border of the light beam every light seemed to be quenched, becoming instead individual spots of shining silver, every detail of fin and eye and "utterly absurd" outline revealed. When he switched off the light their pyrotechnics reappeared, and he looked out into a world of "inky blueness where constellations formed and reformed and passed without ceasing.... With them was a mist of jerking pteropods with their delicate shields, frisking in and out among the hatchet-fish like a pack of dogs around the mounts." [9] Thirteen hundred feet brought twisting, wriggling cyclothones and three squids that shot into the light and out again, changing from black to barred white as they moved.

Anticipating a quarter-mile dive, Will draped flags of both the NYZS and the Explorers Club over the sphere. Now, at 1,426 feet, the two men were curled up tightly on the cold, damp steel, Beebe with his forehead pressed against the glass to see as far as he could, handkerchief tied outlaw-style beneath his eyes to deflect the fog of his breath. Barton watched the dials and spoke continuously over the telephone, assuring the anxious watchers above that the oxygen mix was right. They had agreed to communicate two or three times every minute, to verify that they were neither giddy nor faint. The temperature inside the sphere was 72 degrees, and the slow oscillations of the men's palm-leaf fans kept the air moving. The walls of the sphere were clammy and sweaty from condensation. The window was so exquisitely transparent that there was no distortion or haze, and Will had the feeling that it would be a small thing to just open the hatch and join the incredible parade of life outside. Only with concentration could he make himself believe in the 650 pounds of pressure that pushed against each square inch of the ball's surface—nine tons against the tiny window itself. Yet there, drifting past unconcernedly in the terrible pressure, was another delicate wisp of a jellyfish, pulsating slowly along through a weight of water that would squeeze the life out of a human in an instant.

There came to me at that instant a tremendous wave of emotion, a real appreciation of what was momentarily almost superhuman, cosmic, of the whole situation; our barge, slowly rolling high overhead in the blaz-

ing sunlight, like the merest chip in the midst of ocean, the long cob-
web of cable leading down through the spectrum to our lonely sphere,
where, sealed tight, two conscious human beings sat and peered into
the abyssal darkness as we dangled in mid-water, isolated as a lost planet
in outermost space.[10]

They decided that a quarter mile was enough for the day. The strain
of peering out into the dark was so great that Will felt his strength ebb-
ing after such a long time. They gave the signal, which Gloria relayed to
John, and he ordered the crew to reverse the winch. The return trip
took forty-three minutes, an average of a foot every two seconds. The
men had been down a total of two hours.

After this they went down every day the weather conditions and ship
crew and logistics allowed, although in this first season they never
exceeded the quarter-mile record. Will's appetite for what he was seeing was insatiable. The populousness of the depths amazed him: trawling never produced more than a handful of creatures, so he had not
expected to see more than a few, particularly near a large white sphere
with a searchlight. But the sea life was there in great numbers, gliding
past in the darkness or even coming to the window. He was also surprised at the size of some of the fish they saw, as at those depths it had
been assumed that nothing that large could survive.

Will was able to identify many of the bathypelagic creatures he had
come to know in his deep-sea hauls by their patterns of luminescent
spots, and others by morphology or behavior. But some remained
unclassified: they simply had never been seen before. Else Bostelmann,
a nature painter with accuracy and a decided flair, had joined the expedition to give form and life to the weird creatures Will described
breathlessly from the sphere. Gloria took down each word of description that came up along the telephone line, and Beebe worked with
Bostelmann until her depictions reproduced as exactly as possible what
he had seen. The pictures that resulted revealed creatures so unearthly
that many critics refused to believe in their existence.

After publication of a *National Geographic Magazine* article that
reproduced several of Bostelmann's eerie paintings, Beebe was labeled
an amateur by some hardheaded scientists, and a romantic and a fanta-
sist by many others. But among those who had studied the pitiful remnants of deep-sea fish brought up exploded from the depths, there was
recognition that what he described with such enthusiastic detail was

far from impossible. In fact, in years to come almost every creature
Bostelmann painted from Beebe's descriptions would be seen and identified. "Like an astronomer," he wrote in the *New York Times Magazine*,

> I have had to reach out blindly into an inaccessible region, and, like a
> student of fossils, I have had to work for the most part on dead organ-
> isms. But this new means of exploration changes all this. It is as if the
> astronomer could conquer space, or the paleontologist annihilate time.
> I can see, and photograph, and paint—only I cannot touch them—the
> creatures of the deep abysses. I can be certain at first hand of the tem-
> perature and light in which they live and whether they are solitary wan-
> derers or gathered into schools.[11]

On Gloria's birthday, June 16, Will made an executive decision and
allowed her to go down in the sphere. She and John descended together
to 410 feet—a record for a woman at that time—and emerged elated
with the thrill and with the marvels they had seen. The responsibility
weighed on Will, however, and although he loved sharing the wonders
of the helmet with all and sundry, he seldom allowed anyone but himself, Otis, and occasionally John to dive in the sphere. The dangers that
meant nothing to him were too great to let others take the risk.

He had agreed as a matter of course to let Otis accompany him

*Else Bostelmann painting
a gulper eel dredged from
the depths.*

whenever he wished; he never imagined that he would insist on making every dive. As only two could go at a time, John Tee-Van's presence was sorely missed. John, by now an expert on the fish of the Bermuda area, could have helped identify what they were seeing and, almost more important, add another observer's verification, whereas Otis was preoccupied with the gadgetry. In addition, Barton's frequent spasms of seasickness were a burden in the tiny rolling and spinning sphere. On one particularly trying descent, the crew on board heard Will's exasperated voice come up over the telephone line as Gloria held the receiver away from her ear: "Oh God, Otis—not now!!" Unfortunately for the poor engineer, who already felt left out of the biological excitement, the cry of "Oh God, Otis—not now!" became a gleeful mantra among the crew.

In mid-July of that eventful year, a young woman named Jocelyn Crane joined the staff as soon as she could get there after her graduation from Smith. For as long as she could remember she had dreamed of going on a scientific expedition with William Beebe, whose books had

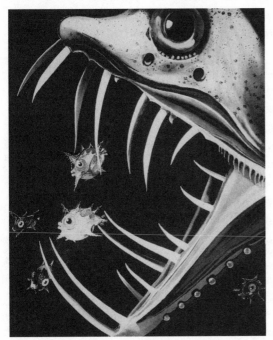

Else Bostelmann's rendering of a saber-toothed viper fish (Chauliodus sloanei) *attacking young ocean sunfish.*

been her companions as she traveled with her much-married socialite mother. From her solitary childhood she had lived a fantasy life through Beebe's tales of birds and people and fish and adventures in exotic locales. She had read and reread every single one of his books, reveling vicariously in his dangers and discoveries, and loving the natural history he canonized. Inspired by a lecture he had given to her class at Smith, she determined to learn what she would need to join Beebe's crew, earning membership in Phi Beta Kappa along the way. By the time she graduated, she had impressed her professors with her insight and energy. As a senior she had researched and written a thorough scientific treatment of the mammals of Hampshire County, Massachusetts, which was published in the prestigious *Journal of Mammalogy*.[12]

Like the college boys who had joined the expedition, Jocelyn worked for free. In addition, her mother, Estelle, contributed a healthy sum to the tropical research station—a sum that would go toward supporting the paid assistants. These helpers, like the young man at Kartabo who had brought a canoe, usually came funded in some way. Each year hundreds of people wrote offering their services, and it was never easy to choose among them. Several through the years were the children of friends; some, like Jocelyn, came recommended by professors Will knew; some were lucky enough to catch his fancy. These summer helpers had to be bright, resourceful, and self-motivated. Many later become doctors, lawyers, and politicians; most remembered their summers with Beebe as rigorous but exciting. Many of the "gang," as Will called them, returned season after season.

With so many young people, there were even more parties than usual. After Gloria's momentous bathysphere descent, birthdays came thick and fast. John's had been celebrated with the usual cake and joke gifts, and a wild ten-clue scavenger hunt that took teams all over the island, competing for conch shells that Will had scavenged furtively from the depths. His own birthday on July 29, however, took the prize for creativity. The planning alone must have required the full-time services of at least half the staff.

> After cocktails, we had a good dinner. Everyone was dressed up as a pirate, rags & knives & bloody smears. . . . They gave me a cedar box with a few gifts. Then two boys got up, came behind me & blindfolded me, and dragged me down the path, & on to the Skink. Everyone got

on, & we pushed off. It seemed to me as if I was going to St. Georges, but we went on, & at last I was pushed out along a plank & ashore at the inside of Castle Island. I was led ashore, & suddenly the St. George Hotel orchestra started up & we marched up to where the moat of the castle was.[13]

William Gregory, the ichthyologist, gave a long-winded mock oration, and the "boys" put on an elaborate pantomime of diving. The boat ladder led into the dry moat, and some of the crew pulled weird fish along on lines. At last one of them pulled a thirteen-foot banner labeled "shark" across, and the helmeted diver, who wore Beebe's bathing suit, panicked and ran.

Then Perkins pulled up a bottle with a note in it, which was tossed to me. It was an old map of the island with a X-marked treasure. I went on & followed the line, & dug in a bed of cactus, & found a big box, & we

Party on Nonsuch: Else Bostelmann, second from left; Gloria Hollister, fourth from left; Beebe (as call girl); Jocelyn Crane; and John Tee-Van. Helen Tee-Van is lying in front.

carried it down to the dungeon, which was lighted with candles and lined with fish net, & here we ate cake & beer & wine & I opened the box & found many gifts & the orchestra enjoyed it and played again & again. (7.29.1930)

Parties on the Bermuda mainland involved all sorts of maneuvering. The long scramble down the rocks to the pier, followed by the bumpy ride across the choppy bay, required waterproof gear and sturdy shoes; conversely, parties usually required formal evening dress. Jocelyn remembered stashing waterproofs and sneakers under rocks at the St. George's dock, slipping on high-heeled pumps and a sequined jacket or stole, smoothing flyaway hair, and walking to the hotel for cocktails or hailing a carriage to take them to whatever party was going on. Peggy Schwab, who grew up in Bermuda, reported that William Beebe and his staff—particularly the good-looking college men—brought more excitement to the stodgy social scene than anything since the foundering of the ship *Sea Venture* on the shore in 1609, an incident that marked the beginning of the Bermuda colony (and that scholars believe may have inspired Shakespeare's *The Tempest*).

Will's agenda for the diving ball was threefold: the experiment itself of descending to great depths in the innovative contraption; observation of deep-sea life forms seen before only as shattered remnants in trawling nets; and mapping and studying the structure of the underwater volcano of which Bermuda forms the top. The dives for this last purpose, which Will called "contour dives," though shallow, were exceedingly dangerous. The ball was essentially towed along beneath the barge, prey to the mercies of both current and underwater geography. On the contour dives Will hoped to map changes in the coral reef fauna as depth increased, providing information on underwater structures that would be valuable to marine biologists and government officials alike. For the contour dives the *Gladisfen* towed the *Ready* with its winches and ball as close to shore as they dared sail on the rare days of perfect calm, preferably with a slight offshore wind. The sphere would be lowered to within two fathoms of the reefs while the *Ready* drifted slowly seaward. Will telephoned his instructions to the crew above as they raised or lowered the ball. But the landmass of Bermuda rose unevenly, and towering ledges and cliffs could loom abruptly out of the darkness, lunging toward them through the small porthole at a sickening rate. At those moments they were keenly aware that the ball had no maneuvering capability.

The first two contour dives were nearly disastrous: the sphere drifted perversely backward, so Will could give only general predictions about the need to be raised or lowered. If the projecting gear had snagged on an unanticipated outcropping, the sphere could have been damaged or helplessly wedged. Once a crag passed two feet beneath, and the men had a "most unpleasant moment" while they were hauled rapidly up thirty feet. Barton solved the problem by affixing two boards to the bottom to act as rudders, which resulted in the whole apparatus pointing forward. Will suggested shifting the shackle that held the cable to the rear, which let them tilt forward slightly, to see the bottom better.

During one shallow dive, Will experienced a brief moment of the purest bliss a scientist can know: he sighted a thin, endlessly long sea serpent never before discovered.

We were drifting slowly along, now lifting over a toothed ridge or set- tling down into a valley of caverns and gorges when, without warning, I saw a long black line undulating over the bottom, clearly visible when

Else Bostelmann's painting of the bathysphere dive that nearly hit the reef.

over a bed of sand, or vanishing behind a mass of giant sea-plumes. A second glance revealed it as the deep-sea transatlantic cable resting quietly on its bed.[14]

The fleeting moment of joyful discovery, followed all too quickly by reality, was an experience familiar to Will, as to most researchers and explorers. "I will never send a cable without thinking of that moment," he wrote, "nor shall I ever forget the breathtaking belief of the first few seconds." It was at once an epiphany, a glimpse of things hoped for and now seen—and a reminder to always look twice before passing judgment.

That first heady season the bathysphere went down fifteen times, four times on unmanned tests. Between the unpredictable wind and weather and the myriad technical complexities, there were only a few days when the sphere and all the concomitant equipage were in working trim, the weather was calm, and the local crew and the two boats were available. In July the huge drum that wound the cable broke, and a new one had to be made and shipped down from New York. The indispensable John Tee-Van developed bronchitis and had to be sent home to recuperate, and Jack Connery slipped and fell between two boats, injuring his back and throwing the camp into a frenzy—just before Dr. W. Reid Blair, who had succeeded Hornaday as zoo director in 1926, was due for a ceremonial visit.

The deepest dive that season had been to a "mere" 1,426 feet—at least 900 feet deeper than man had ever gone. Although Will had not rewritten bathypelagic marine biology, he had demonstrated that the method was feasible and that the depths held wonders. Even in frigid temperatures and at unimaginable pressure the ocean was not empty but teeming with life, with amazing creatures that grew large and swam fast, had functioning vision as well as lights to illumine and lure, and behaved in mysterious, fascinating ways.

The main work of the expedition continued to be studying and classifying the creatures brought up in the dredges. The intrepid *Gladisfen* would blow a blast on its horn as Nonsuch came into view, signaling that they had live specimens. At the sound of the horn, the sleepy camp sprang to life. Jocelyn tossed her books to the floor and pushed away her typewriter while Gloria deliberately stored her slides and microscope where they would be safe from harm. The boys dashed in from tennis or swimming, scooping up armloads of the shallow enamel trays that were used to sort the catch as they went.

Will came running from his writing desk or microscope, John from his fish tanks, Else Bostelmann from her easel, gathering up sketch-book and paints to catch the evanescent colors of the fish before they faded into pallor. The whole camp fell into a straggling line that wended its way down the steep narrow path to the ad hoc marina by the sunken *Sea Fern*.

When the tug pulled in, the staff swarmed aboard, emptying the nets into barrels and cans and tubs of detritus and fish and shell and odd bits. When all the nets had been rinsed carefully into containers so that every bit of living tissue was free, they divided it into trays for sorting. Carrying the awkward buckets and trays, the strange procession made its way back up the cliff path, wet from the inevitable sloshing, muscles tense with the effort of keeping steady.

Then action shifted to the lab. Bright lights illumined the murky trays; hand lenses and microscopes made sense of tiny blobs of jelly or discerned the difference between old Professor Treadwell's worm larvae or Will's precious eel eggs. Each scientist sorted his own species out of the briny soup, and quiet descended as work began in earnest. Will was

Beebe in the lab on Nonsuch.

trying to capture and rear the eggs to see which of the various transparent larval leptocephali became which sort of sea eel. His extensive net
hauls also allowed him to prove that an elusive stalk-eyed fish, classified
in one genus, was actually the juvenile form of the hideous deep-sea
dragonfish *Idiacanthus fasciola,* a very different creature entirely.[15] With
her unique way of staining fish to make their skeletons visible under
ultraviolet light, Gloria was figuring out the complexities of tail structure. John Cannon, one of the college crew, was working on stomach
contents, so he was tossed the guts of all the fish that had them. Patten
Jackson, a good-natured and precocious prep school volunteer, was
studying the structures of the inner ear that allow fish to balance, so he
got the heads.

Patten's outstanding gifts had recommended him to Will despite his
youth. Pat was always the first to volunteer for any difficult or onerous
chore, and Will responded warmly to his eager questions, reminded of
himself as a boy. But Pat, with his cheerful disposition and promising
future, became the Bermuda expedition's first tragedy. When he grew
ill late in September, he attributed it to seasickness. Then old Dr.
Shelly gave him castor oil and laudanum for "gas." He never considered
appendicitis—Pat remembered his appendix being removed years
earlier and had a scar to prove it. When doctors at the hospital in
Hamilton finally operated, they found that the previous surgery had
merely drained the inflamed appendix, a shortcut that had enjoyed a
brief but lethal vogue. The appendix had ruptured and was now gan-
grenous; Pat died on October 2, 1930, bringing the glorious season to a
sorrowful close.

Will, Elswyth, and the staff had to shore up the station for an im-
pending hurricane, and the hard work helped them bear the grief they
all felt. Will especially felt the weight: Patten had been his responsibility. Even if Pat's parents did not blame him, he blamed himself. Only a
few weeks remained of scheduled hauls, and then the station had to be
packed up for the staff's return to New York. Will tried to focus on the
accomplishments of the season—the new worlds he had discovered,
the exciting contour dives and repeated trawls that were slowly painting a picture of Bermuda's oceanographic and geologic history. He
almost looked forward to a winter in New York, where he could rekindle
his excitement by re-creating it on paper for scientists and readers.

CHAPTER 29

On the Air

When you leave the world for which God made you and willfully enter other strange ones, it is reasonable to suppose that your senses and brain have to become readjusted as well as your more physical being.

William Beebe, *The Arcturus Adventure*

IN THE WINTER 1930–31 New York social season, the Beebes had to attend many dinners, dances, cocktail parties, and benefits put on by wealthy patrons. Their own parties, though, catered to the literary and theatrical folks who lived in their building or in other studio buildings nearby. Elswyth lunched with Fannie Hurst, who lived upstairs, and invited friends for cocktails to meet Noël Coward and Gertrude Lawrence. Will still engineered elaborate costume parties and held showings of his films for as many friends as he could shoehorn into the spacious flat. A young Katharine Hepburn was so taken with his movies that she brought friends to the studio to see them and wrote an appreciative letter.[1]

One of Will's major tasks that winter was to write what would be the first of several lengthy articles for *National Geographic Magazine*. With its large readership and commitment to high-quality art and photo reproduction, *National Geographic Magazine* was a perfect venue for his discoveries, and the partnership would be profitable to both; in addition, *Geographic*'s style was pitch-perfect for the intimate personal style T.R. had imprinted on Will. "Round Trip to Davy Jones's Locker" was a stirring account of the bathysphere's first dives into a totally unknown region; it was accompanied by Will's own photos and several of Else Bostelmann's remarkable paintings of the alien life forms Beebe and Barton had seen.[2]

Since Nettie's death, Will had included his father in his social life. His gentle manner, his spirited participation in Will's work and quiet pride in his famous son, and his courtly, gallant demeanor endeared Charles to all Will's friends. He seemed ageless and often joked about his family's longevity. No one was expecting it when, the day after his seventy-ninth birthday on March 13, 1931, he suffered a heart attack, and died eight days later. The funeral took place in Glens Falls, where Charles was buried in the family plot next to Nettie. Obituaries mourned the passing of a warm and sincere man, "the proud father of a very distinguished son," still taking the train to his Brooklyn office a week before his death. The *Pelham Sun* wrote:

> It was only a year or so ago he amazed us with a story of donning a diving suit and walking out from the beach off one of the Bermuda Islands; he boldly kept going until he was in sixteen feet of water below the surface.
>
> His zest for adventure and keen interest in matters relating to his son, his home town and his many close friends, made him a delightful companion and a true friend.[3]

Will's journals are terse about his loss. Early in May he noted sadly that he had left for Bermuda without his father on the dock to wave him off, and when he returned, he wrote that for the first time in his life Dad was not there to meet the boat. Even the customs agents remarked on Charles's absence and spoke of him warmly. They had been familiar with him for years: always neatly dressed and unfailingly polite, he had come many times to collect memorably noisy shipments of living animals or strange-smelling preserved specimens sent by his peripatetic son.

John, Gloria, and Jocelyn went along on the 1931 Bermuda expedition, as well as Howard Barnes, John Cannon, and Jack Guernsey, three of the volunteers from last year, joined this year by Vincent Palmer, an eager freshman from Harvard. Elswyth was in England. It was cold in Bermuda, and the crew immediately put up Celoglass on the open veranda lab so they could unpack there. Will took what comfort he could in the island's stark beauty, the young people's enthusiasm, and the prospect of a season of marine treasure hunts.

Chiriqui—actually the second in what would become a succession of Chiris, each loved more than the last and always named something, like Cheerio or Chiripie, that could be shortened to Chiri—would wake Will at daybreak, stroking his cheek softly on the pillow, and he would shave as the sun rose with the sea in sight, the surf crashing. Chiri was not always so endearing, though: if left out of his cage he was equally likely to tear Will's room to pieces, mixing shaving soap, toothpaste, fossil shells, and talcum together on the bed, or smashing chemicals, specimens, and test tubes in the lab. But Will loved him as much for his mischief as for his affection, and took him swimming every morning on a leash with a belly band.

The bathysphere was not slated to be part of the 1931 expedition. The faltering economy made the Zoological Society leery of overextending, and Barton was making noises about taking back the sphere, which he had publicly donated to the society. Although Will was eager to see more of the depths, he knew that there was more than enough material to keep determined researchers profitably occupied. They continued the deep dredges, and Will and John dissected and classified the denizens of their imaginary cylinder. Gloria studied the evolu-

Chiri wreaking havoc in the lab.

tion of the small differences in skeletal structure that her staining technique revealed, tracing the minute changes that showed evolution at work. The whole troupe partied with fervor in the evenings. On Will's birthday this year the staff tried extra hard to boost his spirits. Elaborate printed programs announced the various entertainments: fireworks, a band hidden in the bushes, a burlesque on the famous wreck of the *Sea Venture* and the discovery of Bermuda, a puppet reenactment of the recent Feipo-Dempsey fight, vaudeville, music, rum punch, and dancing.

The Depression was eating its deadly way through the American and European economies, but its ravening jaws had not yet reached Bermuda's shores. "The pound has crashed," Will wrote on September 25, "but it does not seem to make much difference here." His wealthy friend Edwin Chance grumbled that the Depression was no more than children having to switch from ice cream to oatmeal, and handed over a substantial check. Bermuda, then as now, was a lovely place to be rich and leisured. There were still no automobiles on the island; the quaint horse-drawn carriages and the little train that had been built along the length of the island took no one anywhere in a hurry. A trip to Government House in Hamilton from their St. George's base was an all-day affair.

Despite their relative seclusion from Wall Street, though, even in Bermuda wary investors had begun drawing in their cash, and several choice properties were becoming available. For the first time, Will thought about buying a house of his own. In January 1932, he and Elswyth flew to Bermuda to cement a deal: his friend May Hunter was selling her house in St. George's Parish, and Will was buying it. Elswyth liked the idea of a stylish Bermuda property, and Will was in love with the thought of having a home of his own at last, which he could fix up to suit himself.

The house, a long, low stucco villa in the style popular in Bermuda, sat on a point of land that commanded a sweeping view across the bay to Castle Harbor and on to Nonsuch Island itself, three miles out. From its stately two-level veranda the land fell rapidly off to the water, and Will planned to glass in the lower story to make a long, light-filled lab. He bought a small parcel of land that abutted the property as well, ensuring that there would be plenty of room for outbuildings and field paraphernalia. There was a dock for the small fleet, and a wonderful long cellar that would hold stocks of aquaria and other laboratory apparatus. The new Bermuda Biological Station was nearly complete just around the point, and St. George's harbor, where the bathysphere was docked, was within easy pedaling distance.

Returning to a full schedule of lectures and other commitments in New York, Will was soon fighting his old sinus woes. Infection followed infection, and by early May 1932 he was severely ill—so ill he was unable to attend the funeral of his uncle Clarence, who had survived his twin by just over a year. In June he still was not well, and while he longed to be in Bermuda, his doctors sent him to the Mojave to convalesce in its hot, arid climate. The desert seemed alien and barren after the luxuriance of the tropics, and Will grew restless and bored. Only when he learned where to look—under rocks, beneath decaying carcasses—was he able to find animals to study. He thus leaped at an invitation from Colonel Chance to join his family on a cruise through the Leeward Islands in the Caribbean. He and Gloria spent six weeks working aboard the *Antares*—hard work that was more restful for Will than the forced inactivity of California had been.

By August, Will was back in Bermuda. He and the staff lived on the

New Nonsuch, center, Beebe's home and lab from 1932 to 1941, from the air.

mainland in cottages at the Bermuda Biological Station while the new house was being refurbished. Under the aegis of Princeton professor Edward G. Conklin, the Biological Station had been endowed by a grant from the Rockefeller Foundation and had broad support from several institutions, including the New York Zoological Society and the Smithsonian. Conklin and Beebe were on very good terms, and Will received a standing invitation to house his crew at the station for a modest fee. Even Will admitted that wind and uncertain weather made Nonsuch Island less than ideal as a base of operations, and Elswyth was delighted at the change to the Biological Station's more civilized digs.

For the 1932 season, Otis Barton brought his own photographer. He was determined to get some good footage of the sphere to use in the underwater movies he hoped would make him famous. He and Will painted the bathysphere a vivid turquoise blue, both to cover the rust that streaked the white sides and to camouflage the ball in the water. They hoped to make a record-breaking descent, and this time NBC was planning to broadcast from the bathysphere, sending the explorers' words by radio from the depths of the ocean. There was to be a half-hour broadcast of the preparations for the descent, including the sealing of the two men into the sphere. Then, when they were down as far as seemed prudent, broadcasting would resume for another half hour, during which Will would report in his normal manner on whatever there was to see. It would be the first live U.S. broadcast transmitted from outside the United States, a prospect that made news of its own.

The excitement generated by this unusual broadcast was tremendous. The radio had already accustomed audiences to live entertainment, and to breaking news stories intoned by sonorous announcers. But to hear the voice of the first man ever to penetrate the depths of the ocean—speaking as he actually sank through the water in a spherical steel tomb—lent a thrilling sense of "you-are-there" participation. Everyone had seen the newsreels of the first dives, shown at movie theaters around the globe, and the audience was primed for this participatory experience.

They had also seen news footage of spectacular failure. Although Will had cautioned against it, Barton had replaced the steel plug of the sphere's third eyepiece with the last precious fused quartz window. Otis was not often interested in looking out into the blackness with its tiny constellations of moving lights but wanted to be able to film through one window while shining the searchlight through the other. With three windows, one would be left for observation. Will was uneasy about endangering the one precious quartz lens that remained and had planned to keep it as a spare. In addition, without the factory equipment and know-how, he feared the glass could not be packed in with the same flawless security of the others.

Sure enough, the first time they sent the sphere down for a test dive it came up ponderously, swinging oddly on its massive tether and spraying a fine line of mist across the window. When it landed on the deck they found it was colder than it should be, and a glance in revealed it to be nearly full of water. When they began to cautiously loosen the great nuts, a small hiss and scream grew in intensity until Will called a halt. Realizing that the contents would be under tremendous pressure, the original volume of air being now squeezed into a small pocket at the top, he ordered the crew to stand clear and began taking off the hatch himself.

Suddenly, without the slightest warning, the bolt was torn from our hands, and the mass of heavy metal shot across the deck like the shell from a gun. The trajectory was almost straight, and the brass bolt hur-

The results of an unsuccessful test.

tled into the steel winch thirty feet away across the deck and sheared a
half-inch notch gouged out by the harder metal. This was followed by a
solid cylinder of water, which slackened after a while to a cataract,
pouring out of the hole in the door, some air mingled with the water,
looking like hot steam, instead of compressed air shooting through
ice-cold water.[4]

Anyone who had been in the way, Will realized, would have been
decapitated. Teeth gritted, he ordered the reinstatement of the steel
plate.

Unfortunately, the plate was hand-tightened, just as the quartz had
been, with the same results—and this time Will had a movie camera on
the bridge to record the event. Once again the sphere came up heavy,
and once again they cleared the decks and the door shot free. The ball
was drained and dried; the plug was packed in again, sealed with white
lead, and tightened by the heaviest crew members, pushing in unison
with all their considerable might. On the next test dive, the sphere came
up with a mere gallon of water inside.

By this time, the NBC crew had been milling around waiting for
two weeks. They were hardly impatient—the posh accommodations
on Bermuda, enhanced by abundant rum swizzles, made it much more
like a paid vacation—but Will was feeling increasing pressure to perform. The expedition was receiving no money from NBC directly, but
it was no secret that the zoo expected high-quality publicity from the
event, and Will was not blind to the tangible if indirect benefits that
success would bring to the DTR. The weather, however, steadfastly
refused to cooperate. "We, in our human conceit," Will wrote, "set one
Sunday after another as the appointed time for the dive and broadcast,
and hundreds of thousands of human beings listened in, only to be told
of breaking waves and hopeless conditions." A near-total eclipse of the
sun one day, followed by a startlingly brilliant meteor the next, seemed
ominous, and then a hurricane effectively benched the whole operation
until it passed.[5]

They attempted to set off on September 18, but a threatening gale
caught up with them and they had to put back. The wind did not abate
until the night of the 21st, and the next day they set out with cautious
hope. It was choppy, but there was no swell of the sort that set the barge
to wallowing and Barton's insides to heaving.

To test the communications, they let the bathysphere down to 2,000 feet with an alarm clock as tenant. They could still hear the clock ticking. Quickly they reeled up the sphere and prepared it for a manned descent.

> The staff . . . had gone through the routine so often two years ago and during the last two heart-breaking weeks that preparation was almost instinctive; John Tee-Van was in charge of closing the bathysphere and of the three deck winches and the activities of the crew in lowering and raising it; Gloria Hollister, as before, had the upper ear-phones and was responsible for recording all of my observations; Jocelyn Crane sat by her with charts for recording time, depths, and temperature; Otis Barton and myself were, as usual, to make the dive in company.[6]

After a final survey of all the instruments and a check with the radio crew, they crawled in over the sharp steel bolts and curled up inside the sphere. Will arranged his notebook, pen, and flashlight in a pouch slung around his neck, and tucked his instruments, spectroscope, drying cloths, and assorted wrenches, saws, hammers, and other tools under him "as a hen does her setting of disturbed eggs." Seldom had he been so aware of the force of gravity, he wrote, as every loose thing sought constantly to come to rest at the bottom.

Preparing for the NBC broadcast.

Ford Bard, NBC's most illustrious broadcaster, led up to the dive with a description of the enterprise and, of course, the dangers, in his best measured, portentous tones:

> This is a broadcast of a scientific undertaking. It is <u>not</u> a planned event to make a radio program. There are too many attendant dangers and the exploration is too important, to forecast now just exactly what or how anything may happen. Dr. Beebe . . . will literally enter a new world, from which we will bring you his voice telling of what he sees there. You at your loudspeakers and we, here at the microphones are privileged on-lookers, ex-officio members of the expedition.[7]

Almost immediately, Will lost consciousness of the fact that his every word was being broadcast live to people all over the United States and Britain. His voice, sometimes calm and matter-of-fact, sometimes tense with excitement or curiosity, sometimes full of eager wonder but always with his clipped, slightly nasal New York diction, traveled up through 2,200 feet of telephone cable to receivers on the *Freedom* and was relayed by shortwave to a transmitter on Bermuda. From there it went to AT&T in New Jersey and on to the NBC studios in New York. NBC distributed it over existing networks, which rebroadcast it from the Atlantic to the Pacific, and to England by shortwave to the BBC.

The 1932 dives generated a barrage of headlines, articles, and personal appearances. To short-circuit the hype and misinformation that showed up in the work of even the most well-intentioned reporters, Will wrote the articles that appeared in the *New York Times* himself. His essays "Descent into Perpetual Night" and "First Families of the Deep" have been reprinted many times in anthologies. But the *Times* and other newspapers made up for his restraint with large photos and screaming headlines.

Between dives, Will was writing papers for scientific journals in which he detailed the new species he had seen in the deep dives. He also had the obligatory reports for the *Bulletin of the New York Zoological Society* to produce, and the articles for *Harper's* and *McCall's* that helped pay his considerable rent on West 67th. In addition, the partnership with *National Geographic Magazine* had worked out so well that another article, "The Depths of the Sea," followed in the January 1932 issue,[8] accompanied by Will's photographs of deep-sea fish and a special section of Bostelmann's stunning illustrations, and another that December, "A Wonderer Under Sea." [9] These articles were lengthy and took a great deal of time and effort. But of all the books and articles Will wrote, the *National Geographic Magazine* articles probably did the most to acquaint the world with his work.

Will and John also used the time to complete their labor of love, *Field Book of the Shore Fishes of Bermuda*. The diving helmet allowed them to make unprecedented observations of behavior, and the trawls had brought many new species to light. John's wife, Helen, and Elswyth's mother drew many of the illustrations. With this detailed but readable field guide, Will hoped to cement John Tee-Van's position in the field of ichthyology.

In 1932 Putnam brought out Will's *Nonsuch: Land of Water*, a lyrical appreciation of the little island with chapters on its history and geology, its subservience to the weather, the craggy cedars that covered it then (before disease decimated the population), and the life histories of a few of its shore dwellers, fishes, and migrant birds. As with all of his books, it is a sermon clothed in anecdote and scientific fact. If people would only look, if we would only see what is before us, if we would just change our egocentric perspectives, the world would be a finer place. If prisons and churches had observatories to put us in our place by revealing the vastness and wonder of the universe, there would be fewer prisons and more churches. If people thought about the limitations of the pacifist limpet and periwinkle, safe inside their cement bunkers, we could see that organisms that isolate themselves through geography or conceit do not evolve. Although the rocks should be thick with them, "as safe in the battle of life as a knight in full armor would be in an encounter with half a dozen serfs," the oystercatcher, with its razor-sharp beak and clever technique, can penetrate their defenses with a single well-aimed strike.

And always, he argues the case for evolution. In the snail he sees an evolutionary line that found a comfortable niche early on. Humans took a different path, and he looks back along it.

As my line stretches back my brain contracts, my muscles expand, I drop down on all fours, sprout a tail, develop long ears and snout, my teeth simplify and insects satisfy my hunger; reptilian characters accrue, my ribs increase; I slip into the water, and looking for the last time upon the land, I sink beneath the surface. Gills mark my rhythm of breath, limbs shrink to fins, and even these vanish, while my back-bone, last hold upon the higher life, dissolves to a notocord. At one end of my evolution Roosevelt called me friend—millions of years earlier any passing worm might have hailed me as brother.[10]

If we become smug about our state of sophistication and deride the ignorance or credibility of our ancestors, Will wrote, we should turn the magnifying lens on ourselves and say "Remember Tennessee!" and 1925's ignominious Scopes trial, a wound still smarting in the thirties.

Successful as the bathysphere had been, in terms of both scientific discovery and publicity for the sponsoring organizations, the bleak economic forecast made plans for a 1933 dive season unlikely. In grim recognition of this fact, Will loaned the bathysphere to the American Museum for a major exhibit on underwater exploration. The exhibit opened to great éclat in January 1933, while Will's impacted sinuses sent him once again to the desert, this time to Arizona, where he alternated baking in the desert sun and giving lectures. Even without further dives, he had more than enough material to keep himself and the staff produc-tively occupied. But he champed at the bit still, and readily seized the opportunity for another brief expedition to collect specimens for the

Marquee for Elswyth's play in London.

aquarium and to study the shore fish of the West Indies and Pearl Islands aboard Colonel Chance's *Antares*.

Will and Elswyth sailed together in July for England, where her play *The Tudor Wench* was in rehearsal. After a round of visits with the Milnes and other old friends, Will sailed back in time to join his staff in Bermuda. At his expense they stayed at the Biological Station until the new house, which Will had decided to call New Nonsuch, became habitable. On the sloping lawn they all pitched in to build gardens walled in the native gray coral "Bermuda stone," and Will found vivid ceramic tiles of parrots and flowers to set into the stone matrix. There were ponds for fish, and even a saltwater tank. Jocelyn Crane became expert at managing servants and keeping the house running smoothly. Will moved in officially in August and on August 26 staged a huge housewarming party. There were drinks on the veranda, dinner with cartoon place cards, zany gifts, and mandolin music under an auspicious new moon. Many other parties would follow, complete with charades and practical jokes and Will's trademark rum swizzles. There was always tennis, or deck tennis, and still Will swam every morning, and sunned his lean naked body when he could get to a secluded rock.[11]

The new lab, with its long wall of louvered windows overlooking the harbor, was a marvelous place to work. It was flooded with light and air, and the cellars that lined it along the back of the house accommodated rafts of equipment. The lab's sunlit expanse of glass gave on to a view of the gardens, fish ponds sloping down to the little dock, and on out to Nonsuch Island and beyond. Each researcher had a table by the window, to take advantage of the natural light.

Will was teasing out the details of flying fish development, a mystery that had never been worked out. His addiction to scooping up and studying every patch of sargassum weed that floated by had yielded several odd clumps of weed, held together with thin silklike strands, which he eventually determined were the nests of flying fish. They seemed to be communally built, as the eggs of several different females, in differing stages of development, were found in each. Unable at first to identify the eggs, Will reared them carefully until he could count the number of fin rays. His excitement was tremendous when he could state definitively that flying fish, like birds, had nests.[12]

Between the tide pools, the net hauls, and the parties, the staff of the DTR were kept comfortably busy. Elswyth was still in London, where

The Tudor Wench was enjoying a mild success; Will and Gloria were still enjoying their working relationship, but passion had ebbed. He was becoming increasingly fond of Jocelyn, who worshiped him and was eager to please.

Will and the staff closed New Nonsuch when the weather became dreary at the end of November, and sailed for New York. Elswyth arrived shortly from London, and they enjoyed a festive holiday season before Will began an arduous lecture tour to Philadelphia, Detroit, Notre Dame, Saginaw, Chicago, and Sioux City. Everywhere he was well received. Ever since his first talk to Uncle Clarence's culture club, he had displayed a gift for knowing how to convey information painlessly, and seasoned his public presentations liberally with slides, films, anecdotes, and jokes. At each stop he would be fought over amicably by local scientists eager for information, and wined and dined by potential backers. The high points of the trip for Will, though, were the times he could charm his hosts into going out to the movies. His love affair with the theater had embraced film enthusiastically, and he was never happier or more relaxed than at the cinema. In Sioux City his journal mentions nice people and attentive audiences, but gloats over having seen Zazu Pitts in *The Meanest Gal in Town* and Eddie Cantor in *Roman Scandals,* finding them both "bully."

The New Nonsuch lab: Beebe, John Tee-Van, Jocelyn Crane, Gloria Hollister, Else Bostelmann, and George Swanson.

CHAPTER 30

Half Mile Down

*If one dives and returns to the surface inarticulate with amazement . . . ,
then he deserves to go down again and again. If he is unmoved or disappointed, then there remains for him on earth only a longer or shorter
period of waiting for death.*

William Beebe, *The Arcturus Adventure*

THE DEPRESSION'S CHILLING
effects, filtering through the ranks of the zoo's wealthy benefactors,
not only had beached the bathysphere for the 1933 season but also were
threatening the zoo itself, which responded by cutting salaries and letting go employees. Will offered to take a 50 percent cut in salary to help
the zoo; since his salary went to the DTR, it was like cutting his own
throat. John Tee-Van was able to stay on at a minimum salary thanks to
Helen's family; Gloria was getting by; Jocelyn was still working for a pit-
tance paid by Will, augmented by donations from Herbert Satterlee,
who generously helped with the expenses involved in publishing Will's
technical work on fish. Still, the DTR was in serious danger of being
shut down.

Help again came from the National Geographic Society. Seeing a
great story for its magazine, the society agreed to sponsor the 1934 season in return for another article and a lion's share of the credit. Gilbert
Grosvenor, head of the society, made no demand for a record-breaking
dive, but Beebe was under no illusion about the need to make news.
The society would assign a full-time publicity manager, and Will knew
it was his job to keep him occupied.

The bathysphere had lived for the past year beneath Auguste Piccard's gondola in the Hall of Science of the Century of Progress Exposition in Chicago. Piccard had piloted his balloon up 50,000 feet to study cosmic rays; he and Beebe had lectured and been filmed together as dauntless explorers of alien regions. At the exposition, half a million people had thrust their heads through the bathysphere's narrow door, probably glad they did not have to go farther in. When the Geographic Society agreed to sponsor the expedition, they knew that some refitting of the four-year-old ball would be necessary after its many descents and the year of display; when the technicians at the Watson-Stillman Company in New Jersey, "the place of her birth," checked her out, they discovered several flaws that required major work. The quartz windows were found to have minute fractures, which could cause them to crack under a fraction of the strain they had previously borne. The copper setting of the door and its brass wing bolt had crystallized and had to be replaced.

"When high officials of the Air Reduction Company viewed our old oxygen tanks and chemical trays," Will wrote in *Half Mile Down*, "and saw our palm-leaf fans, they said such things were more or less, contem-

John Tee-Van's diagram of the workings of the 1934 bathysphere.

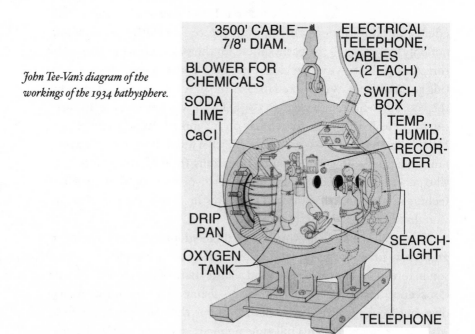

porary with the Stone Age." A new system employed a tiny electric fan that blew over four chemical trays, changing and purifying the air every couple of minutes. New tanks with a sophisticated valve system allowed the oxygen to be regulated more precisely. As before, the companies involved shouldered much of the expense: Bell took the old earphones for their museum and replaced them with new ones; General Electric paid for the new windows. The Burroughs Wellcome Laboratories donated medical supplies that would take care of any contingency "except the possible major one," Will wrote.[1]

John Long, the publicist, turned out to be a real blessing. For one thing, he kept the omnivorous reporters away. The National Geographic Society was footing the bills for the expedition and expected to have exclusive rights to the story. An even greater benefit was that Otis could no longer complain that Will was hogging the limelight, as Long had complete oversight. This had been a sore point ever since the unlikely partnership began: Barton needed publicity desperately to get his infant film production ventures off the ground, but the papers stubbornly clung to Beebe as the popular and charismatic personality the public wanted to see and hear about. Unprepossessing in appearance, Barton came across as gawky and inarticulate. His expertise was in the nuts and bolts of the equipment rather than in the more romantic ideals of the scientific quest for knowledge. Will asked repeatedly for Barton's name to be printed alongside his, and for every description of the expedition to credit him with his part. His name is inserted, in Will's handwriting, into the announcer's script for the 1932 broadcast in several places. But despite these reiterated pleas, the press cut Barton dead.

The Zoological Society took a hit on the publicity as well. National Geographic had stipulated that the operation be called an official National Geographic Society Expedition, and the zoo, unable to fund Beebe themselves, had reluctantly agreed. Out of loyalty Will insisted that the Zoological Society be credited in any articles with all development of the bathysphere and with support of previous and ongoing research. Still, when the publicity associated with the dive earned the Geographic Society a raft of new members, zoo officials wrung their hands.

Once the sphere reached Bermuda, it took a month to get it assembled and tested with the rest of the equipment. The great seven-ton winch Will had salvaged from the *Arcturus* ten years before still worked perfectly, and by August the system was ready for a trial run. To avoid the trouble and expense of domestic staff at the new house, the crew had been living at the Biological Station, walking or cycling the seven-minute commute to the lab they had set up at New Nonsuch. They had been working hard on net hauls, and Will was particularly intrigued by a series of films he and Tee-Van were making of developing eel eggs. The *Skink,* the *Gladisfen,* and the *Ready* were finally primed for a run, and Otis and Will were sealed into the sphere and dropped down an anticlimactic four feet—and promptly had to be hauled up, as water was cascading in around the door.

John, feeling justifiably that for a short test run a few corners could be cut, had fastened only four of the great bolts that held the door. The result, Will wrote, was a joking matter, but still it left behind an unforgettable memory, "a subconscious reaction all its own, which will never quite be eradicated by the instantly succeeding ones of reason and humor." Ultimately, though, he reflected that he had profited by the incident: "Nothing ensures a better seat on a horse than having been bucked or run away with." [2]

On August 7 the weather was calm enough for the dress rehearsal. Carrying only the temperature and humidity gauges and Barton's camera, set—after a characteristic last-moment panic—to expose film by remote signal, the sphere was lowered to a depth of 3,020 feet. It came up holding 400 feet of exposed film, with "no one knew what secrets of fish or the lights of fish concealed in its silver coat." [3] In fact, when developed, the film showed nothing but darkness varied with a few tiny spots of light, but it was at least dry.

A disheartening series of squalls set in to thwart the plans for an immediate dive, but there was always something to be added or set to rights. Will decided that the radical reduction in the rate of oxygen delivery—from two liters per minute to one—needed a dry run, and had himself sealed in the ball on deck for two hours. The new air purification system worked perfectly, but the calcium chloride, which soaked up the moisture from the air, did such a good job that the water it absorbed condensed and dripped weak acid from its tray down into the blower. They shifted the trays so that the blower was on top, and placed a collecting tray underneath. They also discovered that the rubber hose that conveyed the electrical cables had become stiff and brittle and could no longer be packed into its box atop the sphere. This "stuffing box" was the weakest point in the bathysphere's armor, so they replaced the last 600 feet of hose with a new piece.

By August 11 the squalls had died out to swells — daunting to a seasick Barton but not actually dangerous — and Beebe gave the go-ahead. The two men were lowered, as they had been so often in previous years, but Will was astonished all over again at the experience. The golden light of day gave over suddenly to green: faces, tanks, trays, even the blackened walls were tinged with green. After the froth of bubbles the surface became a ceiling, "crinkling, and slowly lifting and settling, while here and there, pinned to this ceiling, were tufts of sargassum weed." The ancient hull of the *Ready,* the bathysphere's home barge, hove into view as they twirled slowly, looking like a coral reef with great streamers of plant and animal life.

The green faded imperceptibly with increasing depth, and Will began to focus on the swarms of tiny copepods and other planktonic life.

> At 320 feet a lovely colony of siphonophores drifted past. At this level they appeared like spun glass. Others which I saw were illumined, but whether by their own or reflected light I cannot say. These are colonial

Preparing for a National Geographic dive, 1934.

creatures like submerged Portuguese men-o'-war, and similar to those beautiful beings are composed of a colony of individuals, which perform separate functions, such as flotation, swimming, stinging, feeding, and breeding, all joined by the common bond of a food canal. Here in their own haunts they swept slowly along like an inverted spray of lilies-of-the-valley, alive and in constant motion. In our nets we find only the half-broken swimming bells, like cracked, crystal chalices, with all the wonderful loops and tendrils and animal flowers completely lost or contracted into a mass of tangled threads.[4]

They encountered yellowtails, pilot fish, blue-banded jacks, and other so-called surface fish at surprisingly low depths. Silvery squid shot past, and lantern fish. At 800 feet they passed through a cloud of copepods, as well as the roundmouth cyclothones. Too often the transient beam of the spotlight frightened away the creatures it was supposed to illumine.

At 1,000 feet they took stock of their surroundings. Stuffing box and door were dry, and the humidity was under such good control that they did not have to use their bandit handkerchiefs to keep the window from fogging. By 1,100 feet, the number of lights from animals increased as the darkness became ever more absolute. Will noted a jelly, its diaphanous folds waving slowly in the terrific pressure, and a transparent four-inch larval eel. Clouds of flying snails sailed past, and small shrimps exploded in a blinding cloud of luminescent anti-ink when they bumped against the glass. A three-inch anglerfish swam past, its grotesque features topped by a pale, lemon-colored light on a slender tentacle, rows of sinister teeth glowing dully.

Just above 1,400 feet, two eighteen-inch black sea eels swam past, and so did a strange sea dragon–like fish. Then Will had the rare treat of seeing a wholly new, unknown fish hang just outside the window for long enough to describe it fully. He grabbed Barton to verify his description, which was of a two-foot-long, pale buff creature, "alive, quiet, watching our strange machine, apparently oblivious that the hinder half of its body was bathed in a strange luminosity."

It was a color worthy of these black depths, like the sickly sprouts of plants in a cellar. Another strange thing was its almost tailless condition, the caudal fin being reduced to a tiny knob or button, while the vertical fins, taking its place, rose high above and stretched far beneath the body. . . . I missed its pelvic fins and its teeth, if it had any, while such things as nostrils and ray counts were, of course, out of the question.[5]

Confident from long experience that his observations were sound, Will named the fish the pallid sailfin, *Bathyembryx istiophasma,* "which is a Grecian way of saying that it comes from deep in the abyss and swims with ghostly sails."

Extremely wary of naming new species, Will ordinarily did not do it without extensive behavioral and anatomical study. But these were not normal conditions, and there would be no second chances. So, with great caution and Otis's inexpert verification, he gave taxonomic names to several of the creatures he was able to see clearly.

Will published his observations and discoveries in the Zoological Society's *Bulletin,*[6] which always had first claim on his output, and in an appendix to *Half Mile Down.* For each order and family, he described in detail how he had made the identification. For the Sternoptychidae, for instance—the skeleton fish and hatchet fish—he reported twenty-eight observations between 650 and 2,800 feet. "During the early dives I did not distinguish between *Argyropelecus* and *Sternoptyx,* but when I began concentrating only upon what I was watching, and refused

Beebe talking to Gloria from the bathysphere.

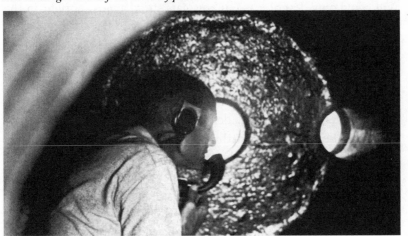

to be distracted by succeeding flares, I could easily tell one form from another, if they were close to the window and side on." Of the cyclothones, "at 2,400 feet I once recognized a pair of ten-inch *Gonostoma elongatum*. The coppery iridescence of their sides shone out clearly, and the serial photophores, with their characteristically large reflectors, were distinct. I probably saw several more of these fish, confusing them with melanostomiatids, but my eye was too slow to make out more than the merest outlines."[7]

Aside from the sheer quantity of life, the most scientifically valuable observation was that there were surprisingly large fish in the depths, and that they moved more quickly than scientists would have predicted. Since they were not often caught in the deep trawling nets, it had been thought that large-bodied animals must be unable to survive in the abyssal depths. Lack of food and light and tremendous pressure on delicate tissues seemed to make the existence of massive creatures unlikely. Will's observations proved that despite the pressure, large fish were common at great depths, a fact that opened new fields of physiological research.

At 2,300 feet on this dive, Hollister told the divers to listen through the earphone to the tug's whistles, saluting the new record depth—100 feet deeper than their deepest dive in 1932. They descended still farther, noting a skeleton fish, a pair of large copper-sided scimitar-mouths, a fish as flat as a moonfish that "entered the beam, and banking steeply, fled in haste." One flying snail, "from among the countless billions of his fellows," flapped back and forth across the glass. And something Will could only describe as rainbow gars stood almost upright in the beam, slender and stiff with long pointed jaws on brilliant scarlet heads, strong blue back of the gills, and clear yellow on the rear and tail. At once he wondered what purpose, in the darkness of the deep, the colors could serve.

In all of this time Will alternated his position at the window with Otis's camera. Despite the inadequate light, Barton was determined to prove that he could take worthwhile photos from the abyss. He had tried a much stronger light, but its terrific heat had threatened the integrity of the window, whose outside was exposed to frigid temperatures and great stress. Nothing resulted from his attempts.

Exhausted from the strain of watching, Will called for an ascent at 2,510 feet. A minute later he saw a new anglerfish, similar to the others but distinctive in its mouth and teeth and its tall, light-topped tentacles. "No pioneer, peering at a Martian landscape, could ever have a greater thrill than did I at such an opportunity."[8] Then, at 1,900 feet, he saw the loveliest specimen yet: a large, almost round brownish fish which, when out of the beam, showed five "unbelievably beautiful lines of light, one equatorial, with two curved ones above and two below. Each line was composed of a series of large, pale yellow lights, and every one of these was surrounded by a semicircle of very small, but intensely purple photophores."[9] Will named this animal *Bathysidus pentagrammus,* the five-lined constellation fish.

They did a contour dive in the afternoon and spent the next day trawling the same waters they had dropped through. As always, they were thrilled with the creatures they hauled from the icy depths, but amazed at the meagerness of the haul compared with what they knew firsthand of the abundance of life below. Either fish were attracted in unusual numbers to the light of the bathysphere, Will wrote in an article for the prestigious journal *Science,* or the number of deep-sea fish caught in their nets was a great underrepresentation of the density of undersea life.

Will, Jocelyn, and John attended the nets as they were brought up and took notes on what they saw. They transferred the catch into tubs of icy salt water and were rewarded by seeing some of the creatures revive. On this haul they brought up a pair of ten-inch scimitar-mouths like the ones Will had seen on the last dive, and for the first time one of the strange black swallowers survived to swim around his jar at full speed. Another treasure was *Dolichopteryx,* or spookfish, living, gay-colored, and almost transparent. As far as Will knew, it was the first ever to be taken alive.

On August 15 the weather held, and the crew set off from St. George's harbor. The *Gladisfen* pulled the *Ready* with its load of steel, cable, and crew back to the same spot where they had descended before, and dropped the sphere over, with Beebe and Barton. On this dive they saw much more life than before: clouds of copepods floating by, including small opalescent *Sapphirina,* blue in the light beam; a huge shadowlike fish at least twenty feet long; and several of the *Lamprotoxus* sea dragons Beebe loved, chin barbels hanging downward, their bow lights shining green. Only sixteen of these had ever been netted, seven of those by Beebe. The record size of these grotesque fish had been eight inches;

but here were four individuals more than twice that, quite possibly a new species.

This time they continued to descend to a depth of 3,028 feet; only a dozen coils of wire were left on the bare winch reel. Concerned that the remainder might fly off spontaneously, Captain Sylvester asked Will to okay the ascent. They had made the half-mile mark he had hoped for without difficulty, and the last few feet had shown more large fish than ever before, so he was content. He cabled Elswyth, who was in London preparing her play *Young Mr. Disraeli* for the theater: DEEP DIVES OVER. 3028 FEET TODAY. NEW WORLD, GIANT FISH. LOVE, WILL.[10]

A subdued controversy continues over some of the fish Beebe described. Taxonomic standards are rigorous and require dissection and verification as well as the procuring of representative or "type" specimens. Will believed devoutly in those standards, but he realized that the unearthly creatures he saw from the bathysphere were unlikely to be caught and classified directly, at least not soon. After much agonizing, he submitted detailed taxonomic descriptions and names for several of the fish he had seen, including the five-lined constellation fish. Will was convinced that he had been accurate and critical in his assessments, and he felt that the assigning of names would help in future deep ocean research.

In retrospect, it is easy to see that this was an unwise move, particularly for someone who was so defensive about his scientific reputation. Although researchers who knew his expertise and his cautious nature accepted his discoveries implicitly, outraged taxonomists scrambled to decry the breach of established scientific etiquette, which they saw as the thin end of a disastrous wedge. Will was both hurt and depressed: he had thought his experience, combined with the unique situation, would speak for itself.

Because some of the creatures have never been seen since, some experts think they never existed. Others, such as Dr. Robert Ballard, who specializes in deep-sea exploration, find it easy to believe that Beebe's exotics remain elusive. With only a tiny fraction—less than 1 percent—of the oceans explored, Ballard thinks the odds against rediscovery are great. The primitive coelacanth was thought to have been extinct for 70 million years until one was brought up by fishermen in 1938, and it was only in 1968 that a new species of huge shark came to light. The sixty-foot giant squid has never been seen alive in its element, and in 2000, the discovery of a new twenty-three-foot deep-sea squid electrified the scientific community. The paper announcing its discovery notes that the existence of such a substantial animal, now known to be common in the world's largest ecosystem yet not previously captured or observed, is an indication of how little is known about life in the deep ocean.[11] Every time a new expedition is launched, new creatures are discovered. Ballard himself has observed dozens of unknown species. Some authorities calculate that the ocean fauna make up 97 percent of the planet's biomass and contain more diversity than even the rain forests.[12] Beebe's scrupulous honesty and his tremendous sense of responsibility to science suggest that unless the oxygen and CO_2 levels made him hallucinate, the fish he saw still live their cloistered lives in the deep Atlantic.

The article Will wrote for *National Geographic Magazine*, "A Half Mile Down," accompanied by sixteen of the eerie paintings Else Bostelmann had made to his specifications, made a gigantic splash. The book of the same name, which came out months later, was on the *New York Times* best-seller list for months, and opportunities to lecture and write articles multiplied. Will accepted many more of these than he wanted, but the grim reality was upon him: the Depression was no bowl of oatmeal, and his sort of big-ticket science was in deadly peril.

Finding his fortune in tatters, Otis Barton knew that he had to make his underwater film addiction pay or "go belatedly into some humdrum business."[13] With characteristic extravagance, he embarked on filming expeditions to Panama and the Bahamas, hiring Hollywood darling Ilya Tolstoy to direct, and local bathing beauties to act the parts of maidens threatened by large, menacing sharks and rays. His stories of these attempts make pathetic reading. The underwater film he pursued to completion, *Titans of the Deep*, included photos of the bathysphere surrounded by starlets dressed as provocatively as diving helmets and wetsuits allowed. Beebe's name figured prominently in the advertising. The clumsy production so embarrassed Will, already beleaguered by accusations of publicity seeking, that he wrote a letter to *Science* renouncing any connection to its production. After this, Barton tried nature photography in Africa and other exotic locales, but he could never get the animals and the equipment to perform at the same time.

Barton never stopped trying to design devices that would go deeper than the bathysphere, and at length succeeded in establishing a depth record in 1949 in his bathysphere-like "Benthoscope." Soon after, however, Auguste Piccard, who had reached great heights in his balloon, entered the competition. Impressed with the bathysphere when he had seen it at the Chicago Exposition, Piccard began toying with the idea of an untethered, maneuverable submersible. In the 1950s Piccard and his son, Jacques, tested a prototype "bathyscaphe"; in 1960, Jacques would descend nearly seven miles in his elegantly designed bathyscaphe *Trieste.*

Even without the bathysphere, the 1935 season in Bermuda was productive, with the tug *Gladisfen,* now well into her dotage, hauling the huge nets up from depths much greater than the bathysphere had plumbed. Will loved the thrill of the hunt, and he and Gloria, Jocelyn, and John described and dissected and stained and studied the strange creatures the nets brought up. The bathysphere had been an important step, but the ecology of the ocean was a vast field of research. They delved into the lives of the organisms they were particularly interested in and sent others, live or preserved, to other specialists. But the bathy-

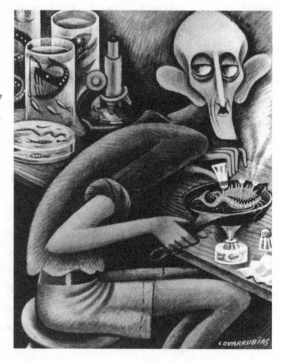

"Professor Beebe, gourmet and ichthyologist, secretly fries his new discovery, instead of pickling it for posterity." Cartoon by Miguel Covarrubias, for his "Lives of the Great" series for Vanity Fair *magazine.*

sphere descents were too egregiously expensive for anyone on a tight budget, and besides, Will had seen what he wished to see. In all, there had been sixteen deep dives over three seasons, and many more contour dives, and he realized how imperfect and inadequate his knowledge of the astonishing fauna of the deep sea would always be. The underwater world was still a mystery, but he had taken the first step toward demythologizing it, and was not unwilling to pass the torch to Barton and the Piccards.

CHAPTER 31

Fishing

*In the depths of the sea, where the sun is powerless to send a single
ray of light and warmth, there live many strange beings, fish and
worms, which, by means of phosphorescent spots and patches, may
light their own way. Of these strange sea folk we know nothing
except from the fragments brought to the surface by the dredge.*

William Beebe, *The Log of the Sun*

WILL WAS NEVER BORED WITH
his writing and the treasure hunts of the dredge, but despite the pretense of "research as usual," the Depression had curtailed his operations drastically; there was no possibility of a Zoological Society–
sponsored expedition, and he needed a forward impetus to work at
peak. With characteristic good luck, he again profited from the seren-
dipitous combination of great wealth and philanthropic spirit. When
in January 1936 Dr. Henry Lloyd, a physician and old friend, proposed a
two-month tuna-fishing expedition with his wife in the West Indies,
Will agreed eagerly.

Lloyd's yacht, the *Hardi Biaou*, was too small to trawl or dredge, but
it was equipped with a laboratory, aquaria, and a glass-bottomed boat.
Will specified the itinerary for the trip and laid out his scientific goals,
which were carefully tailored to provide sport for his hosts as well as
data of interest to science: he would make a thorough study of the
black- and yellow-finned tuna between Martinique and St. Lucia; check
on the struggling colony of birds of paradise that had been established
on Little Tobago; and study the boundary between the West Indian and
South American fish populations.

John and Jocelyn came along. Gloria had won a grant from National Geographic for a research trip of her own to British Guiana on which she hoped, among other goals, to capture more of the intriguing, smelly hoatzins for the zoo. Her position in the DTR was increasingly being filled by Jocelyn as Gloria's own interests and the zoo's kept her away. As a zoo employee in those strapped times, Gloria was called upon to cover a variety of jobs, and not always those she would have chosen. Her Guiana expedition would enhance her scientific credibility as well as take her back to some of the wild familiar places she had loved.

John, Jocelyn, and Will worked diligently to analyze the stomach contents of the tuna caught by the Lloyds, as well as any they could buy from local fishermen. They discovered that the tuna were feeding in the depths and inadvertently bringing creatures familiar from deep ocean dives to the light of day. Sometimes the fish in the tuna's stomachs had been so recently ingested that they revived in aquaria and became objects of investigation in their own right. As they grew, some turned out to be previously unknown larval forms of fish, a discovery that allowed Beebe and Tee-Van to fill in the ontogeny of several species. They scavenged the local fish markets, astounding shopkeepers with requests for entrails and guts, and scoured the shores for samples of sand and odd bits of detritus to complete the picture of the marine environment, from diatoms to killer whales.

Almost immediately after their return from this brief survey, they headed west to join Templeton Crocker, a fabulously wealthy California businessman, on his yacht the *Zaca*. Crocker, who once described his vocation as "multimillionaire," was fascinated by oceanography and had put his sleek black 118-foot, two-masted diesel schooner at the disposal of several scientific expeditions for the California Academy of Sciences. One of these had been to the Galápagos, where scientists had labored to add to the body of knowledge the *Arcturus* expedition had begun. For the current expedition, Will planned to explore the sea and shores of Baja California, trawling, dredging, and helmet diving in an effort to collect and understand the fauna of this little-known area. They would also compare the Pacific forms of fish to those they knew so well from the Atlantic, hoping to find shifts in the populations that would provide clues to their evolution.

The Crocker expedition was a month of intense work for the three scientists as well as for George Swanson, aboard to make scientific renderings, and Toshio Asaeda, a Japanese artist and photographer. Will's plan, subject of course to Crocker's desires, was to concentrate intense dredging, seining, and diving to build up as extensive a body of data on the ecological relationships of the area as possible in such a brief time. They left the somber fleets of warships that surrounded San Diego and sailed south, stopping at Cedros Island, Gorda Banks, and Arena Banks. Across the Gulf of California they stopped at Mazatlán and Guaymas.

Before returning, they made a dogleg to Clarión Island, 360 miles southwest of the tip of Baja California. Beebe thought that a comparison of Clarión's ecology with that of Bermuda, an island similarly isolated and ringed by currents, would be revealing. Crocker and his companion Maurice Willows fished and played cards, lending a hand whenever the big nets came in. Each day they trawled at varying depths, then feverishly sorted and cataloged the hauls. The researchers found intriguing differences in species and distribution and were excited by the abundance. "Fishes, shells, crabs, shrimps, sponges, sea-pens, starfish, urchins—all were piled up in amazing colors and activity. We worked all . . . day and until midnight, and could do no more than list and catalogue them." (4.3.1936)

The ecosystem of the Gulf of California was different from any Will

The Zaca *en route to Baja California, from above. The boom walk juts off to the right.*

had worked in before. Vast kelp beds with tough, straplike leaves lined the coast from shallows to depths. Like the sargassum weed, the kelp supported its own thriving communities of living beings. Darwin had written in the *The Voyage of the Beagle* of the numerous species of fish that are not adapted for life anywhere else. He saw that cormorants and other fishing birds, as well as otters, seals, and porpoises, are dependent on the kelp beds, and that the fragile health of the ecosystem is eternally in danger from human activity. The huge leaves themselves—often hundreds of feet long—provide shelter and support for an infinity of creatures. The specimens Will collected—not without effort, as the tough, rubbery strands were as strong and sinuous as a sea eel, and slippery besides—were covered with whole villages of mollusks, minute shrimps, and crabs.

The bryozoans, or moss animals, that colonized the leaves were particularly fascinating, like corals in many ways but with a slightly more sophisticated body plan. The flat gray white encrustations they formed on the kelp leaves were, under the microscope, exceedingly beautiful and varied.

> I found five species on a single small leaf, each with its individual ground plan and elevation, each with characteristic calcareous ornamentation. Some looked like sculptured, petrified lace-work, and others were plume or feather-like, resembling frost crystals on a windowpane. A moment of quiet under water on the microscope stage, and a hundred circles of eager little tentacles peered out, and slowly expanded into animal flowers.[1]

The *Zaca* docked again in San Diego at the end of May, and the long rail trip back to New York from the kelp forests of the Pacific coast was an opportunity for Will to begin to shape his notes into what would become his nineteenth book, *Zaca Venture*. A month in England with Elswyth leavened a long summer of writing and working out the life histories of the *Zaca* finds. By now very much at home in England, Elswyth introduced Will to the actors and directors who were working on the play that was being made from her book, *Young Mr. Disraeli*. Will still had many friends in England: they had tea again with A. A. Milne, who was as devoted a fan of Beebe's work as Will was of Pooh. Of *Half Mile Down* Milne wrote:

Your book is utterly and entirely fascinating. I don't know which I envy you most: all those moral and physical qualities which you have and I lack, or all that wonder of a new world. No, your book isn't fascinating: it's damnable; I repudiate it utterly. I will not believe that there is anything outside the dry parts of London and Sussex worth living for or looking at. Go and fascinate a younger, un-family man, and take him diving with you.

Accept my homage, my thanks, my apologies. One of the few things in the world of which I am really proud is that I know Will Beebe.[2]

There were no expedition plans for the winter, but January 1937 brought a signal triumph for one of Will's closest colleagues: Jocelyn, diligently striving to turn her crab work into a scientific monograph, was at last granted official status in the DTR and awarded a salary from the zoo, of $1,800 a year. Just when he ought to have been celebrating Jocelyn's promotion, though, Will was hit unusually hard by his old winter sinus complaint, which slipped rapidly into pneumonia. On January 12, Elswyth described him as "all in"; the next day he fainted twice, and was soon under the care of two doctors and two nurses around the clock. For two long weeks he was in mortal danger, and he remained ill and weak long after. Instead of recuperating in the desert, he and Elswyth spent two weeks in Bermuda, where his health improved markedly. By April he was back to his usual overbooked lifestyle and ready to begin the year's Bermuda season.

Will had become more and more certain that the greatest contribution he could make to science would be through the sort of deep and thorough study of a small area that he had pioneered in Kartabo. "One Quarter of a Square Mile of Jungle" was already becoming a classic as professors used it to teach aspiring biologists about the value of the diverse communities that support life in so many habitats, turning out what would become the first generation of bona fide ecologists. To that end, he was investigating his eight-mile cylinder of ocean from top to as close to the bottom as technology and his budget would allow. As in Kartabo's richly populated jungle, the ocean's populations interacted and depended upon one another, and Beebe's new methods of observation promised to make the study of marine ecology practicable. For the first time, population densities and even individual behavior were open to investigation. He and Tee-Van had already published their book on

Bermuda's shore fishes, and he had spent season after season studying the tide-pool creatures of the rocks and beaches.

Despite these worthwhile goals, Will had grown weary of at least some aspects of Bermuda life. He had more invitations to stodgy formal dinners and wacky cocktail parties than ten lifetimes would have absorbed, and more tennis matches than even he could play. So when Templeton Crocker appeared out of nowhere and proposed another, much lengthier expedition on the *Zaca*, he accepted. He did not embrace the idea with unqualified enthusiasm, however. Although Crocker gave him his choice of locations, the list of possibilities was limited to convenient West Coast ports of call with available game fishing, which left out the unspoiled jungle Will craved. And although *Zaca Venture* makes Crocker seem the ideal host, he was in fact a loose cannon: when his friend Willows was away and he was drinking, which was most of the time, he was either artificially hearty, or morose and bullying. When sober he was depressed, defensive, and prone to take offense at the faintest whiff of a slight. The *Zaca*, though beautifully fitted out as a floating laboratory, represented the field biologist's usual deal with the devil: the possibility of brilliant research and discovery, saddled with almost debilitating burdens.

But in 1937 Will had few options; for all its pink sands, Bermuda as a steady diet was cloying, and another winter in New York was certain to

The lab aboard the Zaca.

plunge him into illness once again. November found him, with Jocelyn and John, sailing out of San Diego for the string of bays in Central America south of the Gulf of California, hoping against hope that Crocker's new vocation as amateur shell specialist would carry him through the doldrums of too much money and too little self-esteem. Their chief object was to study the fish, crabs, and mollusks from the tide pools down to 500 fathoms, emphasizing colors, behavior, and ecological relationships.

For five months the *Zaca* cruised down the Pacific coast, this time from Mexico to Colombia. The numerous bays that indent the coast and the few islands that dot it grade southward from desert to tropical rain forest, and the *Zaca* anchored and explored wherever the surf allowed and the shore seemed promising. The Sierras' looming height gave the shore an isolated, uninhabited feeling.

> Never have I seen wild animals so exposed and so easily seen as along this shore. . . . As we row along near the shore, monkeys, tapirs, the largest crocodiles I have ever heard of, boas and birds usually confined to the depths of the jungle are to be seen or heard. There is no commercial fishing here and we find the sea life as it must have been in Cortez's time. John and I have actually volumes of notes and in one day we collected over five thousand specimens. Jocelyn has life histories and development of one hundred and fifty species of crabs.[3]

As they sailed from bay to bay, Will dropped all pretense of being either ornithologist or ichthyologist. He was caught up in the wonder of species and habitat that varied incredibly from one spot to another. He gloried in being once again "that thing barely tolerated by Ultimate Scientists—a Naturalist." Flaunting his intellectual flexibility, Will turned "with unichthyological facility" from a bad day's fishing to the flocks of brilliant birds that rose from the water, banks, and trees as they followed rivers that fed into the bays.

Book of Bays, which came out of this second expedition on the *Zaca,* is a surprising work. Not because of the prose, which is in Will's eulogistic strain, but because of its rare moments of revelation. He actually puts into words the "sublimated conceit" that follows every competition won, whether with water or wind or animal. Every human victory over nature leaves him elated, even "leaping on to a wet, slippery, sea-weeded rock from a bucking skiff, digging in with fingers and toes, clinging with every muscle and nerve" like a limpet while waves crashed over his head. He compares his addiction to noting every bird to the cravings of a dope fiend, and determines to indulge his listmaking urge—whatever its effect on book sales.

His "list," of course, turns out to be more an elegiac essay, its hard-edged facts softened by vivid images. A "dud bomb" of a booby dives and emerges with a shining fish that a frigate bird soon makes it drop, only to catch the fish itself in midair. Brown pelicans stare down their patrician noses at gulls which snap up small fish that drip from the pelicans' great bills. Herons, the true anglers, make human efforts at fishing look second rate. As people go jacklighting for fish at night, he notes the big-eyed night herons waiting in the shadows for dusk, along with the boat-billed heron, with its rounded, pelican-like beak.

> The ibises—white, scarlet, and glossy—probed the mud with their sensitive, curved forceps, and the lovely roseate spoonbills sifted out unfortunate fish and other mud-dwellers with sidewise swathes of their flat mandibles. The wood-ibis—or, better, stork—was the dumbest-looking and the cleverest of all. He stood in the shallows, reached out one great foot and with his toes carefully stirred up the mud and water, with beak poised to seize whatever attempted to escape.[4]

He was fascinated by the sedentary limpets that clung so steadfastly to their rocks in the pounding surf. Many of the larger limpets bore their own colony of barnacles, and whenever a wave covered the limpet even for a few seconds, the barnacles "opened their ivory gates and swept madly at the water with their little curving feet." The limpets were almost impossible to pry off their resting places on the rocks, but when the tide rose they went slowly off foraging, to return at the change of tide, settle into their exact spot, and again become immovable.[5]

The island of Guadalupe provided the trip's most compelling lessons in devastation. Only three decades earlier botanist Edward Palmer had found it a paradise of birds and plants, but it was now barren and dry. Several of the island's plants, as well as its caracaras, juncos, finches, petrels, flickers, and rock and house wrens, had been endemic to the island, the sole representatives of their species. Now nothing was left.

Over the years, cats, rats, and mice had come ashore from the ships and decimated the bird and small mammal populations. Someone with, as Will wrote, "more faith than brains" had turned loose a few goats to provide meat for future ships. By 1937 the goat population had reached 60,000 to 80,000, and every growing thing within reach of the goats had vanished. Every cone or acorn that fell to the ground was eaten, so no new growth was possible.

Researchers were learning from such inadvertent "experiments" that in a predator-free environment the grazers, whether cattle, sheep, deer, goats, or pigs, will consume all tree seedlings. Such a forest will slowly vanish as its individual members die one by one without issue. Like Guadeloupe then, much of the environmentally critical forest cover of the northeastern United States today is being turned into a barren monument to a forest, its progeny cut off by an explosion of predator-free deer.

Wherever they stopped, Jocelyn scavenged for crabs. Over every likely beach, every inch of mud in the rich mangrove swamps, she went after every crab she could find. Catholic in her avidity, she collected them all, but her special interest was in the small fiddler crabs. Male fiddler crabs have one supersized claw, which they wave vigorously,

Jocelyn Crane,
carcinologist.

even while backing rapidly down their burrows to escape. Though they had long been noted as a curiosity, no one had yet studied the many species, with their different patterns of waving and other intriguing behavioral variations. Will, who had watched fiddlers from the northern Atlantic to the South Pacific, had posed the question to Jocelyn: Why should the males carry around one grotesquely large claw, useless for eating, awkward to move, and attractive to all sorts of predators? What hidden advantage does it confer that allows it to persist in the very teeth of natural selection?

At Will's urging, Jocelyn made herself an expert in all aspects of the study of crabs—carcinology—and set about unraveling the mystery of the fiddler's claw with the zeal that had attracted Will's attention from the first. She had become important to him in every aspect of life. Her grit and determination, her love of research against odds, her youth, beauty, and uncomplaining willingness to shoulder the burdens of an expedition—all endeared her to him to the extent that in *Book of Bays,* he uncharacteristically made her a presence in the text—sometimes playfully, as "the carcinologist" who is nearly crushed by the heaving boat in an attempt to clutch a rare crab on the dock; sometimes more seriously, as the avid researcher "kneeling in the midst of the great intertidal sandy expanse, excavating at arm's length some unfortunate crab."[6] Jocelyn Crane is given a personality in Will's prose that John Tee-Van and Gloria Hollister never achieved. It is the carcinologist who demands witheringly, "Who cares?" when he dissects an appealing green parrot to discover its species, and who triumphantly captures and pins a huge fly that had just bitten her, instead of smashing it.

As they had discovered on the last voyage, Templeton Crocker was prone to depression, and when his friend Maurice Willows left, his drinking went out of control. Crocker felt excluded from the research crew, who were so involved with their work that they had little time or taste for socializing. And when they met another yacht and exchanged visits, Will was invariably the focus of attention; he tried to keep a low profile, but his social nature and enthusiasm for his work were too compelling. Crocker was indeed probably as marginalized as he felt. He drank himself into rages every night, repenting in the mornings but nevertheless refusing to apologize. He wrote Will petulant letters detailing his ill treatment, snarled at the crew, and shouted accusations at everyone. He became so unwell that despite his evening rancor, he needed Will to help him shave in the mornings. The doctor who accompanied him worried that he would break down completely and do himself harm.

The atmosphere on the boat became so poisonous that the work, which at first had been rewarding despite Crocker, lost its focus. As soon as Jocelyn had recovered enough from a siege of dengue fever, she, Will, and John left the *Zaca* in Panama and sailed home through the canal—Jocelyn for the first time, Will for the seventh.

When Will chronicled the five months of the second *Zaca* expedition in *Book of Bays,* he dedicated it to Crocker. Several times throughout its pages he mentions Crocker's generosity, his willingness to help, and his wisdom in choosing to spend time in this way. Crocker responded with a wistful note, saying that if Beebe ever found himself in need of a conchologist on another expedition, he would be glad to go. But Will had learned his lesson: he never again accepted an invitation to an expedition, however tempting, that put him and his staff at a rich man's beck and call.

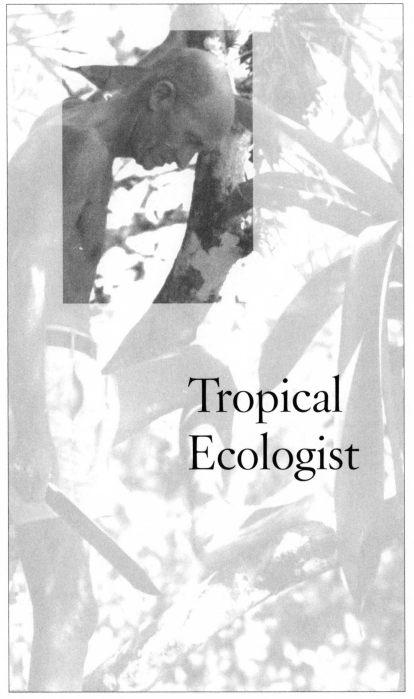

PART IV

Tropical
Ecologist

Ocean to Jungle

All our stories are of the middles of things,—without beginning or end; we scientists are plunged suddenly upon a cosmos in the full uproar of eons of precedent, unable to look ahead. . . .

William Beebe, *Edge of the Jungle*

W ILL, JOHN, AND JOCELYN returned to New York in May 1938. For a year the Depression and the threat of war kept them stateside. In New York Will leavened lab analysis of the hauls from the *Zaca* with writing and with lecture tours that brought in the money he needed. He wrote *Book of Bays* during this year as well as myriad technical papers. He had captured aboard the *Zaca* a record-breaking sailfish—only, unlike Crocker's catch, this was the world's smallest, and Will was delighted. With the tiny fish, combined with younger specimens he had caught over the years in seine nets, he could trace for the first time the ontogeny of the huge game fish from its minute larval form through this diminutive creature whose scales and teeth were perfectly fishlike and ordinary, to the adult, with its thick body armor and slashing, sawtooth swords.[1]

In addition, the entire Department of Tropical Research was drafted to prepare an elaborate exhibit for the World's Fair that was to be held in New York in 1939. The Zoological Society's display would include the bathysphere as well as a diorama of New York's history from the Glacial era, based on the material Will and his staff had found in the Hudson

Gorge. It took a massive effort to design the exhibit, verify all the information that was to go in it, and accumulate and mount the materials.

Elswyth sailed to England on another research trip in August 1938, but found war preparations there unsettling. Bomb shelter drills and gas-mask fittings combined with the dire predictions of friends led her to return after three weeks. She abandoned her popular series of historical novels about England, turning instead to history a little closer to home. Her next books would be centered on George Washington at Mount Vernon and on several generations of a fictional family in colonial Williamsburg, Virginia.

The gathering clouds of war emphasized tensions among the members of the Zoological Society's board. Many of the zoo's most generous and influential patrons, such as the Warburgs and the Schiffs, were Jewish; unfortunately, Madison Grant was becoming increasingly obsessed with the idea of Nordic superiority, and particularly what he saw as the dangers of rampant immigration. He and Henry Fairfield Osborn were both pillars of the nascent eugenics movement; although Osborn's involvement was at least purportedly in the interests of science, Grant made no such claim, and his outspoken books and articles were inflammatory at best. Will was not alone in worrying that Grant's ugly racial politics, shared by many otherwise rational Americans at the time, would alienate people who had long been good friends to the zoo and to Will personally, and create irreparable divisions within the board and trustees.

Desperate to escape the strained atmosphere, Will pulled the financing together—largely from his own pocket—for a short Bermuda season in the fall of 1939. He knew that his picture of the ocean in his eight-mile cylinder was still woefully inadequate, and now there was the added incentive of maybe, just perhaps, dredging up one of the contested fish he had seen from the bathysphere. Anticipation of that vindication added urgency to his usual enjoyment of the net hauls.

John, Jocelyn, and three volunteers accompanied him. They again had rooms at the Biological Station and worked in the sunny veranda-lab at New Nonsuch. Several good deep hauls from the ocean cylinder yielded plenty of interesting work, though no elusive bathysphere creatures. Will and John consoled themselves with studying the eggs of a giant eel and filming their development, a challenging technical feat that opened new vistas of developmental research. Will had met Walt

Disney in California, and their discussions had suggested innovative ways to use film in laboratory settings. The development of the eel was a little-understood phenomenon, and to get it on film would be a fascinating contribution to science. Jocelyn, meanwhile, was absorbed in her crabs.

In September news came first that Germany and Poland were at war, and then that England and France had joined in. Bermuda, being a British colony, was transformed overnight. The Biological Station rented a "wireless," and the news destroyed all pretense of concentration. Groups gathered silently to huddle around the radios and many residents began to leave, closing their houses and heading for Britain while they still could. Will was determined to stay, but on September 10 they heard that the ship that linked Bermuda to New York would make its last run in three days. The threat of submarine attack had become all too real, bringing their series of 1,512 net trawls through Will's eight-mile cylinder to an abrupt end. They packed frantically, trying to close the house and the lab and save what they could of their research, so recently begun. Jocelyn made a last run to Nonsuch and brought back two rare crabs and their eggs, which she carried aboard the *Monarch* in two glass jars and a coffeepot.

In the spring of 1940 Will was laid low by an attack of strep and hospitalized for several weeks. When he recovered, his first thought was to

John Tee-Van's rendering of the eight-mile cylinder, crisscrossed by many seasons of trawls and dives; includes several of the bathysphere dives.

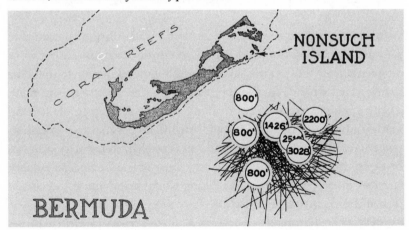

see what changes had been wrought in Bermuda and what they would mean for future research. The first war-induced panic was over, and life—with the addition of hundreds of British officers and soldiers—was returning to some semblance of normal. People had taken heart and moved back, and the social scene was as demanding as ever. When Will finally arrived with his staff in late July—minus John Tee-Van, who had been put in charge of the zoo's exhibit at the New York World's Fair—he found the little island oddly calm. But in Hamilton, he wrote in his journal, soldiers who had been under heavy fire and brought back for R&R wandered the streets with vacant gazes, and Meyer's Wharf, where all the deep-sea trawling equipment was stored, had been turned into a vast coal dump guarded by barbed wire and "boy soldiers."

Although he had never underrated Tee-Van's unique qualities, without him Will became keenly aware of all the things John had done for him so quietly and efficiently, through so many years. Everyone who had ever worked with Will had commented on Tee-Van's invaluable help, whether performing small chores uncomplainingly, taking responsibility for tasks no one else would shoulder, or smoothing over rough encounters between people trapped in close, inescapable contact in confining situations for too long at a stretch. Will also knew well his own inability to deal with the thousands of mechanical details that John had so quietly overseen, so when Bronson Hartley volunteered to help out, it was the beginning of a mutually beneficial relationship. Bronson, a young Bermudian, needed an occupation and was a wonder with any sort of mechanical equipment. Not only was he able to fix anything, he even brought ideas of his own to devising new apparatus for helmet diving.

John was absent again when Jocelyn and Will sailed for Bermuda in May of 1941. In addition to his duties as editor of a multi-author work on Pacific fish, Tee-Van was laying the groundwork for a trip to China to bring back a panda for the zoo, a gift from the Chinese government. The trip, which would have him sailing with the panda en route to Pearl Harbor just under the rope in that terrible December, was an important career move. With the support of his wealthy in-laws, Tee-Van was being considered for an administrative post in the Zoological Society, and this much-publicized undertaking would bring him to the attention of the executive committee.

Meanwhile, Bermuda had become a hub of activity for British forces guarding the Caribbean from German submarines. Will and Elswyth had undergone a scare when the New Nonsuch property was condemned by the Department of War to make a runway for the new American air base. The buildings on Nonsuch Island had also been slated for destruction. But timely intervention by Professor Osborn's son Fair, who had taken over the presidency of the Zoological Society in 1940, and Allyn Jennings, who had succeeded Dr. Blair as the zoo's general director, had saved both.

In May, Will and Jocelyn had hoped to find that all was well with the research station despite the war. From the moment they arrived, however, it was immediately clear that Bermuda had now lost any pretense of calm. An enormous carrier and several other warships floated at anchor just offshore, and large convoys dotted the water. It took them three hours to dock, inching delicately between troop ships. The Biological Station was in the process of being broken up and its furniture and equipment auctioned off, mostly to the American contractors and engineers who had come to Bermuda to build the new base. At every cocktail party or dinner they met more people who were being forced out of their homes, more or less remunerated by the government. W. B. Smith, the contractor who had built most of New Nonsuch, had to see

The garden at New Nonsuch.

his own house standing naked and exposed, every tree having been cut down and burned to clear the airfield. The palatial estate on nearby Long Bird Island was rapidly metamorphosing into military quarters and a mess hall for 600 men.

Although New Nonsuch itself was spared, construction of the airfield was changing the marine environment disastrously. A gigantic dredger sat just offshore, grinding up the coral reefs with its eighty-foot revolving steel ladder, armed with an eight-bladed saw. The ground-up coral was sucked into a pipe and carried back hundreds of feet to the shore, where it was spewed out to fill in between what had once been a chain of small islets. Eventually the fill closed in the gap that had allowed access to Nonsuch Island from the lab.

Many of the reefs were gone completely; the lovely beaches were rimmed with deposits of grease. The once crystalline water, opalescent with oily rainbows, was so turbid from the action of the machinery that nothing was visible, and all life was being choked out. The devastation appalled Beebe; he found an American military contractor to rent New Nonsuch and sailed back to New York.

The winter of 1941 brought the war home to America. When the United States joined the Allies, many of the zoo staff aided the war effort in some way. Eleven trustees and sixty-two staff members had joined up; George Swanson served in the Army, for instance, painting camouflage, and Gloria Hollister resigned from the staff to join the Red Cross. There she met her future husband, Anthony Anable, then serving as a naval officer.

Their Bermuda compromise in ruins, Will and Elswyth gave up on finding a research station that would suit both. Will had never abandoned his dream of understanding more of the ecology of the tropical jungle, and he was weary of the mannered civility of Bermuda and shaken by the excesses of war—even a war he wholeheartedly supported. Elswyth was happiest in a temperate climate, among temperate people. While Will looked for a tropical paradise, she looked for salvation to the New England countryside. She began to look for a country home where she could write, and Will began to put out feelers to donors who had supported the DTR in the past, in hope of founding another Kartabo, which had fallen victim to uncontrollable logging.

If oil, seeping from the vast construction projects on Bermuda, had ruined Will's oceanographic ambitions, it fueled his next field of research. When his need for a new research station became known, Laurance Rockefeller, whose involvement in oil was tempered by a lively interest in science and the environment, brokered a deal between Beebe and Standard Oil of New Jersey. With additional funding from the Guggenheim Foundation, Will was able to mount another expedition to the jungle.

Set to exploit Venezuela's newly discovered wealth of petroleum, Standard Oil had established a refinery and a base camp in Caripito, in the northeast corner about 100 miles west of the island of Trinidad. A small city of North Americans lived in quarters provided by the company, complete with dormitories and mess hall and even a school for the workers' children. Will and his small staff were given a cottage in the spare bachelor quarters as lab and residence. Jocelyn accompanied him, as did the artist George Swanson, assistant Mary VanderPyl, and a new staff member, entomologist Henry Fleming.

Will's object was to continue the jungle work he had started years ago in British Guiana. He was heartsick about Bermuda, but sick too of being typecast as a stuntman rather than a scientist. More than any

The ad hoc lab at the oil encampment at Caripito, Venezuela.

other thrill, he wanted the exhilaration that comes from working a scientific problem through to its solution. He would devote himself to discovering as much about jungle ecology as time and energy allowed. And, of course, he would collect animals for the zoo's ever-growing collections.

At Caripito, everyone worked on a different problem. As usual, Will's interests lay with the entire ecosystem, so his first order of business was a thorough climatic study. It was the dry season when they arrived, and he got a good idea of the desert llano areas to the south that merged into the jungle area. Excavating a dry mudhole, he discovered scores of species in the dried-out mud. Unlikely as it seems, there were thirty-four species of fish alone, waiting out the dry season in patient estivation. In jackets of mucus they kept their bodies moist as the mud solidified around them, lowered their metabolic rate, and slept.

The great spotlights and flares of the refinery were, ironically, a godsend to Fleming's insect collecting: thousands of insects flew to the lights, only to crash against the high walls and lie dead or stupefied in windrows until picked up by eager hands and euthanized, categorized, and preserved. When the dry season of early spring gave way to pouring rains, the change allowed the crew to study the varied ways animals and plants adapted to the dramatically altered conditions. Jocelyn's work with the motion picture camera was producing hundreds of feet of high-quality film of subjects as varied as lizards, snakes, "walking" fish, mouse opossums, and scarlet ibises. They devised a panoply of methods of making animals hold still or perform for their cameras, and constructed an elaborate "island" with a dry moat on which compatible creatures could run free but remain in camera range. Mary VanderPyl studied the frogs that, invisible until the rains, now appeared in droves.

Although Will's reports to the Zoological Society catalog every staff member's work objectively, his journal reveals personalities. Henry Fleming was a particular problem. Instead of fitting into the well-established routine of early rising, hard work, and civilized cocktail and intellectual patter, Henry smoked endlessly, chewed his fingernails, and drank too much. And despite a wife and children back home, he had a tendency to stay out all night dancing—which bothered Will primarily because it left him haggard and late for the morning's labors. Worst of all, Fleming's specimens were pinned hit or miss, with many of the best left out for the ants. His boxed specimens were several decomposing layers deep, and bent at the corners. His desk was filthy, and he could make neither cages nor larval boxes nor fly traps. His love of collecting, though, won Will's grudging respect, and his willingness to work on machinery made him invaluable.

One important study that came out of Caripito was Will's observations of combat behavior in the huge three- to six-inch *Dynastes* beetles *(Megasoma elephas)*, which carry horns on their carapaces. Darwin had theorized that the horns carried by the males ought to be used in combat (male-contest selection), but as the beetles had never been observed to fight, he concluded that they must be ornaments designed to appeal to females (female-choice selection). Alfred Russell Wallace, on the other hand, was equally certain that the horns were defensive, to keep birds and other predators from swallowing them.

Will found several of these rhinoceros beetles at Caripito, drawn to the refinery lights. He made his captive beetles comfortable in capacious terraria and tried pairing males and females, males and males, females with females, to little effect. Instead of fighting, the males he was lucky enough to catch wallowed contentedly in their rotten-banana

Beebe's study of combat among rhinoceros beetles became a classic.

kibble, ignoring both females and other males. But one day he cleaned out the cage and washed off the beetles themselves. As soon as he put them back together, the males faced off and began to heave. Again and again they came together, one levering the other onto his back with the horn until the loser edged sullenly away. When the female was removed, fighting ceased. When she was returned, it began again.

Further inquiry proved that water is a critical releaser for aggression: rhinoceros beetles mate during the rainy season, and the presence of water starts the hormonal onslaught. Dry beetles were phlegmatic, pacifist beetles. The discovery that the horns were actually essential in male-male competitions helped shift the balance in the ongoing debate over sexual selection, and provided what remains a classic insect example.[2]

Although the season at Caripito was productive, as a research station the place was disappointing. For one thing, there were just too many people in the area. And despite having the same yearly rainfall as Kartabo, the forest canopy was lower and not as rich in species. Several weeks there told Beebe why: the dry season was so long and so dry that evergreen trees and shrubs acted like deciduous ones, dropping their leaves. This pseudo-fall had an enormous effect on the epiphytic plants that live in the canopy, which were exposed to the merciless sun year after year, and on the number of habitat niches available for animals. Then, during the wet season, the flat clay soil turned into a swamp; much of the jungle floor, which had been on its way to becoming a parched desert, turned into a shallow, marshy lake, sending hosts of creatures fleeing to low ridges or up tree trunks to keep from drowning. But the bipolar nature of Caripito's climate, while interesting, discouraged long-term, intensive study. Despite Standard Oil's generous offer, Will knew when he left in September that he would not return to Caripito.

The war had cut so radically into both funding and transportation that the DTR was forced to spend 1943 in the States. The Pond Company, the agency that arranged Will's lecture tours, set up talks in cities across the country, where audiences flocked to his mixture of science and adventure. His popularity soared with repeated appearances on the hit radio show *Information, Please,* where he responded to questions that ranged from humorous to arcane. His answers were accurate, deliberate, and often witty.

His contributions to the Zoological Society's annual report for 1943 were triumphs of positive spin. The DTR "explored and exploited the resources of the Laboratory in the Zoological Park." They set up an exhibit on Venezuelan jungles in the old Head and Horns building at the zoo, and Will wrote some chapters for an armed services instructional pamphlet on life in the jungle. The Army and Navy printed and distributed 60,000 copies of *Half Mile Down* for the troops abroad. The department was also in correspondence with dozens of soldiers who wrote in with questions or observations from their remote posts, asking about strange animals they had seen, and even offering to capture specimens and bring them back to the zoo. Making a virtue of necessity, Will and his staff used the occasion of having to consolidate space to organize and preserve the vast collections they had amassed over the years, keeping some for study and sending some to the American Museum.

Although Will had reached sixty-five, the usual retirement age, in July 1942, the zoo's executive committee continued to keep him at his regular salary on a year-by-year basis. The good publicity he generated, together with his long association with the society, kept the balance in his favor. By that time he was the sole remaining member of the zoo's original staff;[3] in addition, he was still doing research and writing books that continued to sell well. His next book, *The Book of Naturalists,* was a departure from earlier works. It was an anthology, a compilation of what he judged to be the best of natural history writing from Aristotle on. The final essay is by a young woman who had been inspired by Beebe to study the oceans. The essay he included on the life of the eel introduced Rachel Carson to the literary world; in her first widely read book, *The Sea Around Us,* she acknowledged the importance of Will's mentorship.[4]

CHAPTER 33

Rancho Grande

One thing is certain, no one can walk for long through a jungle and remain conceited. . . . I know of no better cure for human vanity.

William Beebe, *Book of Bays*

WHILE WILL WAS IN VENEZUELA, Elswyth, exiled from England by war, traveled to Williamsburg to do research for her new series of books on colonial American life. Elswyth had matured from the leggy young woman Will had married into a solid middle age, but she had never changed her image. Her luxuriant wavy hair, which she now wore in a tightly braided coronet, had never been cut, and she dressed in prim velvet Victorian gowns for formal occasions. Although she was a respected part of the literary world, it was Will whose theatrical friends they entertained, and who received invitations to an endless round of cocktail parties and dinners. Will never failed to praise her sense of humor, but her gravity is what people noticed, and her intense gray eyes.

Her health, never robust, continued to plague her. Her diary notes recurrent bouts of "grippe," a blanket term for flu in its many forms. She was often bedridden for two or three weeks at a time. She traced her constant abdominal troubles to her mother's demanding, stressful presence. Edith Ricker had been completely dependent on her daughter since separating from Maurice years before, and the relationship was a rocky one. She had even moved into their apartment, which had been made into a double. Although she had changed her name to Edith Thane, she was a constant reminder of the difficult Iowa past that Elswyth labored to repress. She had been artistic, and Will struggled to find small publishing jobs for her, such as making some illustrations for the field guide to Bermuda's fish that he and Tee-Van had written. What effect her dependence had on Will and Elswyth's relationship is hard to assess, but without the incubus of her presence, they might have found more common ground. Elswyth might even have let her hair down.

In an article for *Today's Woman,* a popular large-format magazine with a progressive slant, writer Helen Worden described Beebe and Thane in an article called "The Marriage That Breaks All the Rules." Worden had lived for years in the apartment beneath Beebe's, and admitted easily that she had been among the many friends of his who were surprised at his choice. She describes Elswyth's gravity, her love of quiet and a retired life. She hints of Beebe's "many women friends" and notes his "flapper-era" sophistication. Was Elswyth jealous? Worden asked coyly.

"Jealousy is an evidence of an inferiority complex, which neither Will nor I have," Elswyth replied. "He says he'd wonder what was wrong with his taste if other men didn't notice me, and I'd feel the same way if

Elswyth in the New York apartment, ca. 1930.

other women didn't notice him."[1] She was uncomfortable on her husband's expeditions, she said, and explained to Worden that they had agreed at the outset to keep their careers separate from their private lives. Elswyth described her fears when Will was diving, and her relief when word came to her in England that he was safe. "'I couldn't have stood it if I'd been there,' she told me with tears in her eyes," Worden reported.

Books, Worden noted, were everywhere in the Beebe apartment. Elswyth waved toward one wall, which housed Will's science, travel, adventure, and mystery collection, along with his first edition of *Alice in Wonderland.* In Elswyth's study were her books—biographies, poetry, and romance. Each had a copy of the *Encyclopedia Britannica,* so they wouldn't have to share. Elswyth said that her chaotic desk caused Will to groan, "but he keeps his hands off. Every pin points the same way in his world."[2]

While she did her best writing at night, she told Worden, Will got up at dawn and thrived on little sleep. They had agreed that he would get up when he pleased, but that she was not to be disturbed until noon. They tried to have tea together in the afternoon when they were both in town, but he was rarely home for dinner, with so many meetings and engagements. Elswyth would always wait up for him, and they would talk companionably of their days. She enjoyed knitting and embroidering while he read, often aloud. And she and her mother made dolls and puppets, some meticulously outfitted to represent the characters in Elswyth's books.[3] She and Will staged elaborate performances for themselves as well as for friends.

When Elswyth was working on a book, she cut out all social engagements and averaged eight to ten hours of writing during the afternoon and evening. In twenty years she had written fourteen books and two plays. "I get down on my knees to my wife for the work she has accomplished," Worden quotes Will as saying. "I thank God it is different from mine. I thank God also that Elswyth likes to play. In that sense I don't believe either of us will ever grow up."[4]

When it became clear that the future of New Nonsuch as a research station was bleak, Will decided that he would help Elswyth buy the country retreat she had been wanting. In October 1942 she found a small farm near Wilmington, Vermont, that was far removed from the city yet accessible to trains. She spent her time increasingly in its quiet isolation, away from the social obligations of New York that Will had always enjoyed more than she. He came up frequently, and they converted the hayloft of the old barn into a spacious aerie for him, with large inviting windows. A local handyman, Elmer LaFlamme, helped with the many major and minor repairs, such as putting in a new heating system to replace the rambling frame house's moody and inadequate boiler. With Elmer's aid, Elswyth was even able to return the property's many sugar maples to modest production. In 1945 she wrote a charming book about reclaiming the place, *The Reluctant Farmer* (later reissued as *The Strength of the Hills*).[5]

Thanks again to Laurance Rockefeller and his associates, Will was able to return to the jungles of Venezuela in 1945. Creole Petroleum, a Venezuelan spin-off from Standard, agreed to foot the bill; to avoid the disappointment of Caripito, Jocelyn had flown down in the spring of 1944 to find a suitable location. Jocelyn, whose reputation as a scientist was becoming formidable, did her research with the thoroughness and zeal that characterized all her pursuits. An article she wrote for the zoo's *Animal Kingdom* magazine described her search for the perfect jungle—a tropical rain forest undisturbed by human beings yet accessible to supplies and transportation. The DTR wanted it to be uninhabited, she explained, "not because we are misanthropes, but because the creatures we study show such strongly anti-human traits."[6]

To the average person, Jocelyn wrote, the word "jungle" conjures up pictures that are "equal parts trees, monkeys, vines, spiders and orchids all tied together with endless loops of boa constrictor." But of the many types of jungle environments, they had tried the deciduous forests of Caripito without success, and the mangrove swamp jungle was impracticable for as many different reasons. What they were really looking for was another Kartabo, the station they had founded in British Guiana. However, British Guiana was not in the gift of Creole Petroleum, so they had to find their Mecca in Venezuela.

On horseback, in small planes, or on foot, Jocelyn trekked from site to site. Several different options, all in good jungle, had fatal flaws: either there were too few animals, or the area was inaccessible, or the soil was riddled with the tricky pits that locals called *tatucos*. In the Andes west of Caracas she found herself at the glorious cliffside ruin of Rancho Grande, which met every one of her criteria.

Rancho Grande was a fortresslike building of concrete and stone that had embodied the dream of Venezuela's erstwhile dictator, Juan Vincente Gomez. He had envisioned a retreat on a grand scale where socialites, high-ranking diplomats, and other dignitaries could fraternize, protected from any hint of the turmoil or revolution that plagued the restive cities. Its view was of cloud forest and bright macaws and toucans rather than the dreary slums of Caracas; its sounds were the howls of monkeys, not blaring radios and automobile horns.

For fifteen years, crews of workmen labored to carve the structure from the rock. But before it could be completed, Gomez died, and with him went his dream. Rancho Grande sat abandoned, its echoing ballrooms papered with lichen and moss, its spiraling halls roofless, the haunt of jaguars, tapirs, and sloths.

The distinguished Venezuelan botanist Henri Pittier, then in his eighties, had identified the area as particularly in need of preservation on account of its unique diversity of plant and animal life. With Pittier's intercession, the Venezuelan government at the time agreed to use Beebe's well-publicized presence to promote the establishment of a science and conservation station of their own, and generously allowed him the use of Rancho Grande.

Jocelyn set up a base camp to assess what would be needed to make some corner of the sprawling edifice habitable. Creole Petroleum finished and outfitted the least dilapidated wing of the giant structure—eighteen rooms worth—for the group. Workers once again commuted

Rancho Grande, surrounded by high cloud forest.

up from nearby Maracay to enclose the vacant windows, seal leaking glass roofs with tar, and pound floorboards onto joists. A whole zoo of jungle creatures was put to flight, but just as many remained, to festoon the walls and naked beams with spiders and frogs, bats and lizards, monkeys, tree snakes, and sloths.

The journey from urban Maracay to Rancho Grande took them from the muddy beaches where Jocelyn's fiddler crabs scurried and waved their way through life, through desert scrub studded with cactus and acacia and looking for all the world like New Mexico. As they went higher, grassy savannah gave way to the green water of Lake Valencia, rich with its own flourishing ecosystem, then through deciduous forest, and gradually to the cloud forest that cloaks the mountain pass where the great ruin sits.

The building perches in a niche carved from the mountainside, curving to fit like an inverted question mark. High ramparts shore it up against the cliff, and the rear disappears into the undisturbed rock face behind. A wide, curving veranda more than 100 feet long fronts a row

Will and Jocelyn on the patio at Rancho Grande.

of cell-like rooms, with high, narrow openings for windows and open drains. The only access to these rooms was from a dank, narrow corridor along the back.

The newly finished wing had vast and echoing—but functional—bathrooms, and the laboratory itself was a wonder, with a wall of windows fronting on the jungle and mountain range, and views clear to Lake Valencia in the distance. Its floors were of elegant terra-cotta and buff tiles, and each researcher had a table with the tools of his trade. But beyond the living area was the labyrinth, where rooms were oval, square, triangular—and often connected only by slippery two-inch girders twenty feet in the air, beloved of looping spider monkeys and bats. The human inhabitants found themselves lost again and again in the dank, echoing corridors.

> The roof is as unexpected as the rest. Some of it is solid concrete ceiling, but over many of the passageways far aloft, are eaves of giant sheets of glass, following yards of nothing but Venezuelan sky. The rain

In Rancho Grande's labyrinth.

drenches, and the fog dims the castellated structure much more than the sun warms it, and the lights and shadows are filled with the varied sounds of flowing or dripping water.[7]

One room that they used as a laundry had an eight-by-thirty-three-foot floor and a fifty-two-foot roof "up in the empyrean." The broad, flat roof of the main building was perfect for sunning, watching birds or the stars, or enjoying cocktails.

Any suitable rooms were converted to cages for a growing menagerie of pets: pacas, agoutis, sloths, monkeys, frogs, and even an ocelot cub lived companionably using girders as perches and aerial roadways, and cavernous floors as dens. Beebe tested his theories of protective coloration by feeding, or attempting to feed, various insects to a motley array of potential predators, confirming to his satisfaction that insects which were cunningly camouflaged were generally edible, while the brightly colored ones were so distasteful that they could flaunt their presence in safety. Chiri, the perennial pet, was honorary chairman of the Rancho Grande Laboratory Tasting Committee.

Only half of the many drains that pierced the ramparts and walls were functional, and the dry ones housed families of violet and silver swallows and bright turquoise tanagers. On the dank interior walls "fungus, lichens, unnamed slimy life, molds, smuts, and mildew" painted tapestries of colorful patterns. A scraping from the walls revealed its own busy world of life under the microscope lens.

Once Will was settled in, his shelves arranged to his specifications, he began the exhaustive survey of the ecology of their research area. Although the annual rainfall of seventy inches was too little for a conventional rain forest, the almost daily clouds and mist kept the humidity reliably above 90 percent. When clouds formed and drifted slowly up the mountainside they gradually enveloped the building in *neblina,* or dense fog. The neblina conspired with the altitude of Rancho Grande to keep the temperature around a chilly 66 degrees, hardly the tropical steam bath Beebe loved. But the flora and fauna thrived on it, so he did too. The wealth of creatures that lived in or flew, ran, or crawled through Rancho Grande's precincts was the kind of riches he could thoroughly enjoy.

CHAPTER 34

High Jungle

*To settle down in a strange country and to study successfully
the wild creatures which inhabit it, demands a few of the elements of real warfare, combined . . . with a large percentage
of luck.*

William Beebe, *Tropical Wildlife in British Guiana*

P ORTACHUELO PASS, THE NOTCH
in the Andes that Rancho Grande commanded, is an age-old migration
route for insects and birds. For whatever reason—perhaps a magnetic
anomaly—flyways from several directions converge on this particular
dip in a forbidding range. By day butterflies by the millions fluttered
through, each individual seemingly aimless, but making astonishing
headway en masse. In one ninety-minute period, for instance, Will and
Henry Fleming calculated that approximately 286,000 small brown
ten-spotted butterflies beat steadily through the sixty-foot-wide pass.[1]
Birds of many species passed through, some on daily trips to favorite
feeding areas and then back, others on arduous journeys of thousands
of miles.

When electricity was finally installed, Rancho Grande's lights became a magnet for night-flying creatures. Confused insects bombarded
the roof and windows, making work difficult and collecting almost too
rewarding. As with the rich net hauls from the depths of the sea, there
was suddenly more to study than time and energy allowed. Even birds
would occasionally blunder into the building, providing fortuitous data

on migrating species. "Years ago," Will wrote in *High Jungle*, "it seemed as if nothing could ever equal the breathless excitement of cascading the contents of a deep-sea net, sea serpents and kraken in miniature, from black, icy depths a mile down. But nights of storm at Rancho Grande, the blackness equaling the opaqueness of ocean's abyss, brought birds as unexpected as they were rare."[2]

One night an unusual white-spotted swift paused at the window long enough to be identified as the third on record anywhere. Seven more crashed down in the course of two weeks, and the bat falcons that nested in a nearby tree caught another, plucked it, and fed it to their nestlings as Will watched aghast through binoculars. Two rails that differed from any in his books appeared out of the dusk with no hint of origin or destination.

In the midst of all this richness, only two weeks after they had moved in officially in early May, Will was holding a swaybacked ladder when it fell along with one of the local assistants, who had been installing a woven twig windbreak on the veranda. Pedro was only slightly concussed, but Will's ankle was broken.

He stayed only one night in the hospital in Maracay, kept fretfully awake by savage mosquitoes and the cries of a woman in labor. Henry brought him back to Rancho Grande frustrated and sore, his foot bruised colorfully enough to put his favorite mildews to shame, and

Collecting night-flying moths from the lighted sheet at Rancho Grande, 1948.

painfully swollen. Pedro fashioned a pair of crutches out of branches, and Will spent the next several weeks reclining in his favorite India "rookha" chair from Calcutta, perched high in Portachuelo Pass with notebook and binoculars. From his station on the ridge he could watch the jungle around him, as well as the butterflies and birds that flew by day through the pass. After he had been there motionless for an hour or so, the animals began to go about their business without regarding him; a hummingbird perched on his hat brim, and a boa slithered over his propped-up leg, tasting and rejecting his cast with its feathery tongue. The chapter he wrote about this time for *High Jungle,* the book that would chronicle this expedition, was ruefully titled "St. Francis of the Plaster Cast."

With this enforced leisure, Will was able to begin the first study of bat falcons—the pair that had so casually fed the rare swift to their nestlings. He watched and recorded their 164-day cycle of courting, nestbuilding, incubation, rearing, and fledging. Through the twenty-two-inch 40x binoculars Herbert Satterlee had given him, Will could identify every prey item by species and sometimes even by sex. His educated observations record important behaviors that were new to science. He was able to note the very different personalities of the two

Watching the bat falcons from the lab at Rancho Grande.

chicks, document for the first time play in birds, and deduce how instinct and experience combine to make the young formidable hunters.

He also had unwonted time for reflection, as he waited more or less patiently for some staff member to help him hobble back from the pass. "As I watched the flying birds and insects," he wrote, "I thought what a grand thing it would be if only the sky trails of volant beings would last for a time like prints in the dew or tracks in new fallen snow."

> Physicists have their cloud chambers wherein atoms and their neutrons and protons register their paths in lines of light. How nice, like vapor sky-writing, if birds could leave at least transitory trails behind them. Think of the sun rising after a night at the height of migration, and what an awe-inspiring sight the sky would present, with its thousands upon thousands of avian trails.[3]

Far from the shore and her fiddler crabs, Jocelyn became fascinated with the small salticid jumping spiders that wave their tiny legs in gestures of courtship or war, much as the fiddlers do. In a terrestrial version of a net haul, she combed the foliage for them, shaking plants furiously "like a small local hurricane," Will wrote, holding an inverted umbrella beneath to catch what fell. She was also interested in the way the orchids and bromeliads that lived high in the trees were pollinated, and was eternally lugging large plants in from the jungle and putting them in pots on the shaded veranda and the lab. Many insects can see wavelengths of light that the human eye is blind to, and many flowers, when viewed through an ultraviolet filter, have markings that point the way to their pollen-bearing organs. Jocelyn dissected the flowers, studied them under ultraviolet light, and monitored the plants for insect pollinators. Many insects, such as the heliconid butterflies, also bear ultraviolet markings on their wings, advertising their species loudly to an exclusive insect market.

Henry Fleming was swamped by the sheer numbers and variety of the insect life, especially around the lamps at night. To be scientifically valid, each specimen had to be meticulously identified, recorded, and described before color could fade. Then they had to be preserved according to their type and eventual destiny, and stored where no insect or mold could destroy them. Though habitually careless, Henry loved

the work, but his gregarious nature rebelled at the monotony. As the
official driver, he managed to escape to Maracay frequently under the
guise of running errands or repairing the notoriously unreliable car.
Considering that he liked to drink with friends and had a bad record
with the police, his late returns caused considerable anxiety. He none-
theless completed and published a thorough study of sixty-seven
species of sphinx moth.

George Swanson, the artist, was at everyone's disposal. Mustered
out of the army by illness, he switched gratefully from painting camouflage to sunsets and vistas, jewel-like beetles, microscopic feather lice,
and spider courtship. The encyclopedic nature of Will's interests combined with the luxuriant jungle life forms made the range of subject
matter daunting. Will was fascinated by color and wanted it recorded
with impeccable accuracy, with no sacrifice of anatomical detail. Swanson rose to the challenge, becoming adept at capturing the life and
spirit of his subjects. Posing live animals often required ingenious solutions, as when Swanson tied a tree snake to a branch by its tail. And

Jocelyn collecting spiders.

always there was the race with death as the bright colors of dead specimens faded into dull grays and browns.

Swanson's favorite days were spent painting plants in the jungle with brilliant birds wheeling past, serenaded by the calls of bellbirds, the eerie cry of red howling monkeys, and the buzzing of gold and purple wasps. "I jumped from spiny rats to impossible caterpillars, to tadpoles of golden froglets found in the wells of bromeliads. Brushes, inks, transparent watercolors, opaque watercolors, oils, sketch pads and slide rules were mixed up with yellow orchids, giant toads, hatching frogs' eggs in dishes, fresh jungle flowers in bottles and dead jungle flowers in jars, all in startling confusion." [4]

The DTR enjoyed two rewarding seasons at Rancho Grande in 1945 and 1946, but was grounded again during 1947. The annual report states confidently that the previous seasons had generated so much material that Will and his associates had to stay home to analyze it. The facts were that funding was tight and Venezuelan politics were in ferment. The Acción Democrática Party, which had seized power in 1945, was disintegrating, and during 1947 a new leader, Rómulo Gallegos, won a fiercely contested election. Will had found it difficult to deal with the chaos before, and dreaded the upheaval and inevitable change of officials both petty and great. When he, Fleming, and Swanson left for Rancho Grande on a frigid day in February for the 1948 season, hoping against hope for sympathetic officials to ease the passage of all their scientific gear, Jocelyn made a side trip to Trinidad and Tobago, hoping to find yet another jungle paradise where the politics would be more secure.

The 1948 season was productive scientifically but full of small irritations, mostly from political appointees, who demanded licenses for lab supplies that had never before been questioned and who expected to be "tipped" for every item passed through customs. Despite these annoyances, Will was able to complete his comprehensive assessment of the ecology of Rancho Grande, coauthored with Jocelyn,[5] as well as technical studies of the migration patterns of several families of birds, butterflies, and moths. On his seventy-first birthday on July 29, 1948, Will remarked that he felt as well as ever. "Physically have not slipped an atom that I can see. Allergies diminished to minimum, and when a hint comes, it dies out at once." His need to compete had not diminished either: he was still able to beat the others in tennis, and could easily run up stairs two at a time. He slept well and, except for a slightly increased frequency of urination and a diminishing sense of taste, was still youthful. His affection for Jocelyn was stable and deep, and he relied on her for the management of all the station's affairs.

Jocelyn, who had come to the DTR fresh from Smith in 1930, had become scientist, research assistant, translator, station manager, bartender, and social director. In 1947 Smith had given her an honorary master's degree, and she had been elected to the Society of Women Geographers, Blair Niles's elite corps of women whose outstanding professional achievements had led them into wild and forbidding places. Other members at that time included such adventurers as Amelia Earhart (who had married Will's friend and publisher George Putnam), Margaret Mead, Gertrude Emerson Sen, and of course Gloria Hollister. In the emerging field of animal behavior, Jocelyn's work on spiders and crabs was attracting considerable attention. She and Will had been lovers for many years now, and in addition to being friend and colleague to Beebe and hostess to phalanxes of visiting scientists, she was turning out scientific research of superlative quality. She handled staff issues, planned the menus, ordered supplies, and dealt with every tedious duty.

On their return to New York in August 1948, the DTR brought the zoo a coral snake, a great male Hercules beetle, and a giant millipede. All of Jocelyn's precious spiders arrived safely, but overzealous Venezuelan customs agents refused to let many of their specimens leave the country without considerable bribes, which Will stubbornly refused to pay. Despite the paradisiacal cloud forest of Rancho Grande—today still a rough but rewarding birding site—he vowed never to return to Venezuela where, as he saw it, political favoritism allowed petty tyrants to make or break a research project.

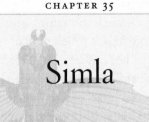

CHAPTER 35

Simla

We must ourselves live among the creatures of the jungle,
and watch them day after day, hoping for the clue as to
the why—the everlasting why of form and color, action
and life.

William Beebe, *Our Search for a Wilderness*

WHEN THE GOVERNMENT
changed as often and as arbitrarily as in Venezuela, Will often said, it
was impossible to keep track of whom to bribe. The book that came out
of the three winters at Rancho Grande, though, was one of his best:
High Jungle. Edwin Way Teale, the respected naturalist who reviewed
it for the *New York Herald Tribune Book Review,* described it as standing
"very high, indeed, on that ridge where literature and natural history
meet."[1] In addition to his chapter on his observations from a chair in
Portachuela Pass, Will brought readers into the world of "marsupial"
frogs, which bear dozens of young from a slit on their backs; of tiny
mouse opossums, jungle wrens, army ants, and leafcutters; of huge bat-
tling Hercules beetles, hummingbirds, and sloths and the living worlds
that colonize their mossy fur. The scientific reports, with their wealth
of tabulated data and experimental results, formed the basis of dozens
of technical papers, most of them published in *Zoologica.*

While Will was in Venezuela, Elswyth was spending more and more
time at the farm in Vermont. She had adopted and reared a baby finch
that had been blown out of its nest, and the slim volume she wrote

about Chewee, *The Bird Who Made Good,* was a surprise success. Chewee lived with Elswyth for several years, riding the carriage of her ever-moving typewriter like a branch in a storm and lovingly grooming her hair. The bird's endearing habits touched many hearts and generated more fan mail than any of her other books. Her Williamsburg series of historical novels had been quite popular as well, and still have a place on many reading lists. Like her novels set in England, they combine accurate historical detail with romantic interest. Beginning with the conflict of loyalties among the residents of colonial Williamsburg, the series follows the descendants of the lead characters on both sides of the Atlantic on and off through the Civil War, the Spanish-American War, and World War I, up to the beginning of World War II.

While she was researching these novels, she would live at an inn for months at a time, leaving Elmer the handyman in charge of the farm. Her mother, whose memory was failing, needed care that Elswyth had neither the time nor the patience to provide, so Florence Wren, a homemaker-companion, moved in and became a beloved family member. Elswyth and Will eventually built her a small house of her own, Wren's Nest, just down Beebe Road from the farmhouse. When Will was in New York Elswyth spent time there too, and he came to Vermont from time to time, writing and watching the birds from his studio above the barn.

At its annual meeting in January 1950, the Zoological Society celebrated Will's fiftieth anniversary with the zoo. During his long tenure

The "marsupial" frog, whose unusual method of giving birth was one focus of study.

he had generated more publicity and produced more scholarly papers than any other employee—possibly any other whole department. That plus his genial humor and unassuming manner earned him the respect and love—if not always the approbation—of all his colleagues. In celebration of his fifty years, testimonials poured in from friends and scientists the world over, and were bound in an illuminated red leather volume complete with illustrations of every phase of his career from all the artists who had worked for him. There were comic strips illustrating Will's favorite tales, like the infamous goldfish story [2] and his comeuppance by a vicious water buffalo. Detailed watercolors showed him in each research site: underwater in a helmet, struggling into the bathysphere's tiny portal, wrestling with an anteater in the river at Kartabo, watching the birth of a volcano in the Galápagos.

A hundred letters chronicled his life, praising his brilliance, his humor, his energy, his unfailing humility in the face of nature. Again and again, the writers attest to the profound impression his life and writings had made on their lives. A note from a prominent geneticist at the New York Aquarium credited Beebe with having made the life of a nat-

Elswyth and Chewee,
"The Bird Who Made Good."

uralist seem "as exciting as the G-man and the movie maker." He wrote that reading *Jungle Peace* and *The Log of the Sun* as a young man had inspired him to ship out on a steamer to see tropical wildlife for himself, and on his return to hang around the bird house at the zoo until Lee Crandall, the curator of birds, gave him a job. Gloria Hollister Anable wrote a reminiscence of the Bermuda days, noting that "through the many intervening years there has been no change in your rare spirit—the enthusiastic spirit of a young naturalist returning from his maiden trip of exploration."

Ernst Mayr of Harvard, dean of the "modern synthesis" linking natural selection with genetics and a curator at the American Museum, wrote of having been inspired by Beebe's jungle books and the grand *Monograph of the Pheasants.* "What could be more exciting than to lead the life that you were leading! With insatiable curiosity you never stopped searching for new worlds to explore." He commends Will for having remained true to the life of a naturalist during the ascendancy of reductionistic biology, when "naturalists were threatened by extinction." "Yours was 'a voice crying in the wilderness' in your insistence on the need for studying animals in their natural surroundings, for studying nature as a whole in order to understand it." He admires Will's transcendent enjoyment of life. "For you there will always be new worlds to conquer."[3]

Other letters spoke of Beebe's rigorous science, his inventive methods, his good nature and abiding friendship. The heartfelt tribute affected Will intensely; despite strep throat, he made a humorous, self-deprecating speech of thanks before sailing the next day for the tropics.

In Trinidad, Jocelyn had found a secluded mountaintop estate that had many of Rancho Grande's advantages without the governmental vagaries that had eroded time and tempers in Venezuela. The small country of Trinidad and Tobago had been colonized first by the Spanish, then the French, and finally the English. African slaves had been imported by the Spanish and French to farm the rich plantations of coffee and cocoa, and when the British outlawed slavery in the mid-1800s, planters brought in thousands of indentured laborers from Malaysia and the East Indies.

In the fifties the people were winning more voice in their own government, but the country was still officially under British rule. Port-of-Spain, its capital, had its Government House, country club, and cricket pitch, as well as its drawling bureaucrats, just like the dozens of colonies Will had visited before. He fit into the hierarchy seamlessly, as he had in colony after colony, his faint Brooklyn accent sounding only a bit exotic and unplaceable to jaded British ears. Always willing to give the Empire its due (and perhaps more), Will credited the British with the stability of Trinidad and Tobago's unique, multifaceted culture—the food, the varied colors and styles of dress, the babble of languages.

Verdant Vale, the estate Jocelyn had found, boasted a house as light and airy as Rancho Grande had been massive. It had been the hill resort of the Siegert family, whose well-guarded recipe for Angostura bitters had made their fortune. A large, low rectangle with the deep overhanging eaves characteristic of Caribbean plantation architecture, the house had wide expanses of windows and French doors that could be closed against rain or heat with wooden shutters. A broad patio looked out over the Arima Valley across lush mountain jungle to their nearest neighbors, the Wrights, at Spring Hill Estate. The patio's fanciful castellations, topped with great jardinieres of exotic plants, recalled the beguiling roof patio at Rancho Grande. Will had reluctantly sold New Nonsuch in 1945 and then used his part of the proceeds to purchase the estate as his gift for the society. He renamed Verdant Vale Simla, after

Simla, Beebe's lab and home from 1949 to 1962.

the lovely and tranquil hill resort in India that in Kipling, and in his own recollections, symbolized ultimate peace. Simla would be for Will the field station and home he had always wanted, the jungle wilderness he had been searching for from his earliest expeditions with Blair. With Jocelyn and Henry to manage the distracting details of the household, the seventy-four-year-old Will set to work exploring its trails and planning new projects with all the enthusiasm of youth.

Beautiful as it was, Simla needed considerable renovation. Henry cleaned and rebuilt the water system, which involved a spring, a cistern, and two temperamental water tanks. The first nights were illuminated solely by candles and fireflies, which gave way to kerosene and gasoline lamps, effective but high-maintenance. Finally the generator, donated by departing army troops at Waller Field, was coaxed into life. It was continually breaking down, but only electricity could power the refrigeration so necessary to the preservation of specimens—and to the inviolable cocktail hour.

As a study site, Simla had less variety than Rancho Grande with its fourteen faunal zones. There was no Portachuela Pass with its endless flow of interesting migrants, but the luxuriant forests provided a constant influx of small mammals, large snakes, and birds for the zoo. Bromeliads in the canopy held entire communities of animals that had never been recorded; the heliconid butterflies and salticid spiders that Jocelyn was studying were around, if not abundant, and colonies of fiddlers were just across the mountains on the shores of the tiny village of Blanchisseuse. Howler monkeys and several varieties of owls joined a dissonant chorus of frogs to enliven the nights. When glassed in, the wide veranda made an excellent laboratory and observation point, sheltered from the unpredictable winds that could whip through, scattering papers, feathers, and moth wings.

Despite its isolation, Simla soon began to attract a steady stream of young "Good Afternoons," the procession of polite local children that had always been a source of interesting creatures. They materialized silently from the jungle, an occasional rare moth or velvety peripatus in their fists, though more often it was a common beetle or a long-dead nestling of a species Will longed to see alive and well. Remembering himself at their age, bringing his treasures to the American Museum, he loved to show the children Simla's creatures. Workers on neighboring plantations, too, would walk miles to bring interesting things to Simla;

Will always paid well if the specimens were alive and healthy. One of his young visitors, Jogie Ramlal, arrived bearing a beautiful live boa. Impressed, Will hired him as an assistant.

Social life at Simla was surprisingly rich. Port-of-Spain had the advantage of Imperial College, a primarily agricultural school whose faculty was passionate about conserving Trinidad's environment, and welcomed Will and his team. And Trinidad, like so many of the British colonies, had long hosted a thriving amateur naturalists' club, made up of British and Creole planters who kept sharp eyes open as they surveyed their holdings, nature-loving East Indian shopkeepers, and government officials, augmented by ornithologists and entomologists from the college.

Ever alert to his self-imposed duty to be an ambassador of the United States and particularly of the Zoological Society, as well as a voice for conservation, Will lectured tirelessly to meetings of the naturalists' club and their wives and children, to bored assemblages at the Officers Club, to riotous cocktail parties at the homes of wealthy planters, and to staid garden parties at Government House. All these audiences enjoyed the film footage Jocelyn had amassed of their various travels and the creatures they studied. The monkey tasting team never failed to draw applause, but so did the time-lapse study of a but-

Simla lab.

terfly emerging from its chrysalis, and the owlet and the porcupine snuggling together for protection from the rain in their outdoor cage.

When the Rockefeller Foundation established a research station on Trinidad to study malaria and other human infections, those scientists visited Simla as well. Beebe's reputation as a clever scientist and genial host also attracted tourists, to whom Will was almost invariably kind in the name of the Zoological Society—though he and Jocelyn often skewered them mercilessly when they left. Will liked to use the garden statues Milne had given him of the characters from *Winnie-the-Pooh* as a sort of litmus test. If guests greeted Pooh & Co. with enthusiasm, they would be treated royally. If not, Will just endured them until they left. Old friends dropped by as well—some to stay and help for a week or two, and others, like Walt Disney and his wife, to reminisce about the old Hollywood days over rum punches.

The Wrights, at neighboring Spring Hill Estate, lived close to their land. Both amateur naturalists, they boarded many of the researchers who came to work in the Arima Valley's lush environment. Henry Newcome Wright, a British lawyer, and his Icelandic wife, Asa, had left England for Newcome's health (he had been gassed during World War I) and were trying their hand at making Spring Hill a self-sustaining coffee, citrus, and cocoa plantation. They welcomed the DTR as a positive force in Trinidad, and embraced Will, Jocelyn, and Henry enthusiasti-

Simla.

cally as neighbors and friends. It was critical to keep in touch with neighbors, since at that time there was no telephone service in the Arima Valley. There was an unpaved "main road" at the bottom of the mountain along the river, but each estate sat on its own hillside, up narrow, twisting roads that clung to sheer cliffs on one side and skirted nearly vertical drop-offs on the other. Through Satterlee's great binoculars on the patio they could see the Wrights' outbuildings, but Spring Hill was four tortuous miles away by road.

Asa's energy and strong views on every subject endeared her to Will, who thought her magnificent. Although most people found her strident manner and loud voice grating, her roughness disguised an affectionate nature, at least where animals were concerned. Asa thought nothing of forgoing social engagements to stay with sick or farrowing pigs, and any hapless worker who knocked a nest out of a tree felt her wrath. While Newcome took care of managing the estate and handling the cash crops, Asa cared for the animals. She raised pigs for market, and usually dressed and often smelled the part. She pelted around the treacherous mountain roads in her truck, bullying the workers and upbraiding anyone who crossed her, often with a sick or injured piglet under each arm.

Will loved the vitality of Spring Hill. On one of his early visits, he described it as "reeking with manure, pigs, burros, goats and chickens. A beautiful peacock perches on top of a peccary's cage, both animal and cage filthy. . . . Yet Asa in her pig apparel is as wonderful as ever." When Asa did attend a dinner party, she appeared in an elegant evening gown she had made herself out of her collection of antique lace, which she loved for its delicacy and beauty, much as she loved butterflies and birds.

The first real expedition to Simla, in 1950, was the DTR's forty-eighth expedition. Will laid out an ambitious research plan that stressed his consuming interest: to study how animals are adapted to successful living in their natural environment and with one another. Social behavior was at the forefront of his mind, an unknown frontier that Will felt was critical to understanding ecology. With Will and Jocelyn on that first trip were entomologist and by now mechanical wizard Henry Fleming as well as Laura Schlageter, a Venezuelan who had painted for them at Rancho Grande. Alcoa, whose huge steamers loaded their holds with pink Venezuelan bauxite to turn into aluminum, gave the party transport on one of its vessels, which took them hopping from island to island in the way Will loved. Thanks to Dramamine, a recent development, "the girls," according to Will, for the first time "saw and enjoyed the following waves from a vertical position." They visited old friends at practically every port, and knew every hidden beach for a good swim and every secluded mud flat where Jocelyn could hunt crabs. The three weeks it took to reach Port-of-Spain from New York were the closest to a vacation that Will ever came.

Once Simla was up and running with Ethel, the Trinidadian cook, and Julia, the housemaid, in place with Ramdial, the caretaker, Will began exploring the property. It was proper rain forest, its high canopy dense with epiphytes and steamy undergrowth beneath. Much of the land had been planted with cocoa, and the orange flamboyant and red immortelle trees planted to shade the young cocoa plants had outpaced almost everything else, their bright scarlet-orange blossoms drowning the valley in the tropical spring, followed by the gold or pink of the poui trees. Ram hacked trails along the narrow ledges, but boulders and sudden drops made walking them rough. Almost immediately Will spotted a tree hung with the pendulous nests of caciques, large noisy oriole relatives whose busy communal life he always liked to observe and record. He heard the calls of bellbirds, an elusive species whose nesting sites and life history remained mysterious. In fact, as they were still unpacking, one of these rare birds crash-landed at a window; without refrigeration, Will had to dissect it at once with his penknife for later study, and store it in a vial of his best rum.

He came upon the dancing arenas of black-and-white manakins, birds whose males divide an area into plots according to rank. They keep their own particular area meticulously clean and perform a courtship dance to attract females. Fourteen species of hummingbird made Simla their home; a mango hummingbird built her tiny nest from lichens and spiderweb just outside his bedroom window, so he could observe the whole life cycle of the jewel-like creatures.

When phalanxes of army ants swept through occasionally, the researchers either fought them fiercely or let them go about their business, leaving the house and lab free of insect pests, depending on whether or not they could secure their specimens and pets from the hordes. Legs of tables and beds were set into cans and jars filled with kerosene, and cages and specimen cases were surrounded with insect tape. After the army passed, carrying with it every insect and rodent that was not able to escape, the compound relaxed, purified and purged.

The research program Will envisioned for Simla was characteristically ambitious. Of primary importance would be the exhaustive ecological profile of the area, of the sort that he had worked out and published for each of his study sites. As with Rancho Grande, he would map the station's climate, rainfall, temperature range, and humidity; its soil and its rock; its topography; and its flora and fauna, from the microscopic to the largest mammals. This mammoth undertaking formed the background of his research, but he was never able to resist following his observations wherever they took him.

His lifelong fascination with animal mimicry made him fascinated by the ctenuchids, chameleons of the moth world, which imitate other creatures so perfectly that many were confused with their alter egos, even by expert taxonomists. These are mostly day-flying moths, often with transparent wings. Some look like wasps, thus protecting themselves from attack by many wasp-fearing predators; some appear moth-like, but when at rest on tree trunks assume strange postures that turn them into leaves or twigs. Every night brought new insects of every sort to the lighted screen, and every night someone made periodic checks to see what was there, to save interesting specimens from the unselective toads who waited, open-mouthed, along the bottom.

A wasp-mimicking fly (left) and moth (center), the inspiration for a study of the function of mimicry.

Will had always been interested in the behavior of the creatures he watched. The way an animal interacted with its habitat, its conspecifics, predators, or prey might be as intricately interwoven with its evolution as its coloration or chemical makeup. Jocelyn's interest focused on the information animals picked up visually about the world, and she launched immediately into the question of what and how much butterflies, with their eye-catching colors and patterns, actually see. If, as lepidopterists had always thought, they were blind to red, why did the local black-and-white heliconids flash bright spots of it on their wings? How did the various nearly identical species of these butterflies tell each other apart for mating?

Henry and a crew of local helpers built a large screened house where they could raise the plants that each species of butterfly and moth used for feeding and laying eggs. Despite generations of eager collectors and greedy poachers, few of the colorful species that brightened the dark tropical jungles were well understood; Jocelyn and Will had to discover for themselves what each species fed on as adults, and which plants provided food for the larvae that hatched from their eggs. Most species will lay eggs only on the particular food plants their caterpillars will

Manicomio, the butterfly house.

need, so when the food plants were known, they had to be located, collected, and coaxed into thriving in the insect house.

The Insectary—which bore a haphazard sign, "Flutter Inn"—was successful from the start. In addition to the heliconids that were Jocelyn's special interest, other species were attracted by the lush flowering plants and vines. Will never met a bug he didn't have questions about, so the Flutter Inn soon had to be joined by another structure, Manicomio ("madhouse" in Spanish).

The latest in the string of Chiris were two capuchin monkeys, Chiripie and Chiribrandy, who formed the heart of the tasting team. The Chiris were tame enough that when they escaped—which was often—they hung about the camp looking for handouts until they were caught. They could have had their freedom if they hadn't raised hell whenever they were out, plucking insects from the work tables, stripping all the orchids from the trees, stealing panties off the line, and raiding every carefully watched bird's nest within sight.

Other pets, transparently disguised as valuable specimens en route to the zoo in New York, began to fill the cages under the accommodating eaves. "The chorus of our caged animals is weird," Will wrote. "The owl Wowllls! The porcupine wails like a lost soul, the rat scampers to get away from the boa, who shows no interest in him, and something knocks, knocks, all night long. Fireflies flash in my room and a vampire hovers around my net. Then at 5:15 the first wren cheerios the almost day." (6.3.1950) Will was up with the wren, eager to begin the work that he had said long ago was better than being a king.

CHAPTER 36

Home to Roost

*I marvel that men can spend whole lives in studying the
life of the planet, watching its creatures run the gamut
from love to hate, bravery to fear, success to failure, life to
death, and not at least be greatly moved by the extremes
possible to our own existences.*

William Beebe, *Pheasant Jungles*

ALMOST IMMEDIATELY, WILL
realized that despite its perfections, Simla's twenty-two acres did not
provide enough scope for research. When the neighboring estate of St.
Pat's came on the market, he agreed to go halves with Frank Storms, a
genial oilman from Caripito who wanted a retreat on Trinidad, to save
the land from developers. The previous owner had allowed Will free
access to all of St. Pat's untouched forest, and he lusted after its huge
saman trees with their immense spread, which housed multiple inter-
locking communities of flora and fauna. And St. Pat's extensive forested
acres, lower on the hillside and nearer the river, were much flatter and
more accessible than Simla's.

The partnership faltered when Storms brought in cattle that fouled
the streams, and as soon as he could afford it Will bit the bullet and
bought Storms out, becoming the laird of St. Pat's 170 acres as well as of
Simla. Elswyth, back home in the Vermont farmhouse with only her
querulous mother, the caregiver Florence Wren, and Elmer the handyman for company, was quietly furious with this expenditure. She and
Will had long ago determined that they would keep their finances

separate, in an unofficial prenuptial agreement designed to keep her pride intact. But she had long felt that despite their warm mutual affection, despite their rational "modern" decision to keep their professional lives separate, he was having more fun. For their entire life together Will had been putting his zoo salary into his research, which she had known would be his practice from the start. But his decision to use his money to purchase St. Pat's, which he openly declared would be his gift to the zoo, rankled despite their agreement. With the difference in their ages, he was almost certain to predecease her; and although by her own estimation she lived simply, and her many books were selling briskly, between her mother and the farm she felt herself hemmed in by care. She couldn't help believing that more money would make her life better.

There is no evidence that she was sexually jealous of Jocelyn. As seemed to have been the case with Gloria, perhaps she didn't suspect the depth of their relationship; or perhaps pride prevented her from acknowledging it. She seems to have preferred the romance of imagination to physical passion, and may have been glad to have Will's needs taken care of elsewhere. But it is hard to believe that she truly didn't

Beebe at eighty, using tree roots to scale a boulder.

care, at least at some level. Will's writing always made it seem as if he and his team were having a wonderful time; Elswyth must have been quizzed unremittingly by acquaintances about why she was in frozen Vermont while he was living his ideal life in the tropics.

By the time Beebe founded Simla in 1949 he was seventy-two, and the Zoological Society had been extending his tenure as director for an unprecedented time. He was still active in research, and his name was still good for publicity. His donation of the substantial Simla and St. Pat's property, ceded to the Zoological Society for $1 in 1951, made him one of the society's elect "Benefactors in Perpetuity." But his continued presence on the list of active staff was a dangerous precedent, and it made every other staff member's retirement harder. After the fiftieth anniversary, even Will knew there had to be a change. Quiet negotiations were under way, and on his seventy-fifth birthday in 1952, his title became director emeritus of the Department of Tropical Research. Jocelyn was promoted to assistant director and received a $500 raise.

Years earlier, millionaire businessman Harrison Williams had taken over the management of the financially feckless Beebe. Williams took $6,000 of his friend's lecturing proceeds and invested the money conservatively; Will added to it gradually, and for several years had been receiving a $200 check every month from the account. In 1952 Williams took him to lunch and dropped the bombshell that in fact, these small investments now amounted to almost $115,000—a considerable sum in those days—and were yielding $6,000 a year, which Williams had been reinvesting. He gave Beebe hell for not having made a will and sent him to a prominent firm of lawyers.

Will's first response to unexpected solvency was panic. How could he possibly deal with so much money? He had seen so many fortunes come and go—what if he lost it all? There had never been enough— there would never be enough—to finance all the expeditions he still wanted to make.

He had always felt himself at a disadvantage among the men of vast wealth he was continually thrown together with—not just the old-moneyed aristocracy of British and American society, but the new-minted millionaires like the pants presser turned utilities baron and pheasant fancier Kuser, or the railroad magnates like Williams, or the land moguls like Wilshire. He had in fact traded on his real-life diffident-boy-from-Brooklyn image when it served him. As the product of

Nettie's "economy equals virtue" ideal, and Emerson and Thoreau's conviction that the well-regulated life requires few amenities, Will had never envied the soft life. He had seen in men like Crocker and Barton the unhappiness too much privilege can bring, and begrudged himself the few luxuries he indulged in.

But here—despite Simla, and St. Pat's, despite the duplex at 33 West 67th and the studio in Hotel des Artistes and his share of the farm in Vermont—here was a grand sufficiency, a nest egg for an ornithologist, and he was determined to keep it safe from predators. He begged his friend Hugh Bullock, a pioneer in the mutual fund industry, for help. "As I understand it (and very vaguely) the money is now in safe keeping, in my own bank; and they collect dividends and deposit them to my account. Shall I leave them as they are, paying no attention to selling or reinvestment? (Of which I am as ignorant as an unborn babe). . . . Whatever business arrangements can be made will be an immense load off my mind. I feel very frightened and exposed with all this money at my disposal. I want some one else to take it away from me!"[1] He fled back to Simla, hungry for the tropics and home. The salesman's son from Brooklyn had weathered the Depression and had enough to secure the life he loved. He could afford to be retired on half pay.

Will's "retirement" didn't faze Simla. He and Jocelyn lived there as an old married couple, with the requisite separate bedrooms but a deepening love mellowed by age and custom. Will pored over moth patterns and bird syringes, wrote endlessly, and kept up his journal. He wrote admiringly of Jocelyn, fondly of Asa and Newcome, and despairingly over Henry, who bit his nails and smoked and still kept his specimens in disarray. And always, he wrote about the new worlds that could be discovered by a simple shift of perspective. Again and again he found himself being caught up in an anthropocentric universe, excited and humbled by a sudden perception of the layers and layers of life existing around him.

Will Beebe's greatest gift—his ability to see through culture's thin and sterile veneer to the astonishing world beneath—sprang from the childish sense of wonder that had never left him. The charm of all his books comes from his irrepressible excitement at the luxuriance, the unexpected beauty, and most of all the cleverness of nature. He embraced the aspects of order and civilization that facilitated his escape into a realm stranger than any fantasy—nature in all its feral extravagance and alien enchantment. Every pin may have pointed the same way in his New York world, as Elswyth saw it, but in his other life, intemperate nature seduced him utterly.

In 1952 Beebe published *Unseen Life of New York,* a charming book that brought the excitement of natural things to the ultra-civilization of New York City. It brought to light creatures that go unseen because they are too small (such as hydras and flatworms), too ancient (the mastodon whose tusks were unearthed near Broadway; the woolly mammoth, ground sloth, and arctic ox that once wandered what would be busy streets and massive skyscrapers), too deep (stalk-eyed dragonfish and fifty-four other species of deep-sea fish from the Hudson Gorge), too dark (bats, moths, opossums, fireflies), too well disguised (crab spiders, walking sticks), too transparent (arrow worms, jellyfish, eel larvae), or even too familiar (cockroaches, English sparrows, red jungle fowl). The humorous drawings made it clear that this was no textbook, but there was an underlying seriousness of purpose beneath the whimsy. Will felt deeply that a simple shift in perspective, whether occasioned via a telescope, a microscope, or the imagination, would have a sanative humbling effect on humans and would in addition eliminate what he saw as one of the worst forms of immorality—boredom.

Except for the particular geology of the New York City area, the stories Will told could be of any city on the continent. He pulls his readers back to look at the city from a space which was just beginning to be explored: "Ninety-two million, nine hundred thousand miles south (or north, east, or west) of the sun, there floats loose in space a pinprick of a planet, which is only one and one third millionth the volume of the sun. On a certain spot of this planet earth is, astronomically speaking, a fraction of a grain of sand which is known as New York City." New York's geologic history shows it careening up and down, "tossed about, raised high in air, lowered until it was at the bottom of a mighty sea; it has more than once been hidden beneath a half mile of solid ice."[2] Excavations reveal the sand and gravel deposited by the retreating glaciers of past ages; the solid bedrock was formed in "the very dawn of the world. . . . Look at it as it crops out here and there in Central Park—quiet, gray, patient—and our individual worries will seem of less importance."[3]

In this world, pollen is not a plague but a marvel, infinitely varied, crystalline but with the potential of life. If only we had ears to hear, the explosions of windborne pollen organs would be a glorious cacophony as startlingly lovely as the grains themselves. The life-and-death struggles going on each second in a bit of pond scum become, under Will's lens, an epic battle.

> Bright moons and stars of diatoms edge themselves along; slender, blue-green ribbons of algae wind about; orange yeast cells just rest and grow; more algae with bright green cells, round spheres rolling gently here and there, barbed-wire tangles of fungus litter the microscopial field. And all about and through the scum move the animal attackers, slipper cells which are super-tanks, for they can turn and twist as they roll onward, forever slaying and sucking in unfortunate cells . . . amoebas quietly but terribly engulfing what comes in their way. [4]

The magic of the lens transformed New York as it had Will's life, and he was as evangelical about its use as he had been about the diving helmet. "No human has fulfilled his manifest destiny of joy and awe in this life if his eyes have never looked through telescope or microscope." [5]

Will Beebe in his Bronx Zoo lab.

The lenses of the ungainly twenty-two-inch binoculars were more or less permanently set up on a tripod in the living room at Simla, pointing either to the heavens or out across the patio toward the jungle-clad hills, where great toucans nested and flocks of seven-color parakeets wheeled, squawking noisily. Will claimed he could tell unerringly if a candidate would be a good coworker by his reaction to the sight of the moon through them. They had been joined by a powerful monocular from a visiting board member, along with a large pair of binoculars slightly more portable than the massive 12/20/40 x 80s. Student volunteers came and went, thrilled to be working with William Beebe but often restless in their mountain isolation. Professor Ellen Ordway credits Beebe with firing her with the determination to become a biologist, but remembers sneaking out at night with Henry and John Cody, another college student who was an artist, to party with neighbors.

Will's dream of a field station that would be a magnet for biologists of all stamps was being realized with a vengeance. Hardly a week went by at Simla that there were not several resident scientists working on assorted problems—at once a blessing and a curse. There were terrific discussions over cocktails, sometimes lasting long into the night, but also terrific disagreements on everything from evolution to tapioca and the recipe for rum swizzles.

Amazingly, Jocelyn was able to manage the ménage and pursue her own interests brilliantly. Her work on the Venezuelan jumping spiders was published to great acclaim—her command of the physiology of the spiders, their relation to their habitat, and the rigorous evaluation of their gestural communication made the study a classic in the nascent field of ethology. Ethology demands at once an understanding of the biology of the creature and of its environment in order to evaluate its behavior. Watching an animal in the sterile environment of a lab is useless, ethologists believe; but its actions in the niche it has evolved to fill,

Signals used by jumping spiders for communication.

and the social system its instinct adapts it to, will tell you everything important about behavior.

Jocelyn had proved that the small, furry salticids have a signaling system involving a repertoire of gestures, and that the minute differences between these gestures are what enable one species to mate only with others of its ilk. The work appeared in five issues of *Zoologica* from 1948 to 1951, at the same time that she was managing Rancho Grande, finding and setting up Simla, and wrangling her ailing and alcoholic mother, Estelle. The articles, which together amounted to a monograph of more than 100 pages, were models of thorough background research, painstaking analysis of the spiders' physical features—she found thirty-six subfamilies—and patient observation. The theories she derived from her observations were both groundbreaking and stunningly obvious to anyone who had watched as carefully as she had. She showed that the waving display of the males differed from species to species, and that the specific patterns released mating behavior in conspecific females.

The salticid communication system inspired her to look more closely at the comical fiddler crabs that she and Will had been watching on every muddy expanse, in every mangrove swamp they had steamed

*Fiddler crab
communication signals.*

past or slogged through for decades. It seemed like high time to test Darwin's idea that "handicaps" such as the male fiddler's claw or the male peacock's eye-catching tail were such liabilities in terms of camouflage that they must serve a function in attracting females.

Jocelyn's first paper on crabs had come out of the Crocker expedition, and she made herself familiar with every aspect of crab physiology. A real interest in the awkward but fast and oddly graceful crustaceans carried her through a study of the literature to real expertise. Her first article on the fiddlers was a nontechnical one in the Zoological Society's *Bulletin* in 1941, "Fiddler's Carnival." [6] It was quickly followed by technical papers on fiddlers — the genus *Uca* — from the west coast of Central America, from Venezuela, and from the northeast coast of the United States. On some beaches she could watch several different species of males signaling simultaneously while the females chose the correct species based on differences in tempo and patterns of the callers. For decades naturalists and biologists had been observing local populations and describing the strange waving patterns. But no one lucky enough to travel extensively had collected data comparing one with another. Jocelyn's work made it clear that a rigorous comparative study might reveal widespread evolutionary patterns. [7]

Until the fifties, funding for science had been largely private. Universities, foundations, or private individuals bore the cost of pursuing research goals, which in the natural sciences could be maddeningly elusive. Beebe had been unprecedentedly lucky in finding support, mostly from wealthy businessmen like Harrison Williams or Laurance Rockefeller. But he had always supplemented the DTR's money from his own pocket, and Simla was entirely his investment. The Zoological Society was having to spread its largesse quite thin because of all the competing needs of an overpopulated, poorly understood planet. Conservation of dwindling resources had become a consuming aim both at home, with captive breeding programs to increase the numbers of rare animals, and abroad, with ambitious projects such as saving California's redwoods, the bison of the Great Plains, and the bird of paradise on Tobago.

After World War II, the important role of science in health, industry, and defense became evident. It was now clear that basic research, projects that might seem irrelevant but that might conceivably have far-reaching effects, was critical to a nation's physical and economic health. In May 1950 Harry Truman signed the National Science Foundation into being, acknowledging the proposition that government sponsorship of science was crucial to the survival of the nation.

The NSF's original budget was laughably small by today's standards: $151,000. It rapidly rose to $3.5 million, and by 1957 stood at $500 million. But Jocelyn's work stood out as research with such concrete and achievable goals, and such broad implications for other fields, that she was one of the NSF's first grant recipients: in December 1954 she received word that she had been awarded a four-year, $18,000 grant to study the fiddler crabs of the Far East. She shocked everyone at the Simla dinner table by bursting into tears.

She had been corresponding wistfully for several years with Michael Tweedie, director of the Raffles Museum in Singapore, about the crabs of the Far East, always ending with the hope of one day coming to study them. When the opportunity came Tweedie was enthusiastic, expressing the hope that Dr. Beebe would come too and give some lectures. Will, for his part, was eager for the trip. He wanted to revisit some of the people and places he had known from the pheasant expedition, and to check on the status of some of the species he had studied then, whose survival had been in question. His health was beginning to fail him, however. He had always been easy prey for respiratory ailments, and some had been life threatening. But most annoying was a gradually worsening condition of the mouth and throat that doctors diagnosed as allergy or "chronic dry pharyngitis."

His symptoms, as he describes them in his journals, sound much like those of Sjogren's syndrome, a type of autoimmune disorder that causes salivary glands to dry up. For Will, it was debilitating. Some days he would be fine, but more and more his mouth would become so swollen and dry that he could barely talk. As the years passed, his voice was often thick, as if he had had a stroke. This, for the garrulous Dr. Beebe, was a horrible thing. And unfortunately for him, doctors diagnosed it as an allergic reaction to "something" in his diet (they referred to it as "mango mouth") and prescribed Dexedrine—a drug whose side effects are now known to include dry mouth. In addition, it is widely used as a dieting aid—something the reed-thin Will needed like stomach flu. He also suffered from severe eczema on his thin legs, which some doctors had him coat with salve to keep them moist while others prescribed calamine or other drying agents. Nothing seemed to help; the eczema came and went mysteriously, on its own malignant schedule.

Because of his mango mouth, Will had become hesitant to accept speaking engagements. He never knew in advance if his voice, always quiet, would become thick and his speech slurred. But his talks were copiously illustrated with motion pictures, and these, together with Jocelyn there to carry the ball when his voice failed, usually got him through. So he accepted Tweedie's invitation and a grant from the Explorers Club (of which he had been an honorary member since 1912), and he and Jocelyn sailed for Europe in May 1955 on the *Andrea Doria*.

Singapore was a paradise of fiddlers. They found three new genera on one peninsula, and the Padang River banks held thousands of crabs, all waving. They filmed each species as it went through its courtship ritual, and sometimes were lucky enough to witness fights between males, which showed that the large claws were not useless for that purpose. They also collected specimens to study and dissect, so that Jocelyn could document each species and get exact measurements of the claws.

The questions they were trying to answer were many, and Jocelyn was nothing if not thorough: over the course of the trip, they collected data and specimens from Pakistan, Penang, Borneo, and Ceylon. She accumulated notebooks crammed with data, 3,400 feet of motion picture film, and 1,500 specimens, carefully preserved in formalin. She and Will reveled in the exotic atmosphere, buying specimen jars and formalin at open-air shops that also sold joss sticks and powdered rhinoceros horn, and where costs were tallied on an abacus. When Jocelyn's crab-digging trowel broke, she had one forged to order by a Singapore smith on the street while they watched.

Will helped her when she needed an extra pair of hands, but spent most of his time making notes on what he saw, rummaging through the bird collections of the scattered museums, and talking to ornithologists about the status of Asian pheasant populations. Still fascinated by the elusive evolutionary origins of flight, he collected specimens of *Draco volans,* a stunning little flying lizard, and several gliding geckos to send back to the zoo, along with two flying snakes (*Chrysopelea ornato*). He was also intrigued by the natural history of a weedy type of heliotrope native to the area, whose wilted or even dried leaves were a potent magnet for some species of butterflies and moths, particularly the day-flying moths that he found so interesting. He had been studying the heliotrope connection since discovering its properties in Venezuela, and hoped that he could discover the origins of the mysterious attraction by observing it where it had evolved along with the moths. The system remained mysterious to Will but is now known to provide the butterflies and moths with alkaloids that make them unpalatable to many predators and even, in some species, act as a sexual attractant.

After two months they had collected thirty-two species of fiddlers, of which two-thirds had been displaying, so Jocelyn could correlate display with specimen. Years more study would eventually generate her mammoth monograph *Fiddler Crabs of the World: Ocypodidae, Genus Uca,*[8] which stands as the definitive analysis of fiddler crab morphology and behavior.

CHAPTER 37

Simla Sunset

Tropical midges of sorts live less than a day—sequoias have felt
their sap quicken at the warmth of fifteen hundred springs.
Somewhere in between these extremes, we open our eyes, look
about us for a time and close them again....

William Beebe, Edge of the Jungle

O NE OF WILL'S FONDEST
desires for his field station was that it become, as he had always
dreamed, a research station for biologists of every stamp. He extended
invitations to scientists studying jungle creatures, plants, or ecology
to visit and study their subjects on the hoof. Among the first to take
advantage of Simla's potential was Ted Schneirla, whose work on army
ants was already world class. The exuberant German researcher Konrad
Lorenz, one of the first to call himself an ethologist, came more than
once to observe and to exchange ideas with Beebe. Like Will, Lorenz
firmly believed that to understand evolution, scientists must know in
detail how animals behave in their native habitats, paying attention to
the experiments that the natural world was constantly making in the
laboratory that is the wilderness. In 1973 Lorenz would win the Nobel
Prize for his groundbreaking work in this field, along with Dutch researcher Niko Tinbergen and Austrian zoologist Karl von Frisch. When
Lorenz and his wife came to Simla in January 1955, his ebullient personality energized everyone as he fished for guppies in the streams and
stalked the difficult, bug-ridden trails clad in his brief European bath-

ing suit. He and Jocelyn and Will talked long into the nights about the need for rigorous scientific methods for evaluating animal behavior.

David Snow, a young ornithologist from Oxford, was also interested in animal behavior. He first came in 1956 to observe the manakins Will had found on Trinidad. Their odd social system, called a *lek,* was just beginning to be studied. In a lek, males compete for females in an odd way: large groups of males, often between fifty and a hundred, gather in a small area to display en masse to females. Attracted by the activity, females arrive and generally mate with the male in the middle of the displaying birds, who is the dominant bird. Males on the periphery of the display get few matings, but may eventually work their way up in the hierarchy and sire offspring of their own. Among the manakins, which are small, active birds, males may clear a dancing arena on the ground of every stick and leaf, or strip the leaves off several adjoining branches and display in the trees, uttering sharp buzzing calls and hopping from branch to branch in a stereotyped sequence; some species also make loud clicks with their wings.

Snow also studied the secretive, cave-dwelling oilbirds, or guacheros, that had long interested Will. Guacheros had established a small colony in a deep chasm on the Wrights' property at Spring Hill. Its entrance, overhanging a swift-running river and covered with a dense growth of air plants, ferns, and orchids, had kept the little population from total extinction, but poachers were pushing it dangerously close throughout its range. As in Venezuela, the practice of gathering the chicks from the nests had grown until there were only a few left in protected or inaccessible caves. The small number that remained had never been studied, and their life habits were a matter of wildly improbable legend. In 1957, the eighty-year-old Beebe hiked to the cave with Snow for the first time in several years. Determined to peer into the nests, he inched slothlike up a slippery tree trunk to look in while the adults wheeled around him, uttering their startling, raucous cries and clicking like typewriters as they gauged their distance with onboard sonar. Will was elated to see the healthy colony, and pleased that his limbs were still strong enough to support him.

Snow returned to Simla for several months each year as resident naturalist, a post Will had long wanted to establish to extend Simla's appeal to researchers. With someone always in residence, the property could be maintained and the facilities kept up and running year-round. Will received an NSF grant in 1958 that allowed Snow to continue his studies of the manakins.[1]

With Henry, Will was still pursuing the cryptic day-flying moths. He was fascinated by their myriad forms, which could mimic wildly different insects or substances, usually something dangerous or distasteful such as wasps or bird feces. The caterpillars of these moths proved to be equally expert in designing elaborate defense systems, and Will would stay up all night to tease out their methods. He illuminated the caterpillars with red light, which they are unable to see, so they would behave normally. One bristly, unprepossessing caterpillar plucked out its hairs one by one and lashed them to its twig with sticky silk, eventually forming a barbed-wire series of whorls behind it and in front,

> denuding its own body of structures and turning them into defense stockades, working cunningly with silk and glue, estimating distance and elevation, stress and strain . . . all directed by an inherited unreasonable knowhow.[2]

Another moth built a spiny protective palisade around its eggs, which the newly hatched larvae could escape only by climbing over the

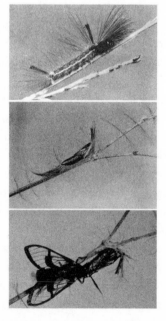

The Colobura dirce *caterpillar, whose defenses Beebe studied; its chrysalis, with protective barbs. It metamorphoses into a clearwinged day-flying moth.*

spiky edge on a matting of silk. Each of these unlikely larvae emerged as a delicate moth with transparent patterned wings and a colorful body designed to elude or deceive enemies. Will watched the caterpillar of *Colobura dirce,* the zebra-striped Dirce Beauty butterfly, take its own fecal pellets as it sat on a branch and mold them into a look-alike twig, resting stretched out on the "twig" tip like a rigid woody extension.

In 1956, the National Geographic Society awarded Will and Jocelyn $3,500 for their work with the colorful heliconid butterflies. No one knew better than Will that taking money from the Geographic Society was cutting a deal with the devil. It entailed writing a long article in lucid and engaging prose, which would be accompanied by wonderful photos that looked effortless but required catering to photographers for weeks, slowing research or speeding it up to fit their schedule, posing for hours as the precious daylight ebbed, taking a day's results with it. But the publicity the articles generated both for a research project and for the zoo was worth the disruption. With the grant money, Jocelyn built more flying cages so that she could fill the needs of more species of butterflies.

Lincoln Brower, who had pioneered the study of chemical toxins in monarch butterflies, came to Simla to work on Jocelyn's heliconids. With his wife, chemist Jane van Zandt Brower—and the helpful Simla animal tasting team, which now included lizards, mantids, and birds— he isolated the plant-produced chemicals that make the brightly patterned heliconid butterflies distasteful to predators. The genetic relationships that Jocelyn had established proved to correspond with the level of toxicity, and her work on the way their colors and patterns are perceived by predators and by other butterflies illuminated the complex interrelatedness of color, chemistry, defense, social systems, and selection. Visiting researchers remember being treated to hors d'oeuvres garnished with bright edible butterflies, and being warned that one of them would be bitter—the toxic original of one of the tasty mimics. It paid to be a butterfly specialist, or at least to have read *National Geographic Magazine.*

With her grant money, Jocelyn was also able to make a research trip to study crabs in the South Pacific in 1956, along with a trip to Africa the following year. Will was unable to go, as his health was uncertain. In addition to his mango mouth, he had suffered from constipation all his life—perhaps resulting from Nettie's overzealous administration of bowel-cleansing stimulants—and it had resolved into a chronic irritable bowel condition that was often debilitating. Will spent much of the time Jocelyn was away in Vermont with Elswyth and Mrs. Thane, who was becoming increasingly difficult. Will and Elmer LaFlamme, Elswyth's handyman, enjoyed each other's company greatly.

Since they were so seldom apart, correspondence between Jocelyn and Will is rare—and what does exist has been heavily edited by Jocelyn.[3] But during these separations they wrote often. "I feel rich in memories," Will wrote from Vermont to Jocelyn in Manila, "for I wake in the morning and remember your voice and then go into the room where I work, the nice barn room, and reread your letters, and am happy. . . . You are so wonderful that I thank heaven that you began to like me while I was still in my 'prime!' Please let it last. . . . I love Manila because it is the start home and Oh! How I want you. All my love, Me." (9.6.1956) "You dearest person," she writes, "Golly how I miss you. It's awful. I'll go along fine, having a grand time and doing a job and for hours only conscious of you in a mildly hungry, unfinished sort of way and then suddenly there'll be a bit of music out a window . . . or without warning start wondering what you're doing and then I think I just can't stick it away from you any longer. Even mangroves set me off because I know you don't like 'em!! Anyway pleeeeese don't send me off again without you. . . . I won't go so you'd better not try." (from Fiji, 8.2.1956) "Dearest person—I do miss and love you so tremendously—Goodness how I want to see you—All my love, J." (from Australia, 9.11.1956)

They also discussed the financing of an apartment for Jocelyn in the Hotel des Artistes that had become available while she was away. Will was helping her to lease it, and he eventually passed the ownership on to her. For propriety's sake, however, nothing was ever acknowledged publicly about their relationship, and Will at least seems to have felt that their attachment remained secret—although he wrote Jocelyn that if it came out, he believed no one would be very surprised, and that their friends would be happy for them. On his eightieth birthday, he wrote that "at my ADVANCED age I am as excited at the thought of you as I was in the lower studio years ago!" (7.27.1957)

In 1957 an American company opened a quarry just downhill from Simla, and its grinding became a constant drone in the jungle, once silent except for animal noises. The upside was that by 1959 it brought electrical power to Simla—freeing everyone, especially Henry, from the all-too-frequent demands of two moody generators. It was no simple project to electrify the tropics: short circuits were a daily occurrence as monkeys and tree porcupines and mouse opossums used the wires as transport networks and nest material, and snakes draped themselves around adjoining wires. The noise of the rock crushers soon slipped into the background, and even the blasting became indistinguishable from the occasional crash of trees borne down by wind, age, and the weight of lianas and epiphytes.

Simla was, as Will had hoped, a productive field station for the Zoological Society. The development of the Pan Am Clipper—an airplane that could land on water—cut the travel time to hours, making Simla a refuge and a prime collection site for zoo personnel. Friends who could afford the plane fare came to visit, and curators stayed while they collected specimens or performed experiments. Jim Oliver, who had succeeded Ditmars as curator of reptiles, visited several times to study the rich snake population, returning home each time with shipments of bushmasters, boas of several varieties, and fer-de-lance, as well as wiry tree snakes and boxes of lizards, frogs, and toads. John Tee-Van, now general director of the aquarium and zoo, and Helen came to visit as often as they could, and Will loved showing them around his "estates," which the Zoological Society would inherit. John had at last been recognized as the eminent scientist he had become under Will's mentorship, and had been awarded an honorary doctorate from Rensselaer Polytechnic Institute in 1955.

William Conway, an intrepid young ornithologist from the St. Louis Zoo who had replaced Lee Crandall as curator of birds at the Bronx Zoological Park in 1956, came to Simla several times to study and collect birds. Beebe had located two nests of the elusive mossy-throated bellbird, identified in 1817 but never trapped alive, whose eerie clanging cry had lured him again and again into the forest on unproductive hunts. With great effort Conway managed to capture a bellbird for the zoo, along with many other interesting specimens, including almost forty manakins and hummingbirds.

Donald Griffin, a Harvard ethologist whose work on bats had uncovered their sonar system of detecting prey, erected a huge flying cage complete with a twelve-foot pond to study fishing bats. Griffin had established that insect-hunting bats send out "clicks" at such a high frequency that they are inaudible to human ears. These signals reflect from any surface they meet, enabling the bats to identify, locate, and follow flying prey; their signal output increases as they close in. Some moths have developed sensors for the clicks, and when they meet a sound of that frequency, they drop like stones to the ground.

These high frequencies can't travel far through water, so no one knew how fishing bats, which skim over the dark surfaces of lakes and ponds at night, catch their prey. With his ingeniously designed "bat detectors," which magnified and dropped the clicks into the range of human perception, Griffin could hear the bats' signals, and in the enclosed area he could listen and watch. He discovered eventually that the bats are able to detect the small ripples made by fish swimming near the surface, zero in, and snatch them with long, hooked claws.

Jocelyn was offended by the strong odor of the caged bats, but Will's only objection was that the bats came out at cocktail hour, seriously depleting the social group. Griffin obligingly moved the cage as far from Jocelyn's sensitive nose as he could, and appeased Beebe by amplifying the sounds picked up by his bat detectors so that they could be heard by the kibitzers on the patio. The bat detectors were wildly popular: many insects communicate in the ultrasonic frequencies, and

Beebe in his Simla lab.

when Griffin and his colleagues were in residence, cocktail hour was enlivened by wild scavenger hunts for the sources of weird new noises.

By 1959 Will's strength and endurance had been gradually lessening for some time, and he began to be unsteady on his matchstick legs. Occasional dizzy spells made tree climbing and long hikes less practical, and his voice was increasingly unreliable. He continued his lab work, following up lines of thought that active fieldwork had crowded out, and tried to content himself with hearing about his associates' discoveries. He made extensive notes on his own observations, which were as keen as ever. Younger staff members brought him bird nests, and he dissected them carefully, stick by stick, grass blade, and liana, teasing out in reverse the mechanics of construction—something he had always wondered about but never taken the time to study. One mockingbird nest had 542 separate elements; on the basis of his intimate acquaintance with every plant within Simla's extensive boundaries, he calculated that the parents had flown 185 miles to collect the material. The primitive bellbird nests, in contrast, were simple platforms of a few interlocked twigs.

He entered actively into the design of experiments with the heliconids, and posed challenging questions about David Snow's manakins, Henry's moths, and Lincoln Brower's monarchs, suggesting new perspectives and avenues of research. He found Jocelyn's work with the fiddlers endlessly interesting, particularly after she turned the lily pond into a crabbery. She installed her own colony of waving fiddlers, marking each individual with paint so she could tell who was doing what. Then she built an observation platform and sat perched like a referee at a tennis match, recording the signals of courtship and aggression in notebooks and on film.

In January and February 1960, Will had several brief periods of confusion, which were diagnosed by his Trinidad friend and doctor Ted Hill as minor strokes. There was no paralysis and the confusion cleared, but he remained unsteady and weak, and the worried staff engaged an English nurse for him. He continued his work at a reduced rate and reigned over visitors and scientists at cocktail hour and at dinner, but his dry mouth and swollen tongue made him increasingly unable to talk. He was lucky that the staff knew all his stories by heart and could supply the punch lines.

He spent more and more time at Simla. Although the Clippers had made the trip an easy one, he could no longer bear the cold of New York or Vermont winters, so visits to Elswyth at the Wilmington farmhouse were curtailed. She never once visited Simla. John and Helen Tee-Van, now retired, had a house in Sherman, Connecticut, that they had named Kartabo, and they made Will welcome there when he came north, but his stays were brief. He maintained his journal-writing habit, but the entries were infrequent. He put his energy into carefully detailed notes on every creature that he noticed, and applied his vast knowledge and experience to understanding the grand scheme of evolution.

Jocelyn Crane was traveling the world in search of fiddlers. Her mission to record and analyze their communication behavior world-wide took her one year to Australia and Fiji, one year to Malaysia, and another to Zanzibar, each time for several months. Her chatty letters were full of her infectious verve, peppered with vivid descriptions of people and food and customs. Will, weak and almost silent, encouraged her to go but lived for her letters, and everyone at Simla, including Jogie Ramlal, the young boa-trapper who had grown up at Simla and become indispensable as general assistant, gathered around the patio to hear them read. From "a Chinese hotel that hasn't any name, being the only one," in the Philippines, for instance, she wrote that it was noon there, with the tide safely in, "so no more crabs under the stilt-legged houses down by the shore. Jeepneys painted like Christmas tops go by my window, and pedicabs decorated with paper flowers and shiny birds that clip on the fronts. Most everybody is having a siesta though, and I can just see all you guys in the Upper Reaches of the Arima Valley with the blankets pulled up under their chins, snoozing away under the moon that was around here some hours ago." (2.19.1962)

Will went north one last time in the spring of 1962, staying briefly in Connecticut. William Conway and the Tee-Vans put him safely on a southbound plane with Jocelyn and a nurse. At Simla on June 2 he was felled by a rapid pneumonia. His weakened lungs gave way in the early hours of June 4 with Jocelyn at his side, in the spare, lofty room he shared with lizards, bats, and hummers, furnished the way he lived: a narrow bed, a washbasin and mirror, a small closet, and his orderly shelves with all he ever needed—Milne, Dunsany, Kipling, and a stuffed bird or two.

According to his wishes, Will was buried in Mucurapo cemetery in Port-of-Spain. His friends at Pan Am provided a plot in their own section of the cemetery, established when a Clipper had gone down in the forties. Conway, now director of the Zoological Society, wrestled a 650-pound boulder from Beebe's beloved Bronx River, and added, at Will's request, a modest bronze plaque reading

William Beebe
1877–1962

He then had it shipped to Trinidad, where today it still stands, partially sunken in the sodden tropical mold, its plaque obscured, between the granite monuments to the Pan Am pilot and a missionary family lost in the crash, amid city blocks of whitewashed family plots and rampant tropical weeds. Memorial services were held at Christ Church in Trinidad and at the Church of the Transfiguration in New York City, so friends from both halves of his world could gather to celebrate Beebe's irrepressible spirit. Obituaries around the globe reminded a forgetful world of the swashbuckling adventures and far-reaching vision of William Beebe, scientist, writer, naturalist, and explorer.

Jocelyn became director of the DTR, and she, Henry, and the staff decided that the best tribute to Will would be the continued research and smooth running of the station. As when young Patten Jackson died

Beebe's grave in Port-of-Spain, Trinidad.

at the Bermuda station, they worked harder than ever. With the help of Jogie Ramlal, Jocelyn was able to keep things going and to maintain her own standards of excellence in research. Researchers continued to come for extended periods, and many, like Donald Griffin, were "regulars."

By 1965, though, Jocelyn's work was taking her north for extended periods. Without her zeal and Will's dynamic energy, it became increasingly difficult to keep Simla running efficiently. No second Beebe arose to reinvigorate it, so the Zoological Society regretfully decided to close it as an official field station. In 1967 Asa Wright, the energetic Icelander, turned neighboring Spring Hill over to a group of Trinidad naturalists. From it they created the first nature center in the West Indies, and the Zoological Society added Simla to it in 1974. A nonprofit institution, the Asa Wright Nature Centre is dedicated to conservation, education, scientific research, and ecotourism. Visitors can watch birds from the wide veranda where Will, Jocelyn, and Asa and Newcome Wright spent so many sociable hours. William Conway is a trustee of the center, and one of its most experienced guides is Jogie Ramlal.

From Spring Hill's vantage point, birders can see the many varieties of hummingbirds, myriad brilliantly colored tanagers, tegu lizards, and countless other animals. Even the oilbirds can still be seen by visitors who don't mind a hike. The long Arima Valley disappears into the far distance, its dense jungle foliage sheltering toucans and parrots, motmots, kiskadees, and dancing manakins, new generations of Schneirla's army ants and the leafcutters that ravaged Jocelyn's carefully tended jardinieres on Simla's patio. The jardinieres have disappeared, along with the statues of Piglet and Eeyore, but a few stalwart researchers still brave Simla's spartan facilities (now the nature center's field station) to study the rich array of plant and animal life that entranced Will.

Jocelyn remained the director of the DTR, joining forces with Rockefeller University to form the Institute for Research in Animal Behavior, which she directed until 1970. Donald Griffin, the bat researcher whose malodorous subjects had disgusted her at Simla, moved to Rockefeller, and in 1965 he and Jocelyn were married. Taking her interest in gestural communication in a direction she and Will had often discussed, Jocelyn Crane Griffin then entered New York University's School of Fine Arts and received her doctorate in 1985. Her thesis was a logical outgrowth of her life work on fiddlers and spiders: it was on the use of gestures as communication in ancient art. She died in Lexington, Massachusetts, in 1998 at the age of eighty-nine.

Blair Niles had died on April 12, 1959, of a cerebral hemorrhage. The obituary in the *New York Times* focused on her 1927 book *Condemned to Devil's Island,* from which a movie had been made, and *The James: From Iron Gate to the Sea,* a historical work about the James River in Virginia. First published in 1939, it had been enlarged to include some informative autobiographical material in 1945. Of her marriage to Will, the *Times* said:

> While the wife of William Beebe, the naturalist, she accompanied him on many scientific expeditions. On these trips she lived the life of the trail, traveled by horseback through wild country, camped in jungles or at the foot of active volcanoes and met with Indians and bandits.
>
> After obtaining a divorce from Mr. Beebe, she was married to Mr. Niles in 1913.[4]

The *Times* also mentioned that she had won the Gold Medal of the Society of Women Geographers in 1944.

The single entry in Will's journal for that day reads in its entirety, "Word from John [Tee-Van] of the Death of Blair on Monday." (4.18.1959)

Upon Will's death, Elswyth cleared out the elegant apartment on 67th Street. She sold his extensive scientific library to Princeton University, and the rest went to dealers in New York and Vermont. Her own papers, carefully edited, she placed in the Mugar Library of Boston University and at the University of Iowa. Elmer LaFlamme, her hardworking handyman, moved into the downstairs bedroom in the Vermont farmhouse, and after Elswyth's mother's death in 1964 they lived companionably with a succession of cocker spaniels.

Will had been obsessively worried that Elswyth would write a biography of him after his death. To prevent the possibility, he left his scientific papers as well as the journals he had been keeping since childhood to Jocelyn. Elswyth contested this bitterly, as she did his will, which divided his assets between her and Jocelyn. After Elswyth's death in 1984, Jocelyn placed Beebe's papers in the Department of Rare Books and Special Collections at Princeton University's Firestone Library.

Will's office at the Bronx Zoo still stands, and the well-worn rolltop desk presented to him by William Hornaday is now occupied by George Schaller, who with a large group of explorers, ecologists, and researchers continues the tradition Will pioneered of scientific research teamed with appreciation and aggressive conservation of the world's dwindling natural resources.

William Conway, who had joined the Zoological Society in 1956 in Beebe's old position as curator of birds, replaced John Tee-Van as director on his retirement in 1962, taking the society in directions both old and new, and vastly increasing its field research to fifty-three nations. As it had been since Hornaday's time—indeed, since it had first been envisioned by Teddy Roosevelt and his Boone and Crockett Club—the New York Zoological Society was a pioneer in the conservation of the world's fast-vanishing wildlife.

Under Conway's leadership, the Zoological Society became the Wildlife Conservation Society, dedicated to the conservation of natural resources worldwide. In the zoo itself, some of the grand old buildings remain much as they were when Will Beebe first walked through the gates in 1899. But the Bronx Zoo is no sterile monument to past glories. Animal care specialists work continually to raise the quality of life of the creatures that are sheltered on its acres. The naturalistic enclosures that Hornaday and Grant envisioned keep growing to fit the needs of the species that range there. And ongoing research into the minds of the animals themselves, spurred by Donald Griffin's crusade to make animal thinking a legitimate field of scientific study, inspires new approaches to making their lives interesting, challenging, and more like life in the wild.

Field researchers working with the Wildlife Conservation Society fan out across the United States as well as remote parts of the globe to study animals and their habitats. No organization in the world is so thoroughly engaged in field research and practical conservation. As Beebe prophesied, intense study of the animal in its world is the only way to protect and preserve, and even to bring species back from the brink of extinction, as the zoo did with the bison in the early 1900s.

The effects William Beebe had on science, particularly on deep ocean biology and neotropical ecology, are enormous and lasting. He made an effective transition between the Victorian natural historian, content to collect and classify the natural world, and the modern experimental biologist, alive to the effect of its natural habitat and mode of living on every aspect of a creature's life. His early conviction of the truth of Darwin's theory of natural selection shaped his inquiry into the lives of pheasants, the embryology of fish, and the phenomenon of mimicry, and led him to make pioneering studies of selection on the Galápagos. He was an early advocate of the idea that evolution often proceeds in a radial pattern, rather than in a strictly linear way, as the pheasants had shown him in 1910.

Beebe's overarching belief that there can be no real knowledge of an organism without a concomitant understanding of its whole ecosystem was wholly new. In his first intense study of the quarter mile of jungle at Kartabo, his prescient analysis of an eight-mile cylinder of ocean off Bermuda, the dissection of the fourteen faunal zones of Rancho Grande, Venezuela, and his study of the ecology of the Arima Valley in Trinidad, he raised the level of understanding of what is now known as community ecology to a new standard. With his diving helmet, trawling nets, and bathysphere, he broke down the physical and mental barriers to studying the deep ocean and made the first serious studies of the world's richest faunal communities, the coral reefs. The dozens of new species he introduced to the world, many named in his honor by other researchers, stand as a lasting tribute to his thoroughness and his informed imagination.

Ironically, the very breadth of Beebe's knowledge and vision have worked against him in the recent era of specialization. His unprecedented study of pheasants set a standard of avian research, but his greatest contributions were not in ornithology. The groundbreaking analysis at Kartabo gave tropical ecology its most powerful research tool, but the field had yet to be born. His study of behavioral adaptations foreshadowed behavioral ecology decades before it became a recognized field of study. His work on moths and butterflies and beetles and mantids and a host of other insects was often brilliant, but he never claimed to be an entomologist. His daring venture into the depths inspired Jacques Cousteau and opened the doors to the fertile field of deep-ocean studies, but he was no marine biologist. Out of loyalty to the zoo, he published all his research papers in *Zoologica,* whose proud generalist mandate lost its authority as specialist journals multiplied with the exponential increase of scientific fields.

The impact of Beebe and Barton's descent into the alien world of the abyss can hardly be overstated. Every scientist who had longed to explore its depths—even Will—had thought of the ocean as the ultimate impenetrable mystery. Once he and Barton had been there and back, the "glass floor" was shattered; suddenly a whole field of research was open for discovery.

Throughout his life, it was Will's joy to shepherd young scientists into active service. In an era in which women were not accepted as equals physically or intellectually, he never wavered in his assumption that the secondary sexual characteristics that set the sexes apart had nothing to do with the primary character that made people smart or talented or good company. Gloria Hollister Anable, whose years with Beebe sharpened her keen understanding of the natural world and the dire effects human development could inflict, became interested in the threatened Mianus River watershed in Connecticut. Her effort to save the watershed would evolve into what is now The Nature Conservancy, a nonprofit organization with more than a million members and a workable blueprint for protecting the world's most threatened ecosystems.

Ruth Rose, who became a scriptwriter, Isabel Cooper the artist, Rachel Carson, and Jocelyn Crane, whose career in serious science became dearer to him than his own, were all mentored and encouraged by Beebe. Many women were among the visiting researchers who used Kartabo or Rancho Grande or Simla to test their mettle and crystallize their ideas.

Though his talks were riveting and his personal charisma unprecedented in science, it was through his books that Beebe influenced generations of nature lovers. Thirteen of his books were translated into a total of twelve languages, and for decades his beautifully crafted essays could be found in school and college texts. One of a phalanx of Beebe's scientific stepchildren, Edward O. Wilson, ant specialist and father of sociobiology, counts Beebe as an important early factor in his choice of career. In his autobiography *Naturalist,* Wilson remembers that as a child he read Beebe's books and hungered for the world of the "naturalist-explorer of the Venezuelan jungles." Wilson went to Simla to work on ants during the last year of Beebe's life; he remembers dining with Jocelyn and Will, "admiring the silver candlesticks given to Beebe by his friend Rudyard Kipling and talking tropical natural history here in the place where so much of the best research on it had been conceived." When marine biologist and pioneering diver Sylvia Earle launched her own submersible vehicle, Deep Rover, she packed her treasured copy of *Half Mile Down,* whose awed descriptions of undersea life had first awakened her desire to study marine ecology. In what William Conway called the far-reaching web of Beebe's intellectual offspring, his broad vision and contagious enthusiasm live on.

Most important to Will, and to the wilderness he respected and loved, his passionate commitment and challenging perspectives endure in the great work of conserving what is left of the natural world. His melancholy descriptions of Kalacoon's luxuriance, destroyed irretrievably by human greed for rubber; his angry outbursts about whole populations of gorgeous birds or helpless sea lions being destroyed for vanity or self-indulgence; his aching nostalgia for a world of decency and respect for human and animal alike—all reached audiences of many nationalities, at many levels of understanding.

Most of all, the driving force of his unquenchable enthusiasm for the natural world, his exhilaration in its wonders, inspired a generation of ardent nature lovers. Whether they went on to become field biologists or teachers or naturalists in any walk of life who support conservation efforts, his ideals and his enthusiasm live in the progeny of his scientific life and in his pen. "Who can be bored for a moment," wrote Beebe in 1921, "in the short existence vouchsafed us here; with dramatic beginnings barely hidden in the dust, with the excitement of every moment of the present, and with all of cosmic possibility lying just concealed in the future?"

In researching and writing this book I discovered a vast subculture of people who had loved and been inspired by William Beebe, either personally or through his writing. So many have given generously of their time and knowledge that I can never repay the debt. Jocelyn Crane, of course, was the driving force behind the work; I fervently wish I had moved quickly enough for her to have seen the result. Her husband, Donald Griffin, knew Beebe as well. His recollections, gentle corrections, and interest made him a treasured mentor as well as a friend. Our trip to Bermuda together, to scatter Jocelyn's ashes over the coral reefs she had loved, was a pilgrimage as well as research.

William Conway, director emeritus of the Wildlife Conservation Society, has been invaluable as a source of information both about Beebe and the Bronx Zoo, and about the creatures Beebe studied. He has in every way facilitated my interactions with the zoo, which was Beebe's employer throughout his adult life.

Two members of the Wildlife Conservation Society's board, Robert G. Goelet and Howard Phipps Jr., generously funded research trips to Beebe's field stations in Bermuda, Trinidad, and Nova Scotia and to several other locations important to Beebe; they also underwrote the less glamorous but necessary trips to the zoo's archives in New York. At the archives, Steven Johnson, head librarian, was an endless source of information, providing me with research materials and good advice. Diane Shapiro, image archivist, steered me through the labyrinth of photographic resources, devoting precious hours to ensure that this book's photographs would present the most accurate portrait possible of the zoo's early years and of Beebe's life.

The Rare Books and Special Collections staff of Princeton University have been unswervingly friendly and helpful throughout this arduous process, which they must have thought would never end. Don Skemer facilitated my access to the material both before and after Jocelyn Crane's death. Margaret Sherry Rich, AnnaLee Pauls, Charles Greene, Jane Snedeker, and Linda Bogue provided assistance in many ways, even stretching the rules to procure for me a blessedly nonregulation chair.

Howard Gotlieb, of the Howard Gotlieb Archival Research Center of Boston University, graciously allowed me the use of Elswyth Thane's collection of Beebe's correspondence, and Sean D. Noël facilitated my work. The Gotlieb collection houses a fascinating array of Beebe's early memorabilia, including even his baby clothes.

In Bermuda, Bob and Jean Flath helped me locate people and places Beebe knew. David Wingate and Jeremy Madeiros, zealous guardians of the ecological wealth of Nonsuch Island, kindly allowed me to visit the old field station there. John Barker, who has made several faithful copies of the bathysphere for museums, offered his considerable expertise, and Greg Hartley—son of Bronson Hartley, who worked with Beebe—helped me experience the world of the coral reef in an updated version of Beebe's diving helmet. Phyllis Schwab, who lives in the house Beebe called New Nonsuch, was full of tales of Bermuda's social life.

In Trinidad, Ian Lambie, who heads the Asa Wright Nature Centre on the old Spring Hill and Simla estates, was courteous and helpful; Mary Alkins-Koo, of the University of the West Indies, was well informed about Beebe's time at Simla and proved to be an eager accomplice in our lively search for Beebe's mysteriously elusive grave.

Closer to home, the Fitzgerald family graciously allowed me to tour their Vermont home, which was Elswyth Thane and William Beebe's farmhouse retreat, and lent me photographs of Elswyth in her later years. Lucie Palmer—whose husband, Vincent, had worked closely with Beebe, and who made a career of painting underwater—told wonderful stories of an exciting life. John Cannon, who also worked with Beebe in the thirties, and John Cody and Ellen Ordway, who worked with Beebe at Simla, provided much interesting information and many illuminating stories.

Professor Andrew Dobson, of Princeton's Department of Ecology and Evolutionary Biology, deserves credit for encouraging me to take on this project. Beebe was an early inspiration to him, as well as to many other ecologists and conservationists today, and he encourages his students to study Beebe and his work. He also good-naturedly read and commented on the manuscript throughout its lengthy gestation. My long-suffering editor at Island Press, Jonathan Cobb, has been patient and kind throughout; he has been a strong advocate for Beebe, and for clarity and terseness in telling his story.

My heartfelt appreciation goes to the memory of my dear friend Janet Stoltzfus, whose energy and spirit quickened the early stages of this project. Her inspiring love of research and sense of history infuse the entire manuscript. She did not live to see the fruition of her vision, but I hope she would have approved and enjoyed the results.

My husband and colleague, James Gould, I can never acknowledge adequately. He inspired, organized, encouraged, and supported me, buoyed my spirits, corrected my science, accompanied me on research trips, and gave the project far more time and energy than he could spare. Without him, this book would never have come to be.

Key to Acronyms Used in Notes

AA: *The Arcturus Adventure*
APC: Author's personal collection
BB: *Book of Bays*
BU: Howard Gotlieb Archival
  Research Center, Mugar Library,
  Boston University
EJ: *Edge of the Jungle*
GWE: *Galápagos: World's End*
HJ: *High Jungle*
HMD: *Half Mile Down*
JP: *Jungle Peace*
LS: *The Log of the Sun*
NLW: *Nonsuch: Land of Water*
OSW: *Our Search for a Wilderness*
PJ: *Pheasant Jungles*

PLH: *Pheasants: Their Lives and Homes*
PM: *A Monograph of the Pheasants*
PU: William Beebe Papers (CO661),
  Manuscripts Division, Department
  of Rare Books and Special
  Collections, Princeton
  University Library
TB: *The Bird: Its Form and Function*
TBLM: *Two Bird Lovers in Mexico*
TWBG: *Tropical Wildlife in British
  Guiana*
ULNY: *Unseen Life of New York*
WCS: Wildlife Conservation Society
  Archives, Bronx Zoo, New York City
ZV: *Zaca Venture*

PROLOGUE
1.   *HMD*, 171–72.

CHAPTER 1. COUNTING CROWS
1.   From William Beebe (WB) to his father, 5.31.1889. A copy of this letter is in the author's personal collection.
2.   4.14.1891; William Beebe Papers (CO661), Manuscripts Division, Department of Rare Books and Special Collections, Princeton University Library. These journals are arranged chronologically. Further references to these journals will be given in the text.
3.   1.18.1890; Elswyth Thane Collection, Howard Gotlieb Archival Research Center, Boston University, Box 78. Further references to these archives will be cited as BU, followed by the box in which the material is located.
4.   7.14.1888; BU 80.

CHAPTER 2. FLEDGING
1.   4.9.1891; BU 78.
2.   11.19.1890; BU 82.
3.   4.16.1891; BU 78.

4. 1.2.1891; BU 82.
5. 4.19.1891; BU 78.
6. 5.29.1891; BU 78.
7. Frank M. Chapman, *Autobiography of a Bird-Lover* (New York: D. Apple-
    ton, 1933), 71–72.
8. Daniel Beard, *The American Boy's Handy Book* (New York: Scribner, 1882;
    repr., Jaffrey, NH: Godine, 1983), 206.
9. Beard, *Handy Book,* 232–33.

CHAPTER 5. THE SCIENTISTS' APPRENTICE
1. Holograph letter from WB to Henry Holt, 6.11.1902; Correspondence
    series, Box 7, Folder 11, Henry Holt Archives, Manuscripts Division, De-
    partment of Rare Books and Special Collections, Princeton University
    Library.

CHAPTER 6. BRONX ZOOLOGICAL PARK
1. The ensuing description of the beginning of the New York Zoological
    Society is taken largely from William Bridges, *Gathering of Animals: An
    Unconventional History of the New York Zoological Society* (New York: Harper
    & Row, 1974), 1–98.
2. Bridges, *Gathering of Animals,* 16.
3. Bridges, *Gathering of Animals,* 31.
4. Quoted in Bridges, *Gathering of Animals,* 34.
5. Bridges, *Gathering of Animals,* 60.

CHAPTER 7. WIDENING HORIZONS
1. Bridges, *Gathering of Animals,* 125.
2. Undated journal entry for mid-June 1900, PU.
3. 9.2.1901; BU 82.
4. 9.13.1901; BU 82.
5. 9.17.1901; BU 78.

CHAPTER 8. NESTBUILDING
1. 4.20.1902; BU 82.
2. 3.14.1902; BU 78.
3. 8.5.1902; BU 78.
4. "Sweet Birds Sang Wedding Hymn," *Richmond Times,* 8.6.1902.
5. 8.17.1902; BU 78.

CHAPTER 9. FLYING SOUTH
1. 2.18.1903; BU 78.
2. William Beebe, "Five Days among the Birds on Cobb Island, Virginia,"
    Eighth Annual Report of the New York Zoological Society for 1903, 162.

CHAPTER 10. MIGRATION
1.   *TBLM,* 52–53.

CHAPTER 11. THE NATURALIST AS AUTHOR
1.   These and the following reviews can be found in Beebe's scrapbook in the
      Beebe Papers, WCS: *Review of Reviews,* 12.1906; *New York Herald,* 2.3.1907;
      Hartford Courant, 11.12.1906.
2.   *TB,* ix.
3.   *TB,* 18.
4.   *TB,* 187.
5.   *LS,* 117.

CHAPTER 12. RAIN FOREST AT LAST
1.   *OSW,* 73–74.
2.   3.18.1908; BU 78.
3.   *OSW,* 78.
4.   Box 6, Folder 6, PU; p. 2.

CHAPTER 13. IN SEARCH OF WILDNESS
1.   *OSW,* 195–96.
2.   *OSW,* 144–45.
3.   *OSW,* 167.
4.   *OSW,* 285.
5.   *OSW,* 295.
6.   *OSW,* 295.
7.   *OSW,* 317.
8.   *OSW,* 385.

CHAPTER 14. CITY LIGHTS
1.   *OSW,* 348–49.
2.   *Zoologica* 1 (1909): 443–66.
3.   *Zoologica* 1 (1909): 67–114.
4.   *Zoologica* 1 (1910): 139–49.
5.   *PM,* xx.
6.   A new species, the Congo peacock *(Afropavo),* has since been discovered
      in Africa, a finding that has generated a great deal of interest for the light
      it casts on the history of the pheasants. Its existence was unknown at this
      time, however.
7.   2.2.1910; BU 82.

CHAPTER 15. PHEASANT JUNGLES
1.   2.28.1910; BU 82.
2.   2.28.1910; BU 82.
3.   *PJ,* 31.

4. 3.31.1910; BU 82.
5. 3.19.1910; BU 82.
6. 3.15.1910; BU 77.
7. 4.1.1910; BU 77.
8. *PJ*, 46–47.
9. *PJ*, 47.
10. *PJ*, 49.
11. Letter from WB to Colonel Kuser, 4.30.1910. Copied in Beebe's journal, pp. 174–76, in Box 5, Folder 1, PU.
12. 4.17.1910; BU 82.

CHAPTER 16. WESTERN HIMALAYAS
1. *PJ*, 83.
2. Now frustratingly relabeled *Anadenus altivagus* by taxonomists.
3. *PJ*, 85.
4. *PJ*, 81–82.
5. WB to Mrs. Rice, 5.10.1910; BU 82.
6. 6.05.1910; BU 77.
7. 7.14.1910; BU 82.
8. 7.15.1910; BU 82.
9. WB to Kuser from Fort Kapit, 7.20.1910; BU 77.
10. 8.12.1910; BU 82.
11. 9.29.1910; BU 77.

CHAPTER 17. BEARING EAST
1. 9.20.1910; BU 77.
2. WB to Nettie, 9.20.1910; BU 82.
3. 10.24.1910; BU 77.
4. *PJ*, 108.
5. WB to Charles, 11.25.1910; BU 82.
6. WB to Kuser, 12.26.1910; BU 77.
7. *PM,* vol. 1, xliv.
8. *PJ*, 130.
9. *PJ*, 115.
10. WB to Nettie, 2.28.1910; BU 82.
11. Blair to Nettie, 12.29.1910; BU 82.
12. WB to Kuser, 3.29.1911; BU 77.
13. 12.28.1910; BU 82.

CHAPTER 18. RECALIBRATION
1. *PM,* xxv.

CHAPTER 19. BETRAYAL
1. Quoted in Leslie Curtis, *Reno Reveries: Impressions of Local Life* (Reno: Chas. E. Weck, 1912), 60.

2. Case No. 9928, Second Judicial District Court, Washoe County, Reno, Nevada, p. 4.
3. Depression had claimed the life of Blair's brilliant uncle Theodorick Pryor, a Princeton graduate, who walked off a Brooklyn pier at the age of twenty.
4. *New York Times*, 9.07.1913; Box 14, Folder 32, PU.
5. 11.29.1915; BU 78.
6. 4.11.1915; BU 78.
7. 4.18.1915; BU 78.
8. 4.11.1915; BU 78.
9. 5.6.1915; BU 78.

CHAPTER 20. WILDERNESS FOUND
1. 2.2.1916; BU 78.
2. Theodore Roosevelt, "A Naturalist's Tropical Laboratory: Beebe's Laboratory in Demerara," *Scribner's Magazine* 61 (1917): 46–64.
3. 10.13.1916; draft of note from WB to T.R. in Box 16, Folder 10, PU.
4. *NLW,* 130.
5. WB to Charles Beebe, 3.10.1916; BU 78.
6. 3.22.1916; BU 78.
7. William Beebe, "Fauna of Four Square Feet of Jungle Debris," *Zoologica* 2 (1916): 107–19.
8. Beebe, "Four Square Feet," 117.
9. *EJ,* 39.
10. *TWBG,* 81.
11. *TWBG,* 87.
12. *TWBG,* 93.
13. WB to W. T. Hornaday, 7.23.1916; Director's Office file, WCS.
14. 11.29.1916; BU 78.

CHAPTER 21. JUNGLE PEACE
1. 12.5.1917; BU 118.
2. 1.11.1918; BU 78.
3. See William Beebe, "A Naturalist in Paris," *The Atlantic Monthly* 121 (1918): 721–32; "A Red Indian Day," *The Atlantic Monthly* 122 (1918): 23–31; "Animal Life at the Front," *Bulletin of the New York Zoological Society* 21 (1918): 1614–16; "Battle Photography," *Scribner's Magazine* 64 (1918): 399–409.
4. 2.15.1918; BU 78.
5. 11.18.1918; H. F. Osborn file 2-29, American Museum of Natural History Archives, New York City.
6. *Twenty-third Annual Report of the New York Zoological Society for 1918,* 49.
7. *JP,* 4.
8. *JP,* 64–65.
9. In William Beebe, *Jungle Peace* (2nd printing, New York: Henry Holt & Co., 1922), ix–x.
10. 3.4.1919; BU 78.

11. WB to W. T. Hornaday, 4.10.1919; Director's Office file, WCS.

12. William Beebe, "The Three-Toed Sloth *Bradypus cuculliger cuculliger* Wagler," *Zoologica* 7 (1926): 1–67.

13. 7.20.1920; BU 78.

CHAPTER 22. THE ENCANTADAS

1. Quote from "Log of the Noma," APC. Excerpts in Box 6, Folder 6, PU.

2. These may have been the much-publicized penguins WB sent to the title character in *The Man Who Came to Dinner,* the 1939 Kaufman and Hart stage hit inspired by WB's eccentric neighbor Alexander Woollcott.

3. *GWE,* 180.

4. In the twenties the islands of the Galápagos Archipelago were still known by their English names. Albemarle is now Isabela, Narborough has become Fernandina, Tower is Genovesa, and Chatham is San Cristóbal.

5. *GWE,* 174.

CHAPTER 23. NEW YORK AERIE

1. As Grant wrote Osborn, "I am very anxious to demonstrate to others the value of his publicity work for the Society, but he is an erratic genius, and he forgets sometimes that he is receiving a salary from the Society, small as it may be." 6.24.1928; Osborn Papers, Box 8, Folder 38, WCS.

2. *Zoologica* 6 (1925): 5–193.

3. Nicholas Roosevelt, "By Sea to the End of the World," *New York Times Book Review,* 2.24.1924, 1, 25.

4. WB to Nettie from Kartabo, 4.8.1919; BU 78.

CHAPTER 25. FIRE AND WATER

1. *AA,* 126–27.

2. One fathom = six feet.

3. *AA,* 180–81.

4. *AA,* 373, 375.

5. *AA,* 377.

6. *AA,* 379.

7. *AA,* 381.

8. *AA,* 382.

CHAPTER 26. ELSWYTH

1. Elswyth Thane, *Riders of the Wind* (New York: Frederick A. Stokes Co., 1926; republished Mattitack, NY: Aeonian Press, 1976), 63.

2. Charles Shaw, *Bookman* 66 (1928): 635–37.

CHAPTER 27. BERMUDA DIARY

1. Otis Barton, *The World Beneath the Sea* (New York: Thomas Crowell, 1953), 12.

2. *HMD*, 91–92.

3. Barton, *World Beneath*, 14–15.

4. Barton, *World Beneath*, 17.

5. Elswyth Thane, *The Reluctant Farmer* (New York: Duell, Sloan and Pearce, 1950), 4.

6. Helen Worden, "The Marriage That Breaks All the Rules," *Today's Woman*, 86; BU 122. I have been unable to date this article closer than 1945.

7. 5.27.1930; loose journal pages, APC. Will was lucky to have experts working on the technical aspects of the dive: the actual oxygen flow rate was one liter per minute per person, not one per hour. This journal entry was written in the excitement of the first evening after the dive; in everything he wrote about it later, the flow rate is correct.

CHAPTER 28. OUT OF THE DEPTHS

1. *HMD*, 103.

2. *HMD*, 104.

3. *HMD*, 107.

4. *HMD*, 111.

5. *HMD*, 113.

6. Barton, *World Beneath*, 33.

7. *HMD*, 121.

8. *HMD*, 122.

9. *HMD*, 127.

10. *HMD*, 134.

11. William Beebe, "Down into Davy Jones's Locker," *New York Times Magazine*, 7.13.1930, 19.

12. Jocelyn Crane, "The Mammals of Hampshire County, Massachusetts," *Journal of Mammalogy* 12 (1931): 267–73.

13. Castle Island, a favorite DTR picnic spot, is a small rocky island about a quarter of a mile from Nonsuch, with the picturesque ruins of a 1612 English fortification.

14. *HMD*, 144.

15. William Beebe, "New Data on the Deep Sea Fish *Stylophthalmus* and *Idiacanthus*," *Science* 78 (1933): 390.

CHAPTER 29. ON THE AIR

1. Undated holograph letter from Katharine Hepburn to WB, BU 67:

    My dear Mr. Beebe, this is just a note to thank you for [the] interesting evening. It was lots of fun & it was also charming of you to really keep your promise & let us see the movies which we all enjoyed tremendously. *Katharine Hepburn*

2. William Beebe, "Round Trip to Davy Jones's Locker," *National Geographic Magazine* 59 (1931): 653–78.

3. *Pelham Sun*, 3.24.1931.

4. *HMD*, 154.

5. *HMD,* 178.
6. *HMD,* 157–58.
7. Script for Ford Bard's narration, holograph copy; APC.
8. William Beebe, "The Depths of the Sea," *National Geographic Magazine* 61 (1932): 65–88.
9. William Beebe, "A Wonderer Under Sea," *National Geographic Magazine* 62 (1932): 740–58.
10. *NLW,* 207.
11. Will would pay for his trust in the healthful effects of the sun in later life, when he had to have several small malignant growths removed from his hairless scalp.
12. *NLW,* 69.

CHAPTER 30. HALF MILE DOWN
1. *HMD,* 183–84.
2. *HMD,* 190.
3. *HMD,* 194.
4. *HMD,* 198–99.
5. *HMD,* 205.
6. See for example *Bulletin of the New York Zoological Society* 35 (1932): 175–77; 37 (1934): 190–93; *Zoologica* 13 (1932): 47–107.
7. *HMD,* 317–18.
8. *HMD,* 211.
9. *HMD,* 212.
10. Cable from WB to Elswyth Thane, 8.15.1934; BU 28.
11. Michael Vecchione et al., "Worldwide Observations of Remarkable Deep-Sea Squids," *Science* 294 (2001): 2505.
12. Robert Ballard, *The Eternal Darkness: A Personal History of Deep-Sea Exploration* (Princeton, NJ: Princeton University Press, 2000), 28.
13. Barton, *World Beneath,* 41.

CHAPTER 31. FISHING
1. *ZV,* 20–21.
2. 1.1.1937; BU 104.
3. WB to Fair Osborn, 2.11.1938; Beebe-Osborn Correspondence, WCS.
4. *BB,* 106–7.
5. *BB,* 125.
6. *BB,* 218.

CHAPTER 32. OCEAN TO JUNGLE
1. *BB,* 85–86.
2. William Beebe, "The Function of Secondary Sexual Characteristics in Two Species of Dynastidae (Coleoptera)," *Zoologica* 29 (1944): 53–58. Also described in "Battles of the Beetles," *HJ,* 300–18.

3. Of the original executive committee, H. F. Osborn and W. W. Niles had died in 1935, and Madison Grant in 1937. Hornaday, the original general director, had also died in 1937.

4. "My absorption in the mystery and meaning of the sea have been stimulated and the writing of this book aided by the friendship and encouragement of William Beebe." Rachel Carson, *The Sea Around Us* (New York: Oxford University Press, 1951), vi. In 1949 Beebe had arranged for Carson to dive in a helmet, to give her an understanding of the sea that he felt was invaluable for an oceanographic researcher.

CHAPTER 33. RANCHO GRANDE
1. Worden, "Marriage," 86.
2. Worden, "Marriage," 88.
3. Under the name Edith Thane, Elswyth's mother published a book of her own, *Marionettes Are People Too,* with Elswyth's publisher (New York: Duell, Sloan and Pearce, 1948). George Swanson did the illustrations.
4. Worden, "Marriage," 89.
5. New York: Duell, Sloan and Pearce, 1945. Elswyth had been publishing with Harcourt, Brace, & Co., but after a charge that she had reprinted some of Disraeli's letters without proper credit in her book *Young Mr. Disraeli,* she changed publishers.
6. Jocelyn Crane, "Shopping for a Jungle," *Animal Kingdom* 47 (1945): 3–13. The Zoological Society's *Bulletin* had morphed in 1942 into *Animal Kingdom,* a slick, chatty, well-illustrated magazine for society members that featured articles by staff members about their adventures. Competitive as ever, Will had won $10 in a contest to name the new magazine.
7. *HJ,* 62.

CHAPTER 34. HIGH JUNGLE
1. *HJ,* 237.
2. *HJ,* 232.
3. *HJ,* 128.
4. George Swanson, "Jungle Studio," *Animal Kingdom* 48 (1946), 175.
5. William Beebe and Jocelyn Crane, "Ecology of Rancho Grande, a Subtropical Cloud Forest in Northern Venezuela," *Zoologica* 32 (1947): 43–60.

CHAPTER 35. SIMLA
1. Edwin Way Teale, *New York Herald Tribune Book Review,* 5.29.1949.
2. On his way back to the train late one night after a lecture, Will pulled his handkerchief from his pocket for a sneeze. Along with it came a dead goldfish, which dropped into a deep snowdrift. He had been too polite to refuse a bereft woman (and wealthy zoo donor) who wanted her departed pet autopsied by the great Dr. Beebe. Instead of helping Will, who on hands and knees was digging through the snow for what he claimed was a

goldfish, a skeptical policeman tried to take him to the police station to "sleep it off, sir." Will told this story often and with great relish; the DTR presented him with a large toy goldfish, which became a team mascot.

3.   Ernst Mayr to WB, 50th Anniversary Volume, holograph; APC.

CHAPTER 36. HOME TO ROOST

1.   Undated holograph letter from WB to Hugh Bullock; APC.
2.   *ULNY,* 4.
3.   *ULNY,* 12.
4.   *ULNY,* 25.
5.   *ULNY,* 19.
6.   Jocelyn Crane, "Fiddler's Carnival," *Bulletin of the New York Zoological Society* 44 (1941): 188–125.
7.   See Jocelyn Crane, "Crabs of the Genus *Uca* from the West Coast of Central America," *Zoologica* 26 (1941): 145–208; "Display, Breeding and Relationships of Fiddler Crabs of the Genus *Uca* of the Northeastern United States," *Zoologica* 28 (1943): 217–223; "On the Color Changes of Fiddler Crabs (Genus *Uca*) in the Field," *Zoologica* 29 (1944): 161–168; "Intertidal Brachygnathous Crabs from the West Coast of Tropical America, with Special Reference to Ecology," *Zoologica* 32 (1947): 69–95; "Combat and Its Ritualization in Fiddler Crabs (Ocypodidae), with Special Reference to *Uca rapax*," *Zoologica* 52 (1967): 49–75.
8.   Princeton University Press, 1975.

CHAPTER 37. SIMLA SUNSET

1.   See David Snow, "The Displays of the Manakins *Pipra pipra* and *Tyranneutes virescens*," *Ibis* 103a (1961): 110–13.
2   William Beebe, *Animal Kingdom* 57 (1954): 5.
3.   All the correspondence in this chapter between Jocelyn and WB is in APC.
4.   "Blair Niles, 71, Writer, Is Dead; Author of 'Condemned to Devil's Island' Recounted Travels in Many Lands," *New York Times,* 4.15.1959.

BOOKS AND ARTICLES BY WILLIAM BEEBE

Adventuring with Beebe. New York: Duell, Sloan and Pearce, 1955.

"Animal Life at the Front." *Bulletin of the New York Zoological Society* 21 (1918): 1622–24.

The Arcturus Adventure. New York: G. P. Putnam's Sons, 1926.

"Battle Photography." *Scribner's Magazine* 64 (1918): 399–409.

Beneath Tropic Seas. New York: G. P. Putnam's Sons, 1928.

"The Bird Called 'Brown Creeper.'" *Harper's Young People* 16 (1.15.1895): 79.

The Bird: Its Form and Function. New York: Henry Holt & Co., 1906. Reprint, New York: Dover Publications, 1965.

"Blue Jay Caches the Nut." *Science* 89 (1939): 366.

Book of Bays. New York: Harcourt, Brace & Co., 1942.

The Book of Naturalists. New York: Alfred A. Knopf, 1944. Reprint, Princeton, NJ: Princeton University Press, 1988.

"A Contribution to the Life History of the Euchromid Moth *Aethria carnicauda.*" *Zoologica* 38 (1953): 155–60.

"The Depths of the Sea." *National Geographic Magazine* 61 (1932): 65–88.

"Down into Davy Jones's Locker." *New York Times Magazine* (July 13, 1930): 19.

"Ecology of the Hoatzin." *Zoologica* 1 (1909): 443–66.

"Ecology of Rancho Grande, a Subtropical Cloud Forest in Northern Venezuela" (coauthored with Jocelyn Crane). *Zoologica* 32 (1947): 43–60.

Edge of the Jungle. New York: Henry Holt & Co., 1921.

"Exploration of the Deep Sea." *Science* 76 (1932): 344.

Exploring with Beebe. New York: G. P. Putnam's Sons, 1932.

"Fauna of Four Square Feet of Jungle Debris." *Zoologica* 2 (1916): 107–19.

Field Book of the Shore Fishes of Bermuda (coauthored with John Tee-Van). New York: G. P. Putnam's Sons, 1933.

"Five Days among the Birds on Cobb Island, Virginia." *Eighth Annual Report of the New York Zoological Society for 1903,* 161–81.

"The Function of Secondary Sexual Characteristics in Two Species of Dynastidae (Coleoptera)." *Zoologica* 29 (1944): 53–58.

Galápagos: World's End. New York: G. P. Putnam's Sons, 1924. Reprint, New York: Dover Publications, 1988.

"Geographical Variation in Birds with Especial Reference to the Effects of Humidity." *Zoologica* 1 (1907): 1–41.

Half Mile Down. New York: Harcourt, Brace & Co., 1934.

"A Half Mile Down." *National Geographic Magazine* 66 (1934): 661–704.

High Jungle. New York: Duell, Sloan and Pearce, 1949.

"High World of the Rain Forest." *National Geographic Magazine* 113 (1958): 838–55.

"Home Life of the Bat Falcon, *Falco albigularis albigularis* Daudin." *Zoologica* 35 (1950): 69–86.

"Insect Migration at Rancho Grande in North-Central Venezuela." *Zoologica* 34 (1949): 107–10.

"Introduction to the Ecology of the Arima Valley, Trinidad, BWI." *Zoologica* 37 (1952): 157–83.

Jungle Days. New York: G. P. Putnam's Sons, 1925.

"Jungle Night." *The Atlantic Monthly* 120 (1917): 69–79.

Jungle Peace. New York: Henry Holt & Co., 1919.

"Jungle Sluggard." *Ladies' Home Journal* 42 (1925): 12ff.

The Log of the Sun. New York: Henry Holt & Co., 1906.

"Mojave." *The Atlantic Monthly* 150 (1932): 395–402.

A Monograph of the Pheasants (4 vols). London: H. F. Witherby & Co., 1918, 1920, 1921, 1922. Reprinted as 2 vols., New York: Dover Publications, 1990.

"A Naturalist in Paris." *The Atlantic Monthly* 121 (1918): 721–32.

"New Data on the Deep Sea Fish *Stylophthalmus* and *Idiacanthus.*" *Science* 78 (1933): 390.

"Nineteen New Species and Four Post-larval Deep Sea Fishes." *Zoologica* 13 (1932): 47–107.

Nonsuch: Land of Water. New York: Brewer, Warren & Putnam, 1932.

"Note on the Humboldt Current and the Sargasso Sea." *Science* 63 (1926): 91–92.

"Notes on the Hercules Beetle *Dynastes hercules* (Linn) at Rancho Grande, Venezuela, with Special Reference to Combat Behavior." *Zoologica* 32 (1947): 109–16.

"Oceanographic Work at Bermuda of the New York Zoological Society." *Science* 80 (1934): 495–96.

Our Search for a Wilderness (coauthored with Mary Blair Beebe). New York: Henry Holt & Co., 1910.

Pheasant Jungles. New York: G. P. Putnam's Sons, 1927.

Pheasants: Their Lives and Homes (2 vols.). New York: Doubleday, Page & Co., 1926.

"Physical Factors in the Ecology of Caripito, Venezuela." *Zoologica* 28 (1943): 53–59.

"Preliminary Report on an Investigation of the Seasonal Changes of Color in Birds." *American Naturalist* 42 (1908): 34–38.

"Round Trip to Davy Jones's Locker." *National Geographic Magazine* 59 (1931): 653–78.

"Scale Adaptation and Utilization in *Aesiocopa patulana* Walker." *Zoologica* 32 (1947): 147–53.

"Second Eastern Pacific *Zaca* Expedition of the New York Zoological Society." *Science* 87 (1938): 522–23.

"Studies of a Tropical Jungle: One Quarter of a Square Mile of Jungle at Kartabo, British Guiana." *Zoologica* 6 (1925): 5–193.

"Submerged Beach off Bermuda." *Science* 74 (1931): 629.

"Success of the Indoor Flying Cage." *Sixth Annual Report of the New York Zoological Society for 1901,* 128–36.

"The Three-Toed Sloth *Bradypus cuculliger cuculliger* Wagler." *Zoologica* 7 (1926): 1–67.

Tropical Wildlife in British Guiana (coauthored with G. Inness Hartley and Paul G. Howes). New York: New York Zoological Society, 1917.

Two Bird Lovers in Mexico. Boston: Houghton, Mifflin & Co., 1905.

Unseen Life of New York as a Naturalist Sees It. New York: Duell, Sloan and Pearce, 1953.

"Vertebrate Fauna of a Tropical Dry Season Mud-hole." *Zoologica* 30 (1945): 81–88.

"A Wonderer Under Sea." *National Geographic Magazine* 62 (1932): 740–58.

"A Yard of Jungle." *The Atlantic Monthly* 117 (1916): 40–47.

"Young Sailfish." *Science* 94 (1941): 300–301.

Zaca Venture. New York: Harcourt, Brace & Co., 1938.

BOOKS AND ARTICLES BY OTHER AUTHORS

Allen, Thomas B. "William Beebe." In *Into the Unknown: The Story of Exploration.* Edited by Jonathan B. Tourtellot, 296–301. Washington, DC: National Geographic Society, 1987.

Austin, Elizabeth S. *Frank M. Chapman in Florida: His Journals and Letters.* Gainesville, FL: University of Florida Press, 1967.

Ballard, Robert. *The Eternal Darkness: A Personal History of Deep-Sea Exploration.* Princeton, NJ: Princeton University Press, 2000.

Barton, Otis. *The World Beneath the Sea.* New York: Thomas Crowell, 1953.

Beard, Daniel C. *The American Boy's Handy Book.* New York: Scribner, 1882. Reprint, Jaffrey, NH: Godine, 1983.

Berra, Tim M. *William Beebe: An Annotated Bibliography.* Hamden, CT: Archon Books, 1977.

Blake, Nelson Manfred. *The Road to Reno: A History of Divorce in the United States.* New York: Macmillan, 1962.

Brands, H. W. *T.R.: The Last Romantic.* New York: Basic Books, 1997.

Bridges, William. *Gathering of Animals: An Unconventional History of the New York Zoological Society.* New York: Harper & Row, 1974.

Broad, William J. *The Universe Below: Discovering the Secrets of the Deep Sea.* New York: Simon & Schuster, 1997.

Brooke, Margaret. *My Life in Sarawak.* London: Methuen, 1913. Fifth printing, New York: Oxford University Press, 1990.

Brooke, Sylvia. *Queen of the Head-hunters: An Autobiography.* London: Sidgwick & Jackson, 1970.

Burgess, Thomas. *Take Me Under the Sea.* Salem, MA: The Ocean Archives, 1994.

Burroughs, John. *The Art of Seeing Things: Essays by John Burroughs.* Edited by Charlotte Z. Walker. Syracuse, NY: Syracuse University Press, 2001.

———. *Bird and Bough.* Boston: Houghton Mifflin, 1906.

Carrington, J. Cullen. *Charlotte County, Virginia: Historical, Statistical, and Present Attractions.* Richmond, VA: The Heritage Press, 1907.

Carson, Rachel. *The Sea Around Us.* New York: Oxford University Press, 1951.

Chapman, Frank M. *Autobiography of a Bird-Lover.* New York: D. Appleton, 1933.

———. *Camps and Cruises of an Ornithologist.* New York: D. Appleton, 1908.

———. *My Tropical Air Castle: Nature Studies in Panama.* New York: D. Appleton, 1928.

Coues, Elliott. *Handbook of Field and General Ornithology: A Manual of the Structure and Classification of Birds. With Instructions for Collecting and Preserving Specimens.* London: Macmillan, 1890.

Crandall, Lee. "In Memoriam: Charles William Beebe." *The Auk* 81 (1984): 36–41.

Crane, Jocelyn. *Fiddler Crabs of the World: Ocypodidae, Genus Uca.* Princeton, NJ: Princeton University Press, 1975.

———. "Keeping House for Tropical Butterflies." *National Geographic Magazine* 112 (1957): 193–218.

———. "The Mammals of Hampshire County, Massachusetts." *Journal of Mammalogy* 12 (1931): 267–73.

———. "Shopping for a Jungle." *Animal Kingdom* 47 (1945): 3–13.

Darwin, Charles. *On the Origin of Species by Means of Natural Selection.* New York: D. Appleton, 1895.

Delacour, Jean. *The Pheasants of the World.* New York: Scribner, 1951.

Earle, Sylvia. *Sea Change: A Message of the Oceans.* New York: G. P. Putnam's Sons, 1995.

Ellis, Richard. *Deep Atlantic: Life, Death, and Exploration in the Abyss.* New York: Alfred A. Knopf, 1996.

Gartner, Carol B. *Rachel Carson.* New York: Frederick Engar, 1983.

Goddard, Donald. *Saving Wildlife: A Century of Conservation.* New York: Harry N. Abrams, 1995.

Hellman, Geoffrey. *Bankers, Bones & Beetles: The First Century of the American Museum of Natural History.* New York: Natural History Press, 1968.

Holzman, Robert S. *Adapt or Perish: The Life of General Roger A. Pryor, C.S.A.* Hamden, CT: Archon Books, 1976.

Hooker, Joseph D. *Himalayan Journals: Notes of a Naturalist.* 2nd ed. London: John Murray, 1855.

Hornaday, William Temple. *The Minds and Manners of Wild Animals: A Book of Personal Observations.* New York: Scribner, 1923.

Hoyt, Erich. *Creatures of the Deep: In Search of the Sea's "Monsters" and the World They Live In.* Buffalo, NY: Firefly Books, 2001.

Mazuchelli, Nina. *The Indian Alps and How We Crossed Them.* London: Longmans, Green and Co., 1876.

Miller, Char. *Gifford Pinchot and the Making of Modern Environmentalism.* Washington, DC: Island Press, 2001.

Murray, John, and Johan Hjort, *The Depths of the Ocean.* New York: Macmillan, 1912.

Niles, Blair. *Condemned to Devil's Island: The Biography of an Unknown Convict.* New York: Harcourt, Brace & Co, 1928.

———. *The James: From Iron Gate to the Sea.* New York: Farrar & Rinehart, 1939.

———. *Strange Brother.* New York: Arno Press, 1975.

Preston, Douglas J. *Dinosaurs in the Attic: An Excursion into the American Museum of Natural History.* New York: St. Martin's Press, 1986.

Pryor, Mrs. Roger A. *My Day: Reminiscences of a Long Life.* New York: Macmillan, 1909.

Putnam, David. *David Goes Voyaging.* New York: G. P. Putnam's Sons, 1925.

Raby, Peter. *Bright Paradise: Victorian Scientific Travellers.* Princeton, NJ: Princeton University Press, 1996.

Rainger, Ronald. *An Agenda for Antiquity: Henry Fairfield Osborn and Vertebrate Paleontology at the American Museum of Natural History, 1890–1935.* Tuscaloosa, AL: University of Alabama Press, 1991.

Robison, Bruce, and Judith Connor. *The Deep Sea.* Monterey, CA: Monterey Bay Aquarium Press, 1999.

Roosevelt, Theodore. "A Naturalist's Tropical Laboratory: Beebe's Laboratory in Demerara." *Scribner's Magazine* 61 (1917): 46–64.

Thane, Elswyth. *The Bird Who Made Good.* New York: Duell, Sloan and Pearce, 1947.

———. *The Reluctant Farmer.* New York: Duell, Sloan and Pearce, 1950.

———. *Riders of the Wind.* New York: Frederick A. Stokes Co., 1926. Republished Mattitack, NY: Aeonian Press, 1976.

———. *Young Mr. Disraeli.* New York: Harcourt, Brace & Co, 1936.

Tracy, Henry Chester. *American Naturalists.* New York: Dutton, 1930.

Verteuil, Anthony de. *Great Estates of Trinidad.* Port-of-Spain, Trinidad: Litho Press, 2000.

Welker, Robert Henry. *Natural Man: The Life of William Beebe.* Bloomington, IN: Indiana University Press, 1975.

Wilcove, David S. *The Condor's Shadow: The Loss and Recovery of Wildlife in America.* New York: Freeman, 1994.

Wilson, Edward O. *Naturalist.* Washington, DC: Island Press, 1994.

Yanni, Carla. *Nature's Museums: Victorian Science and the Architecture of Display.* Baltimore: The Johns Hopkins University Press, 1999.

(**AU** = Author's collection; **BU** = Elswyth Thane Collection, Howard Gotlieb Archival Research Center, Boston University; **JC** = Jocelyn Crane Griffin papers, in author's collection; **PU** = William Beebe Papers (CO661), Manuscripts Division, Department of Rare Books and Special Collections, Princeton University Library)

Location	Source	Location	Source
iii	WCS	85	WCS
Frontis	BU	87	Charles Livingston
v	Charles Livingston		Bull, JC
	Bull, JC	94	WCS
1	WCS	95	WCS
3	WCS	96	WCS
5	WCS	98	WCS
7	BU	102	WCS
9	WCS	110	AU
10	WCS	112	WCS
11	BU	114	WCS
12	BU	118	JC
13	BU	120	WCS
22	PU	121	WCS
29	JC	123	WCS
31	PU	124	*Two Bird Lovers in*
37	JC		*Mexico,* WCS
39	AU	125	WCS
42	AU	126	WCS
47	BU	132	C. R. Knight, WCS
49	BU	134	C. R. Knight, WCS
53	BU	136	AU
55	WCS	140	WCS
59	WCS	142	WCS
63	WCS	143	WCS
64	WCS	145	WCS
68	BU	146	WCS
70	BU	152	WCS
74	AC	153	WCS
77	BU	158	WCS
83	WCS	163	WCS

Location	Source	Location	Source
166	WCS	272	WCS
167	H. Grönvold, WCS	273	WCS
172	WCS	275	WCS
175	AU	277	JC
177	AU	279	PU
182	*New York Times,* WCS	282	WCS
183	WCS	285	WCS
186	WCS	286	WCS
189	WCS	289	WCS
191	WCS	291	Else Bostlemann, WCS
193	WCS	294	WCS
196	WCS	295	Else Bostlemann, WCS
200	WCS	297	JC
205	PU	299	Else Bostlemann, JC
209	JC	301	WCS
211	WCS	305	WCS
215	WCS	307	WCS
216	WCS	309	WCS
219	WCS	311	WCS
220	AU	314	JC
221	WCS	316	WCS
223	WCS	318	WCS
225	WCS	321	WCS
229	WCS	323	WCS
230	WCS	328	Miguel Covarrubias,
233	AU		*Vanity Fair,* WCS
237	Ralph Barton,	332	WCS
	New York Times, WCS	335	WCS
239	JC	338	JC
240	PU	341	WCS
244	WCS	343	WCS
245	WCS	345	WCS
248	WCS	347	JC
250	WCS	349	WCS
251	WCS	351	WCS
254	WCS	355	JC
255	WCS	358	WCS
257	WCS	359	WCS
260	WCS	360	WCS
263	BU	363	WCS
264	BU	364	WCS
268	BU	366	WCS
270	WCS	370	WCS

Location	Source	Location	Source
371	JC	387	BU
373	WCS	388	JC
375	WCS	389	JC
376	JC	396	WCS
379	WCS	400	JC
380	JC	403	JC
383	JC	409	WCS